SIGILLVM · VNIVERSITATIS · SCOLARIVM · STVDII · PRAGENSIS

Pavel Cejnar

A Condensed Course of Quantum Mechanics

Charles University
Karolinum Press 2017

Reviewed by:

Prof. Jiří Hořejší (Prague)
prof. Jean-Paul Blaizot (Paris)

CATALOGUING-IN-PUBLICATION – NATIONAL LIBRARY
OF THE CZECH REPUBLIC

Cejnar Pavel
A Condensed Course of Quantum Mechanics / Pavel Cejnar. –
1st ed. – Prague : Karolinum Press, 2013
Published by: Charles University
ISBN 978-80-246-2321-4

530.145.6
– quantum mechanics
– handbooks, manuals, etc.
– kvantová mechanika
– příručky

530 – Physics [6]
53 – Fyzika [6]

ISBN 978-80-246-2321-4
ISBN 978-80-246-2349-8 (pdf)

Contents

HEALTH AUTHORITY WARNING
THINKING ABOUT
QUANTUM PHYSICS
CAUSES INSOMNIA

Preface

This book was conceived as a collection of notes to my two-semester lecture on quantum mechanics for third-year students of physics at the Faculty of Mathematics and Physics of the Charles University in Prague. It was created in 2011-12.

At first, I just wanted to write down the most important facts, formulas and derivations in a compact form. The information flew in a succinct, "staccato" style, organized in larger and smaller bits (the ■ and ▶ items), rarely interrupted by wordy explanations. I enjoyed the thick, homogeneous mathematical form of the notes. Calculations, calculations, calculations... I thought of a horrified historian or sociologist who finds no oasis of words. This is how we, tough guys, speak!

However, I discovered that the dense form of the notes was hardly digestible even for tough guys. I had to add some words. To create a "storyteller" who wraps the bare formulas into some minimal amount of phrases. His voice, though still rather laconic, may help to provide the proper motivation and clarify the relevant context. I also formed a system of specific "environments" to facilitate the navigation. In particular: Among crowds of calculations there appears a hierarchy of highlighted formulas:* important essential 1 essential 2 crucial

Assumptions or foundational concepts, irreducible to other statements/concepts, appear in boxes:† Answer to ultimate question of life, universe & everything = 42

Here and there come some historical notes:‡ ◀ 2013: *Condensed Course* issued

Handmade schemes (drawn on a whiteboard) illustrate some basic notions.

In this way, the notes have turned into a more serious thing. They almost became a *textbook*! The one distinguished from many others by expanded mathematical derivations (they are mostly given really step by step) and reduced verbal stuffing (just necessary comments in between calculations). This makes the book particularly well suited for conservation purposes—acquired knowledge needs to be stored in a *condensed*, dense enough form, having a compact, nearly tabular structure.

However, as follows from what has been said, this book *cannot* be considered a standard textbook. It may hardly be read with ease and fluency of some more epic treatises. One rather needs to proceed cautiously as a detective, who has to precisely fix all objects on the stage (all symbols, relations etc.) before making any small step forward. This book can be used as a teaching tool, but preferably together with an

*Such formulas are highly recommended to memorize! Although all students of physics & mathematics seem to share a deep contempt for any kind of memorization, I have to stress that all results cannot be rederived in reasonable time limits. There is no escape from saving the key formulas to the memory and using them as quickly reachable starting points for further calculations.

†However, these assumptions do not constitute a closed system of axioms in the strict mathematical sense.

‡I believe that knowledge of history is an important part of understanding. The concepts do not levitate in vacuum but grow from the roots formed by concrete circumstances of their creation. If overlooking these roots, one may misunderstand the concepts.

oral course or a more talkative textbook on quantum mechanics. Below I list some of my favorite candidates for additional guiding texts [1–8].

I have to stress that the notes cover only some parts of *non-relativistic* quantum mechanics. The selection of topics is partly fixed by the settled presentation of the field, and partly results from my personal orientation. The strategy is to introduce the complete general formalism along with its exemplary applications to simple systems (this takes approx. one semester) and then (in the second semester) to proceed to some more specialized problems. Relativistic quantum mechanics is totally absent here; it is postponed as a prelude for the quantum field theory course.

Quantum mechanics is a complex subject. It obligates one to have the skills of a mathematician as well as the thinking of a philosopher. Indeed, the mathematical basis of quantum physics is rather abstract and it is not obvious how to connect it with the observed "reality". No physical theory but quantum mechanics needs such a sophisticated PR department. We will touch the interpretation issues here, but only very slightly. Those who want to cultivate their opinion (but not to disappear from the intelligible world) are forwarded to the classic [9]. The life saving trick in this *terra incognita* is to tune mind to the joy of thinking rather than to the demand of final answers. The concluding part of the theory may still be missing.

Before we start I should not forget to thank all the brave testers—the first men, mostly students, who have been subject to the influence of this book at its various stages of preparation. They were clever enough to discover a lot of mistakes. Be sure that the remaining mistakes are due to their generous decision to leave some fish for the successors.

In Prague, January 2013

Recommended textbooks:

[1] J.J. Sakurai, *Modern Quantum Mechanics* (Addison-Wesley, 1985, 1994)
A modified edition of the same book:
[2] J.J. Sakurai, J.J. Napolitano, *Modern Quantum Mechanics* (Addison-Wesley, 2011)
[3] G. Auletta, M. Fortunato, G. Parisi, *Quantum Mechanics* (Cambridge University Press, 2009)
[4] L.E. Ballantine, *Quantum Mechanics. A Modern Development* (World Scientific, Singapore, 1998)
[5] A. Peres, *Quantum Theory: Concepts and Methods* (Kluwer, 1995)
[6] A. Bohm, *Quantum Mechanics: Foundations and Applications* (Springer, 1979, 1993)
[7] W. Greiner, *Quantum Mechanics: An Introduction* (Springer, 1989)
W. Greiner, *Quantum Mechanics: Special Chapters* (Springer, 1998)
W. Greiner, B Müller, *Quantum Mechanics: Symmetries* (Springer, 1989)
[8] E. Merzbacher, *Quantum Mechanics* (Wiley, 1998)

Further reading:

[9] J.S. Bell, *Speakable and Unspeakable in Quantum Mechanics* (Cambridge University Press, 1987)

Rough guide to notation (succinct and incomplete)

symbol	meaning
	Spaces, state vectors & wavefunctions
$\underline{\mathcal{H}}$, $\mathcal{H}$, $\overline{\mathcal{H}}$	Gelfand's hierarchy of spaces (rigged Hilbert space)
ℓ^2, $\mathcal{L}^2(\mathbb{R}^3)$, $\mathbb{C}^d$	specific separable or finite Hilbert spaces
$\|\psi\rangle$, $\langle\psi'\|$; $\langle\psi'\|\psi\rangle$	"ket" & "bra" forms of state vectors; scalar product
$\|\|\psi\|\| = \sqrt{\langle\psi\|\psi\rangle}$	vector norm
$\alpha\|\psi\rangle+\beta\|\psi'\rangle$	superposition $\equiv$ linear combination of state vectors ($\alpha,\beta\in\mathbb{C}$)
$\|\phi_i\rangle$, $\|\Phi_{ij}\rangle\equiv\|\phi_{1i}\rangle_1\|\phi_{2j}\rangle_2$	general basis vector in $\mathcal{H}$; separable basis vector in $\mathcal{H}_1\otimes\mathcal{H}_2$
$\|\psi\rangle_1\|\psi'\rangle_2$	general separable vector in $\mathcal{H}_1\otimes\mathcal{H}_2$
$\|a\rangle$, $\|a_i\rangle$, $\|a_i^{(k)}\rangle$	eigenvector of $\hat{A}$ with eigenvalue a or a_i (degeneracy index k)
$\|E_i\rangle$, $\|E\rangle$	energy eigenvectors
$\|\uparrow\rangle$, $\|\downarrow\rangle$	up & down projection states of spin $s=\frac{1}{2}$
$\substack{\|lm_l\rangle \\ \|sm_s\rangle}$, $\|jm_j\rangle$	states with ($\substack{\text{orbital} \\ \text{spin}}$,total) ang. momentum ($\substack{l \\ s}$, j), projection $m_\bullet$
$\psi(\underbrace{\vec{x},m_s}) \equiv \boldsymbol{\Psi}(\vec{x})$	single-particle wavefunction in single/multicomponent forms
$\Psi(\boldsymbol{\xi}_1\ldots\boldsymbol{\xi}_N)$	N-particle wavefunction
$R_{nl}(r) = \frac{u_{nl}(r)}{r}$	radial wavefunction
$\text{Span}\{\|\psi_1\rangle\ldots\|\psi_n\rangle\}$	linear space spanned by the given vectors
$\mathcal{N}$, $d_{\mathcal{H}}$	normalization coefficient & dimension of space $\mathcal{H}$
	Observables & operators
$\hat{O}$, $\hat{O}^\dagger$, $\hat{O}^{-1}$	operator, its Hermitian conjugate & inverse
$\hat{I}$, $\hat{U}$	identity operator & unitary operator
$\hat{P}_a$, $\hat{\boldsymbol{\Pi}}_{(a_1,a_2)}$	projectors to discrete & continuous eigenvalue subspaces
$\|\|\hat{A}\|\|$	operator norm
$\hat{A}_1\otimes\hat{A}_2$	tensor product of operators acting in $\mathcal{H}_1\otimes\mathcal{H}_2$
$\hat{H}$, $\hat{T}$, $\hat{V}$; $\hat{H}'$	Hamiltonian, its kinetic & potential terms; perturbation
$\vec{\nabla}$, Δ	gradient & Laplace operator (or also an interval, gap...)
$\hat{\vec{x}}$, $\hat{\vec{p}}$, $\hat{P}$	coordinate, momentum vectors & spatial parity operator
$\hat{\vec{L}}$, $\hat{\vec{S}}$; $\hat{\vec{J}}$, $\hat{J}_\pm$	orbital, spin & total angular momentum, shift operators for $\hat{J}_3$
$\hat{\vec{\sigma}}$	the triplet of Pauli matrices
$\hat{T}_{\Delta o}$	$\|o\rangle\rightarrow\|o+\Delta o\rangle$ eigenvector shift operator for general operator $\hat{O}$
$\hat{G}_i$, $\hat{C}_{\mathcal{G}}$	generator & Casimir operator of a group $\mathcal{G}$
$\hat{b}$, $\hat{b}^\dagger$; $\hat{a}$, $\hat{a}^\dagger$; $\hat{c}$, $\hat{c}^\dagger$	annihilation, creation operators for bosons, fermions, or both
$\hat{N}$, $\hat{N}_k$	total number of particles & number of particles in k^{th}state
$\hat{R}_{\vec{n}\phi}$, $\mathbf{R}(\alpha\beta\gamma)$	rotation operator in $\mathcal{H}$ & rotation matrix in 3D (Euler angles)
$\hat{U}(t)$, $\hat{U}(t_1,t_0)$	evolution operator for times $t_0 \xrightarrow{t} t_1$
$\hat{\mathcal{T}}$, $\mathfrak{T}$	time reversal operator & time ordering of operator product
$\hat{G}(t)$, $G(\vec{x}t\|\vec{x}_0t_0)$	Green operator & propagator

$\hat{O}_{\rm S}$, $\hat{O}_{\rm H}(t)$, $\hat{O}_{\rm D}(t)$	Schrödinger, Heisenberg, Dirac representations of operator
$[\hat{A}^{\lambda_1}{\times}\hat{B}^{\lambda_1}]^{\lambda}_{\mu}$	tensor coupling of spherical tensor operators $\hat{A}^{\lambda_1}_{\mu_1}, \hat{B}^{\lambda_1}_{\mu_2}$
$[\hat{A},\hat{B}]$,$\{\hat{A},\hat{B}\}$	commutator & anticommutator of operators
$\{A,B\}$	Poisson bracket of classical observables A,B
${\rm Tr}\,\hat{O}$, ${\rm Tr}_1\hat{O}$	trace of operator/matrix, partial trace over $\mathcal{H}_1$ in $\mathcal{H}_1\otimes\mathcal{H}_2$
${\rm Det}\,\hat{O}$, ${\rm Def}(\hat{O})$	determinant of matrix/operator, definition domain of operator
	Statistics, probabilities & densities
$p_{\psi}(a)$	probability to measure value a of observable A in state $\vert\psi\rangle$
$\langle A\rangle_{\psi}$, $\langle a\rangle_c$	average of A-distribution in $\vert\psi\rangle$, average of a for a parameter c
$\langle\langle A^2\rangle\rangle_{\psi}\equiv\Delta^2_{\psi}A$	dispersion of A-distribution in $\vert\psi\rangle\equiv$ squared uncertainty
$p_c(a\vert b)$	conditional probability of a given b (depending on parameter c)
$\rho(\vec{x},t)$, $\vec{j}(\vec{x},t)$	probability density & flow at point $\vec{x}$, time t
$\hat{\rho}$, $W_{\rho}(\vec{x},\vec{p})$, S_{ρ}	density operator/matrix, Wigner distribution function, entropy
$\varrho(E)$	density of energy eigenstates
	Functions
$j_l,n_l,h^{\pm}_l(kr)$	Bessel, Neumann & Hankel functions
$L^j_i(\rho)$, $H_n(\xi)$	$\left\{\begin{smallmatrix}\text{associated}\\ \text{generalized}\end{smallmatrix}\right\}$ Laguerre polynomials & Hermite polynomials
$P_{lm}(\cos\vartheta)$,$Y_{lm}(\vartheta,\varphi)$	associated Legendre polynomial, spherical harmonics (sph.angles)
$D^j_{m'm}(\alpha\beta\gamma)$	Wigner matrix function $\equiv D^j_{m'm}(\mathbf{R})$ (Euler angles of rotation $\mathbf{R}$)
$\delta(x)$, $\delta_{\epsilon}(x)$; $\Theta(x)$	Dirac δ-function, sequence of functions $\xrightarrow{\epsilon\to 0}\delta$; step function
$Z(\beta)$, $Z(\beta,\mu)$	(grand)canonical partition funcs. (inv.temperature,chem.potential)
$\left\{\begin{smallmatrix}S[\vec{x}(t)]\\ S(\vec{x},t)\end{smallmatrix}\right\}$, $\mathcal{L}(\vec{x},\dot{\vec{x}})$	classical action (functional & function forms), Lagrangian
$V(\vec{x})$, $\vec{A}(\vec{x})$	scalar & vector potentials
$S_{ji},P_{ji},W_{ji}(t)$	$j\to i$ transition amplitude, probability & rate (time)
F_l, S_l, $\delta_l(k)$	partial wave amplitude, S-matrix & phase shift (|wavevector|)
$f_{\vec{k}}(\vec{k}')\equiv f_{\vec{k}}(\vartheta,\varphi)$	scattering amplitude (direction/angles)
$\frac{d\sigma}{d\Omega}(\vartheta,\varphi)$	differential cross section ($\sigma\equiv$ integral cross section)
	Miscellaneous
$(1,2,3)\equiv(x,y,z)$	indices of Cartesian components
$\vec{n}$, $\left\{\begin{smallmatrix}(\vec{n}_x,\vec{n}_y,\vec{n}_z)\\ (\vec{n}_r,\vec{n}_{\vartheta},\vec{n}_{\varphi})\end{smallmatrix}\right\}$	unit vector, $\left\{\begin{smallmatrix}\text{Cartesian}\\ \text{spherical}\end{smallmatrix}\right\}$ orthonormal coordinate vectors
δ_{ij}, ε_{ijk}	Kronecker & Levi-Civita symbols
$C^{jm}_{j_1m_1j_2m_2}$	Clebsch-Gordan coefficient $\equiv\langle j_1j_2jm\vert j_1m_1j_2m_2\rangle$
$\hbar,c,e$	Planck constant, speed of light, elementary charge
$M,\mathcal{M}$; q	particle mass & two-particle reduced mass; particle charge
$\vec{k}$, ω, λ	wavevector, frequency, wavelength (or perturbation parameter)
ε_k, n_k	energies & occupation numbers of single-particle states
$\{X_i\}_{i\in\mathcal{D}}$,$\{X(c)\}_{c\in\mathcal{C}}$	discrete/continuous set of objects
Min,Max,Sup$\{X_i\}_i$	minimum, maximum, supremum of a set of numbers
$\bullet$; *iff*	blind index denoting objects from a given set; if and only if

INTRODUCTION

Before sailing out, we encourage the crew to get ready for adventures. Quantum mechanics deals with phenomena, which are rather unusual from our common macroscopic experience. Description of these phenomena makes us sacrifice some principles which we used to consider self-evident.

■ Quantum level

Quantum theory describes objects on the atomic and subatomic scales, but also larger objects if they are observed with an extremely **high resolution**.

► Planck constant

The domain of applicability of quantum mechanics is determined with the aid of a new constant: $\boxed{\hbar \doteq 1.05 \cdot 10^{-34}\ \mathrm{J{\cdot}s} \doteq 0.66\ \mathrm{eV{\cdot}fs}}$ (units of action)

► Consider 2 classical trajectories $\mathbf{q}_1(t)$ & $\mathbf{q}_2(t)$ (in a general multidimensional configuration space) which (in the given experimental situation) are on the limit of distinguishability. The difference of actions: $\Delta S = |S[\mathbf{q}_1(t)] - S[\mathbf{q}_2(t)]|$

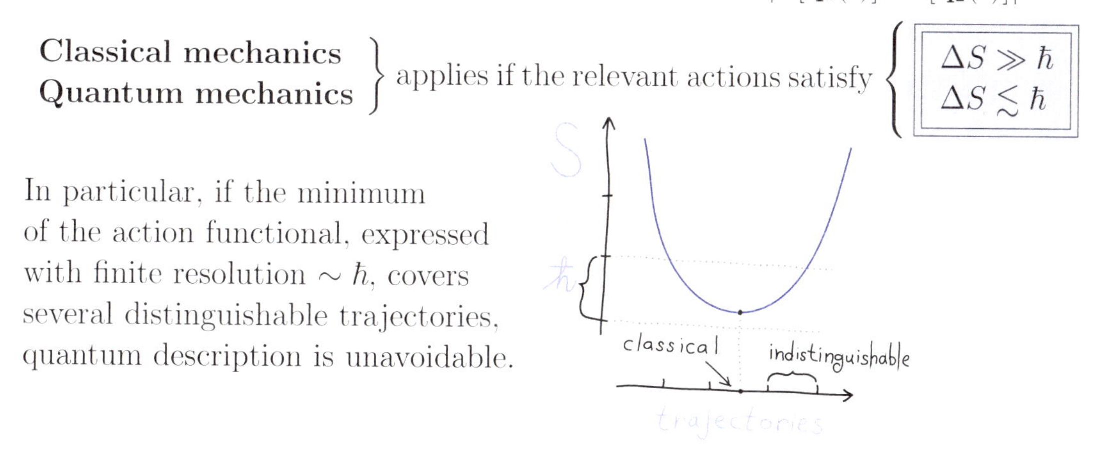

Classical mechanics / **Quantum mechanics** } applies if the relevant actions satisfy $\left\{ \begin{array}{l} \Delta S \gg \hbar \\ \Delta S \lesssim \hbar \end{array} \right.$

In particular, if the minimum of the action functional, expressed with finite resolution $\sim \hbar$, covers several distinguishable trajectories, quantum description is unavoidable.

◄ **Historical remark**

1900: Max Planck introduced $\hbar$ along with the quanta of electromagnetic radiation to explain the blackbody radiation law

1905: Albert Einstein confirmed elmag. quanta in the explanation of photoeffect

1913: Niels Bohr introduces a quantum model of atoms ("old quantum mechanics")

■ Double slit experiment

According to Richard Feynman & some others, this is the most crucial quantum experiment that allows one to realize how unusual the quantum world is.

► Arrangement

Emitter E of *individual* particles, shield with slits A and B, screen S

Both trajectories $\vec{x}_{\mathrm{A}}(t)$ and $\vec{x}_{\mathrm{B}}(t)$ from $\vec{x}_{\mathrm{E}}$ to $\vec{x}_{\mathrm{S}}$ minimize the action

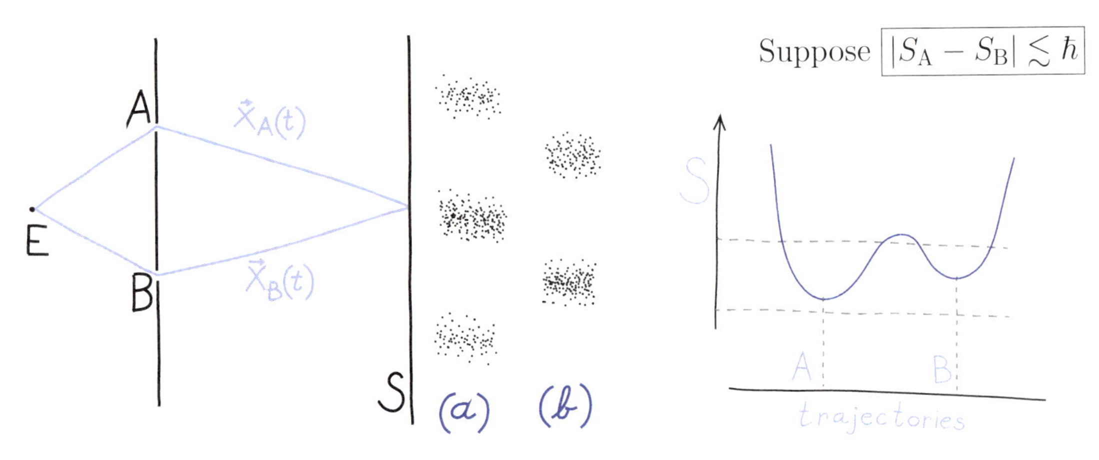

▶ **Regimes of measurement**

(a) Interference setup: particle position measured only at the screen
⇒ interference pattern with individual particle hits

(b) Which-path setup: prior the screen measurement, the particle position measured immediately after the slits ⇒ no interference pattern

Delayed-choice experiment: The choice of setup (a)/(b) is made *after* the particle passed the slits. The same outcome as if the decision was made before.

Paradox: The outcome of the interference setup indicates a wave-like behavior of particles (passage through both slits simultaneously). The outcome of the which-path setup shows a corpuscular behavior (passage through one slit only). The outcome of the delayed-choice experiment invalidates the possibility that the particle "changes clothes" according to the setup selected.

◀ **Historical remark**
1805 (approx.): Thomas Young performed double-slit experiment with light
1927: C. Davisson & L. Germer demonstrate interference of electrons on crystals
1961: first double-slit experiment with massive particles (electrons)
1970's: double-slit experiments with individual electrons
1990's-present: progress in realizations of which-path setup & delayed-choice exp.

■ Wavefunction & superposition principle

To explain the outcome of the interference setup of the double-slit experiment, one has to assume that particles possess some wave properties.

▶ Particle attributed by a wavefunction: $\boxed{\psi(\vec{x},t) \equiv \sqrt{\rho(\vec{x},t)}\; e^{i\varphi(\vec{x},t)}} \in \mathbb{C}$

Squared modulus $|\psi(\vec{x},t)|^2 = \rho(\vec{x},t) \geq 0$ is the **probability density** to detect the particle at position $\vec{x}$. Normalization: $\boxed{\int |\psi(\vec{x},t)|^2\, d\vec{x} = 1}$ $\forall t$

Phase $\varphi(\vec{x},t) \in \mathbb{R}$ has **no "classical" interpretation**

$\psi(\vec{x},t) \equiv$ instantaneous density of the **probability amplitude** for finding the particle at various places (**particle** is inherently a **delocalized object!**)

▶ **Superposition of wavefunctions**

The outcome of the interference setup depends on the fact that waves can be summed up. Consider 2 wavefunctions $\psi_{\rm A}(\vec{x},t)$ & $\psi_{\rm B}(\vec{x},t)$

$$\int |\psi_{\rm A}|^2 d\vec{x} < \infty, \int |\psi_{\rm B}|^2 d\vec{x} < \infty \Rightarrow \boxed{\int |\alpha\psi_{\rm A}+\beta\psi_{\rm B}|^2 d\vec{x} < \infty} \; \forall \alpha,\beta \in \mathbb{C}$$

$\Rightarrow$ any linear combination of normalizable wavefunctions is a normalizable wavefunction $\Rightarrow$ these functions form a linear vector space $\mathcal{L}^2(\mathbb{R}^3)$

▶ **Interference phenomenon**

Probability density for a superposition of waves is not the sum of densities for individual waves

Choose $\left\{ \begin{smallmatrix} \alpha=|\alpha|e^{i\varphi_\alpha} \\ \beta=|\beta|e^{i\varphi_\beta} \end{smallmatrix} \right\}$ such that $\int |\alpha\psi_{\rm A}+\beta\psi_{\rm B}|^2 d\vec{x} = 1$ (with $\left\{ \begin{smallmatrix} \psi_{\rm A} \\ \psi_{\rm B} \end{smallmatrix} \right\}$ normalized)

$$\Rightarrow \boxed{\underbrace{|\alpha\psi_{\rm A}+\beta\psi_{\rm B}|^2}_{\rho_{\alpha{\rm A}+\beta{\rm B}}} = \underbrace{|\alpha\psi_{\rm A}|^2}_{|\alpha|^2\rho_{\rm A}} + \underbrace{|\beta\psi_{\rm B}|^2}_{|\beta|^2\rho_{\rm B}} + \underbrace{2|\alpha\beta\psi_{\rm A}\psi_{\rm B}|\cos(\varphi_{\rm A}+\varphi_\alpha-\varphi_{\rm B}-\varphi_\beta)}_{\text{interference terms}}}$$

▶ **Description of the interference setup** in the double slit experiment

1) Initial wavefunction between emission (t=0) and slits (t=$t_{\rm AB}$): $\psi(\vec{x},t)$

2) Wf. at $t \gtrsim t_{\rm AB}$ (right after the slits): $\psi(\vec{x},t_{\rm AB}^+) \approx \alpha\delta_{\rm A}(\vec{x}-\vec{x}_{\rm A}) + \beta\delta_{\rm B}(\vec{x}-\vec{x}_{\rm B})$ with $\delta_\bullet(\vec{x}-\vec{x}_\bullet) \equiv$ wf. localized on the respective slit ($\delta_\bullet$=0 away from the slit) and $\alpha,\beta \equiv$ coefficients depending on the "experimental details"

3) Wf. at $t_{\rm S}$=$t_{\rm AB}+\Delta t$ (just before screen): $\psi(\vec{x},t_{\rm S}) \approx \alpha\psi_{\rm A}(\vec{x},\Delta t)+\beta\psi_{\rm B}(\vec{x},\Delta t)$ with $\psi_\bullet(\vec{x},\Delta t) \equiv$ the wf. developed from $\delta_\bullet(\vec{x}-\vec{x}_\bullet)$ in time Δt

$\Rightarrow$ Distribution on screen: $\boxed{\rho(\vec{x}_{\rm S}) \approx |\alpha\psi_{\rm A}(\vec{x}_{\rm S},\Delta t) + \beta\psi_{\rm B}(\vec{x}_{\rm S},\Delta t)|^2}$

▶ **Dirac delta function** (mathematical intermezzo)

$\delta(x) \equiv$ a generalized function (distribution) $\equiv$ limit of a series of ordinary functions: $\boxed{\delta(x) = \lim_{\epsilon\to 0}\delta_\epsilon(x)}$ with, e.g.: $\delta_\epsilon(x) \equiv \begin{cases} \frac{1}{\epsilon} & \text{for } x\in[-\frac{\epsilon}{2},+\frac{\epsilon}{2}] \\ 0 & \text{otherwise} \end{cases}$

$$\Rightarrow \boxed{\text{Support}\,[\delta(x)] \equiv \{x=0\} \quad \& \quad \int_{-\infty}^{+\infty} \delta(x)\,dx = 1}$$

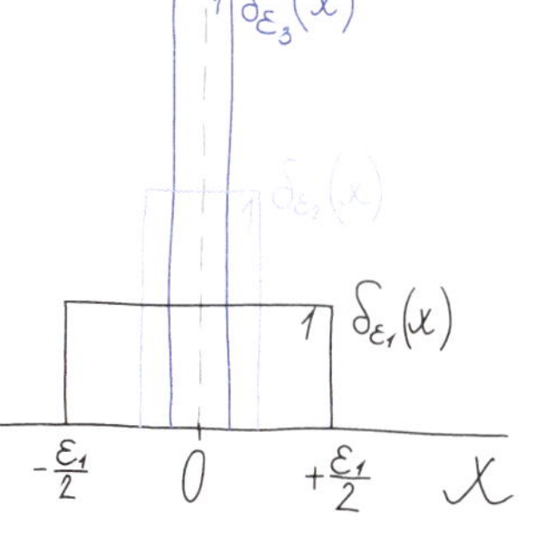

Other limiting realizations of δ-function:

$\delta_\epsilon(x) = \frac{1}{\pi}\frac{\epsilon}{\epsilon^2+x^2}$ (Cauchy or Breit-Wigner form)

$\delta_\epsilon(x) = \frac{1}{\sqrt{2\pi\epsilon^2}}e^{-\frac{x^2}{2\epsilon^2}}$ (Gaussian form)

$\delta_\epsilon(x) = \frac{1}{\pi}\frac{\sin(x\epsilon^{-1})}{x} = \frac{1}{2\pi}\int_{-\epsilon^{-1}}^{+\epsilon^{-1}} e^{iqx}dq$ (Fourier transformation of unity)

In 3D space:

$$\overbrace{\delta_{\epsilon_1}(x_1-x_1')\delta_{\epsilon_2}(x_2-x_2')\delta_{\epsilon_3}(x_3-x_3')}^{\delta_{\vec{\epsilon}}(\vec{x}-\vec{x}')} \xrightarrow{\vec{\epsilon}\to 0} \overbrace{\delta(x_1-x_1')\delta(x_2-x_2')\delta(x_3-x_3')}^{\delta(\vec{x}-\vec{x}')}$$

Defining property in terms of distribution theory: $\boxed{\int f(\vec{x})\delta(\vec{x}-\vec{x}')\,d\vec{x} = f(\vec{x}')}$

▶ **Delocalized wavefunctions**

Any wavefunction can be expressed as: $\boxed{\overbrace{\psi(\vec{x},t)}^{|\psi(t)\rangle} = \int \psi(\vec{x}',t)\,\overbrace{\delta(\vec{x}-\vec{x}')}^{|\vec{x}'\rangle}\,d\vec{x}'}$

$\Rightarrow$ general state $|\psi(t)\rangle \equiv$ **superposition of localized states** $|\vec{x}'\rangle \equiv \delta(\vec{x}-\vec{x}')$ with coefficients equal to the respective wavefunction values $\psi(\vec{x}',t)$

But note that $\delta(\vec{x}-\vec{x}') \notin \mathcal{L}^2(\mathbb{R}^3) \quad \Leftarrow \quad$ no sense of $|\delta(\vec{x}-\vec{x}')|^2$

◀ **Historical remark**

1800-10: Thomas Young formulates the superposition principle for waves
1924: Louis de Broglie introduces the concept of particle wavefunction
1926: Erwin Schrödinger formulates wave mechanics
1926: Max Born provides the probabilistic interpretation of wavefunction
1926-32: John von Neumann formulates QM through linear vector spaces
1927-30: Paul Dirac includes into the formulation the δ function
1940's-60's: L. Schwarz, I.M. Gelfand, N.Y. Vilenkin work out proper mathematical background for the generalized functions (distribution theory, rigged Hilbert spaces)

■ **Quantum measurement**

To explain the outcome of the which-path setup of the interference experiment, one has to assume that in quantum mechanics the measurement has a dramatic effect on the system: "**reduction**", "**collapse**" of its wavefunction

▶ **Change of the wavefunction in measurement**

Example: position measurement detecting the particle (in time t_0) within the box $(x_1' \pm \frac{\epsilon_1}{2}, x_2' \pm \frac{\epsilon_2}{2}, x_3' \pm \frac{\epsilon_3}{2}) \Rightarrow$ the wavefunction changed:

$$\psi(\vec{x},t_0)_{\text{ delocalized}} \xrightarrow{\text{reduction}} \psi(\vec{x},t_0+dt) \propto \delta_{\vec{\epsilon}}(\vec{x}-\vec{x}')\psi(\vec{x},t_0)_{\text{ localized}}$$

In an *ideal* ($\epsilon\to 0$) measurement:

$\boxed{\psi(\vec{x},t) \to \delta(\vec{x}-\vec{x}')}$ or $\boxed{|\psi(t)\rangle \to |\vec{x}'\rangle}$

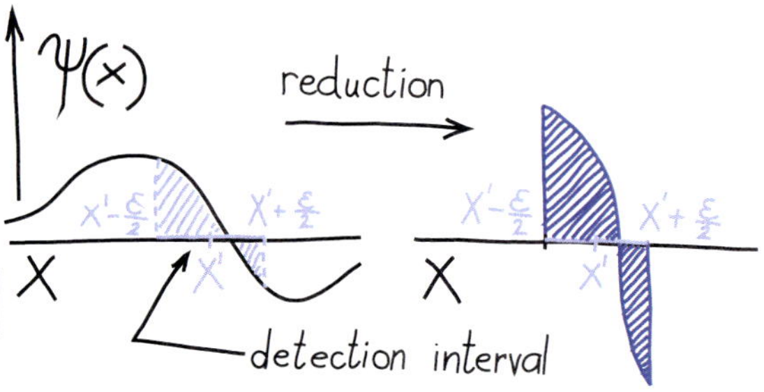

▶ **Description of the which-path setup** in the double slit experiment

1) Initial wavefunction: $\psi(\vec{x},t)$

2) After the slits: $\psi(\vec{x},t_{\rm AB}^{+}) \approx \alpha\delta_{\rm A}(\vec{x}-\vec{x}_{\rm A}) + \beta\delta_{\rm B}(\vec{x}-\vec{x}_{\rm B})$

3) After which-path measurement: $\psi(\vec{x},t_{\rm AB}^{++}) \approx \begin{cases} \delta_{\rm A}(\vec{x}-\vec{x}_{\rm A}) \text{ probability } \approx|\alpha|^2 \\ \delta_{\rm B}(\vec{x}-\vec{x}_{\rm B}) \text{ probability } \approx|\beta|^2 \end{cases}$

4) Before screen: $\psi(\vec{x},t_{\rm S}) \approx \begin{cases} \psi_{\rm A}(\vec{x},\Delta t) \text{ probability } \approx|\alpha|^2 \\ \psi_{\rm B}(\vec{x},\Delta t) \text{ probability } \approx|\beta|^2 \end{cases}$

⇒ Distribution on screen: $\boxed{\rho(\vec{x}_{\rm S}) \approx |\alpha|^2|\psi_{\rm A}(\vec{x}_{\rm S},\Delta t)|^2 + |\beta|^2|\psi_{\rm B}(\vec{x}_{\rm S},\Delta t)|^2}$

The interference pattern destroyed! This is a direct consequence of the wavefunction collapse caused by the which-path measurement.

◀ **Historical remark**

1927: the first explicit note of wavefunction collapse by Werner Heisenberg
1932: inclusion of collapse into the math. formulation of QM by John von Neumann
1930's-present: discussions about physical meaning of the collapse

■ Some general consequences

Already at this initial stage, we can foresee some general features of the "quantum world", which seem counterintuitive in the classical context.

▶ **Contextuality**

Particles show either wave or corpuscular properties, in accord with the specific experimental arrangement. One may say—in more sweeping manner—that the observed "reality" emerges during the act of observation. The actual result depends on a wider "context" of the physical process that is investigated.

▶ **Quantum logic**

An attempt to assign the strange properties of the quantum world to a non-classical underlying logic. In the double slit experiment it can be introduced via the following "propositions":

$$\left.\begin{array}{l} A \equiv \text{ passage through slit A} \\ B \equiv \text{ passage through slit B} \end{array}\right\} \to S \equiv \text{detection at given place of screen}$$

Different outcomes of interference & which-path setups indicate the inequality:

$$\boxed{\underbrace{(A \vee B) \wedge S}_{\text{interference setup}} \neq \underbrace{(A \wedge S) \vee (B \wedge S)}_{\text{which-path setup}}} \quad \Rightarrow \text{violation of a common logic axiom}$$

▶ Rule for **general branching processes** with alternative paths A & B:

Probability that the system passed through the branching (real or "logical") while its path has not been detected depends on whether the paths can/cannot, *in principle*, be distinguished (e.g., by a delayed or more detailed measurement):

Indistinguishable paths ⇒ sum of amplitudes $\boxed{\psi_{A\vee B} \propto \psi_A + \psi_B}$

Distinguishable paths ⇒ sum of probabilities (densities) $\boxed{\rho_{A\vee B} \propto \rho_A + \rho_B}$

◀ **Historical remark**
1924-35: Bohr (Copenhagen) versus Einstein debate. Niels Bohr defends a "subjective" approach (with the observer playing a role in the "creation" of reality)
1936: Garrett Birkhoff and John von Neumann formally introduce quantum logic
1920's-present: Neverending discussions on the interpretation of quantum physics

1. FORMALISM ⇝ 2. SIMPLE SYSTEMS

Quantum mechanics has rather deep mathematical foundations. Such that the interpretation of abstract formalism in terms of "common sense" becomes a nontrivial issue. This may lead some of us to philosophical meditations about the link of physical theory to reality. Here we focus mostly on mastering the theory on a technical level. Elements of the abstract formalism are outlined in Chapter 1, while their simple concrete applications are sketched in Chapter 2. To keep a link between the *Geist* and *Substanz*, we present these chapters in an alternating, entangled way.

1.1 Space of quantum states

Any theory starts from identification of the relevant attributes of the system under study which are necessary for its unique characterization. In physical theories, these attributes represent specific mathematical entities which fill in some spaces.

■ Hilbert space

The formalism of quantum theory is based on mathematics matured at the beginning of 20th century. The essential idea turned out to be the following: to capture quantum uncertainty, distinct states of a system cannot be always perfectly distinguishable. The states must show some "overlaps". This is exactly the property of vectors in linear spaces.

▶ State of a physical system

State ≡ a "complete" set of parameters characterizing the physical system. The set does not have to be exhaustive (determining all aspects of the given system), but it has to be *complete* in the sense of *autonomous determinism*: the knowledge of state at a single time (t=0) suffices to uniquely determine the state at any time ($t \gtrless 0$).
Let $|\psi\rangle$ denote a *mathematical entity* describing an arbitrary physical state ψ of a given quantum system (shortcut: $|\psi\rangle \equiv$ "a state"). Let $\mathcal{H}$ be a system-specific *space* of all such entities.

▶ **Requirement 1**: $\mathcal{H}$ supports the **superposition principle**

$|\psi_1\rangle, |\psi_2\rangle \in \mathcal{H}$ and $\alpha, \beta \in \mathbb{C} \Rightarrow$ $\boxed{|\psi\rangle = \alpha|\psi_1\rangle + \beta|\psi_2\rangle \quad \in \mathcal{H}}$

$\Rightarrow \mathcal{H}$ is a *complex vector space*

Scaling $|\psi'\rangle = \alpha|\psi\rangle$ has no physical consequences: states = rays of vectors

▶ **Requirement 2**: $\mathcal{H}$ supports a **scalar product** $\langle\psi_1|\psi_2\rangle \in \mathbb{C}$

Properties: $\langle\psi_1|\psi_2\rangle = \langle\psi_2|\psi_1\rangle^*$, $\langle\psi_1|\alpha\psi_2+\beta\psi_3\rangle = \alpha\langle\psi_1|\psi_2\rangle + \beta\langle\psi_1|\psi_3\rangle$, $\langle\psi|\psi\rangle \geq 0$

Norm: $||\psi||^2 \equiv \langle\psi|\psi\rangle$

$\Rightarrow$ Distance: $d^2(\psi_1, \psi_2) \equiv ||\psi_1 - \psi_2||^2 = \langle\psi_1|\psi_1\rangle + \langle\psi_2|\psi_2\rangle - 2\mathrm{Re}\langle\psi_1|\psi_2\rangle$

$\Rightarrow$ Normalized state vector: $\langle\psi|\psi\rangle = 1$

Schwarz inequality: $|\langle\psi_1|\psi_2\rangle|^2 \leq \underbrace{\langle\psi_1|\psi_1\rangle}_{1} \underbrace{\langle\psi_2|\psi_2\rangle}_{1}$

Why we need scalar product:

Outcomes of measurements on a quantum system are in general **indeterministic** (described in the probabilistic way, see Sec. 1.2). A single measurement does not allow one to uniquely determine the state. Quantum **amplitude** & **probability** to identify state $|\psi_2\rangle$ with $|\psi_1\rangle$ or vice versa (for $||\psi_1||=||\psi_2||=1$) in an "optimal" single measurement: $\boxed{\underbrace{A_{\psi_2}(\psi_1) \equiv \langle\psi_1|\psi_2\rangle}_{\text{amplitude}} \quad \underbrace{P_{\psi_2}(\psi_1) \equiv |\langle\psi_1|\psi_2\rangle|^2}_{\text{probability}}}$

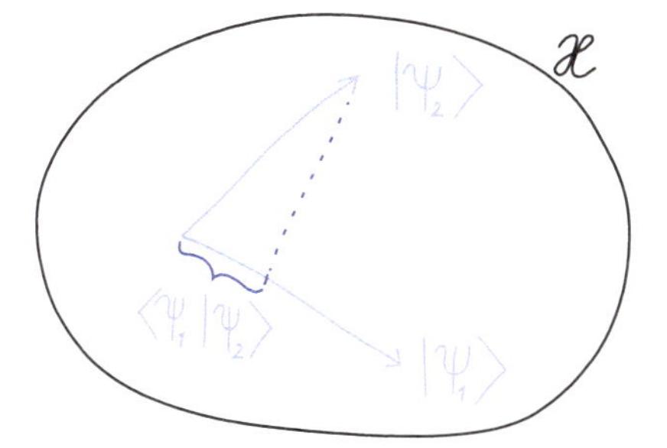

Consequence: States $|\psi_1\rangle, |\psi_2\rangle$ are perfectly distinguishable *iff* orthogonal

General QM terminology:

amplitude $\quad A \in \mathbb{C}$

probability $\quad |A|^2 \equiv P \in [0, 1]$

▶ **Requirement 3**: $\mathcal{H}$ is *complete* (for "security" reasons)

$\forall$ converging (in the sense of distance d) sequence of vectors the limit $\in \mathcal{H}$

▶ 1)+2)+3)$\Rightarrow$ **Postulate**: space of physical states $\mathcal{H}$ = **Hilbert space**

▶ $\mathcal{H}$ is **separable** if $\exists$ countable (sometimes finite) basis of vectors

Systems with finite particle numbers, subspaces of selected degrees of freedom

$\{|\phi_i\rangle\}_i \equiv$ an orthonormal basis $\langle\phi_i|\phi_j\rangle = \delta_{ij} \quad \Rightarrow$

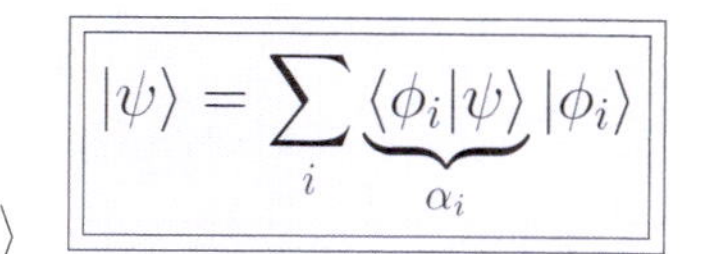

$$|\psi\rangle = \sum_i \underbrace{\langle\phi_i|\psi\rangle}_{\alpha_i} |\phi_i\rangle$$

Each state $|\psi\rangle$ can be expressed as a complex superposition of an enumerable set of basis states $|\phi_i\rangle$

▶ $\mathcal{H}$ is **nonseparable** if it has *no countable basis*

Systems with unbounded particle numbers, quantum fields, continuum

▶ **Any separable $\mathcal{H}$ is isomorphic with $\boldsymbol{\ell}^2$**

Definition of the $\boldsymbol{\ell}^2$ **space**: infinite "columns" $|\psi\rangle \equiv \begin{pmatrix} \alpha_1 \\ \alpha_2 \\ \vdots \end{pmatrix}$ with $\sum_{i=1}^{\infty} |\alpha_i|^2 < \infty$

Mapping $\mathcal{H} \to \ell^2$: components α_i associated with expansion coefficients $\langle\phi_i|\psi\rangle$ of $|\psi\rangle$ in a given basis

Superpositions $a|\psi\rangle + b|\psi'\rangle$ mapped onto $\begin{pmatrix} a\alpha_1 + b\alpha'_1 \\ a\alpha_2 + b\alpha'_2 \\ \vdots \end{pmatrix}$

Scalar product represented by: $\langle\psi|\psi'\rangle \equiv \sum_i \alpha_i^* \alpha'_i = (\alpha_1^*, \alpha_2^*, \ldots) \begin{pmatrix} \alpha'_1 \\ \alpha'_2 \\ \vdots \end{pmatrix}$

◀ **Historical remark**
1900-10: David Hilbert (with E. Schmidt) introduces the ∞-dimensional space of square-integrable functions and elaborates the theory of such spaces
1927: John von Neumann (working under Hilbert) introduces abstract Hilbert spaces into QM (1932: book *Mathematische Grundlagen der Quantenmechanik*)

■ Rigged Hilbert space

Although the standard Hilbert space is sufficient for consistent formulation of QM, we will see soon (Sec. 2.1) that its suitable extension is very helpful.

▶ Hierarchy of spaces based on $\mathcal{H} \equiv \ell^2$

$\underline{\mathcal{H}} \equiv$ sequences $|\psi\rangle$ with $\sum_i |\alpha_i|^2 i^m < \infty$ for $m = 0, 1, 2, \ldots$ (dense subset of ℓ^2)

$\overline{\mathcal{H}}$ (conjugate space to $\underline{\mathcal{H}}$) $\equiv$ sequences $|\psi\rangle$ for which $\langle\psi'|\psi\rangle < \infty \ \forall |\psi'\rangle \in \underline{\mathcal{H}}$

$\Rightarrow \sum_i \alpha_i'^* \alpha_i < \infty \Rightarrow \sum_i |\alpha_i|^2 \frac{1}{i^m} < \infty \Rightarrow |\alpha_i|^2$ may polynomially diverge

These are linear vector spaces but not Hilbert spaces:

$\underline{\mathcal{H}}$ is not complete

$\overline{\mathcal{H}}$ does not have scalar product

The smaller is $\underline{\mathcal{H}}$, the larger is $\overline{\mathcal{H}}$

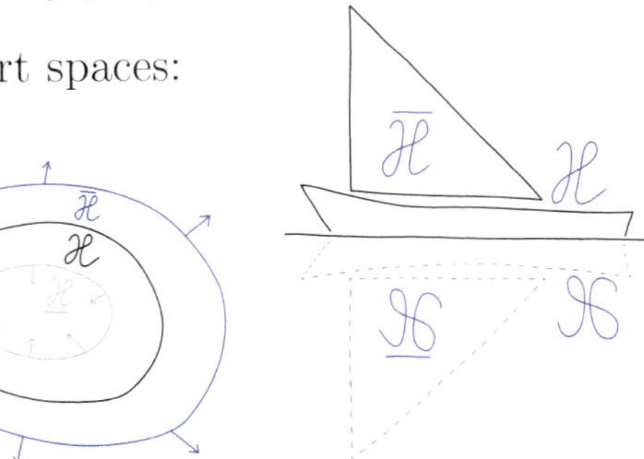

▶ **Gelfand triple** ("sandwich")

$\boxed{\underline{\mathcal{H}} \subset \mathcal{H} \subset \overline{\mathcal{H}}}$ $\equiv$ "rigged Hilbert space"

It turns out that solutions of some basic quantum problems $\notin \mathcal{H}$ but $\in \overline{\mathcal{H}}$, while the definition domain of some quantum operators is not $\mathcal{H}$ but $\underline{\mathcal{H}}$

■ Dirac notation

Physicists are proud to master a symbolic technique that makes some involved mathematical reductions much easier to follow. Although the "bra-ket" formalism is not always fully rigorous, it is extremely efficient especially when dealing with the action of linear operators in Hilbert spaces.

▶ Kets & bras

$\forall$ vector $|\psi\rangle$, called "**ket**", there $\exists$ a linear functional $F_\psi \equiv \langle\psi|$, called "**bra**", such that the value assigned to a vector $|\phi\rangle$ is: $\boxed{F_\psi(\phi) \equiv \langle\psi|\phi\rangle}$ ("bra-c-ket")

Superposition principle for bras: $\alpha\langle\psi_1|+\beta\langle\psi_2| \equiv \langle\alpha^*\psi_1+\beta^*\psi_2|$

The space of bras is isomorphic with the space of kets $\equiv \mathcal{H}$

Matrix form: $\langle\psi| \equiv (\,\alpha_1^*, \alpha_2^*, \ldots\,)$ $\quad \begin{pmatrix}\alpha_1\\ \alpha_2\\ \vdots\end{pmatrix} \equiv |\psi\rangle$

▶ Linear operators

Linear operators play a very important role in QM. They will be subject to systematic study from Sec. 1.2. Here we just introduce basic notions.

Linear operator $\boxed{\hat{O}|\psi\rangle = |\psi'\rangle}$ $\equiv$ mapping $\mathcal{H}\to\mathcal{H}$ such that:

$$\hat{O}(\alpha|\psi_1\rangle+\beta|\psi_2\rangle) = \alpha|\psi_1'\rangle+\beta|\psi_2'\rangle$$

$\Rightarrow \hat{O}$ defined through its action on any basis: $\{|\phi_i\rangle\}_{i=1}^{d_\mathcal{H}} \xrightarrow{\hat{O}} \{|\phi_i'\rangle\}_{i=1}^{d_\mathcal{H}}$

$$\Rightarrow \hat{O}|\psi\rangle = \sum_{i=1}^{d_\mathcal{H}} \underbrace{\langle\phi_i|\psi\rangle}_{\alpha_i}\underbrace{\hat{O}|\phi_i\rangle}_{|\phi_i'\rangle} = \sum_{i=1}^{d_\mathcal{H}} |\phi_i'\rangle\langle\phi_i|\psi\rangle \quad \Rightarrow \quad \boxed{\hat{O} \equiv \sum_{i=1}^{d_\mathcal{H}} |\phi_i'\rangle\langle\phi_i|}$$

Each term $|\phi_i'\rangle\langle\phi_i|$ represents action of $\hat{O}$ on $|\phi_i\rangle$

Expression via $|\phi_j\rangle\langle\phi_i|$ obtained by using the $\{|\phi_i\rangle\}\xrightarrow{\hat{I}}\{|\phi_i\rangle\}$ operator:

$$\boxed{\sum_{i=1}^{d_\mathcal{H}} |\phi_i\rangle\langle\phi_i| = \hat{I}} \equiv \text{unity operator} \Rightarrow \boxed{\underbrace{\hat{O}}_{\hat{I}\hat{O}\hat{I}} = \sum_{i=1}^{d_\mathcal{H}}\sum_{j=1}^{d_\mathcal{H}} \underbrace{\langle\phi_j|\hat{O}\phi_i\rangle}_{\langle\phi_j|\hat{O}|\phi_i\rangle\equiv O_{ji}} |\phi_j\rangle\langle\phi_i|}$$

Matrix form: $\hat{O} \equiv \begin{pmatrix} O_{11} & O_{12} & \cdots \\ O_{21} & O_{22} & \\ \vdots & & \ddots \end{pmatrix}$

▶ Projectors

Linear operators satisfying $\boxed{\hat{P}^2 = \hat{P}}$ (repeated projection is redundant)

Let $\{|\phi_i\rangle\}_{i=1}^{d_0} \equiv$ orthonormal basis of a subspace $\mathcal{H}_0 \subset \mathcal{H}$

$\langle\phi_i|\phi_j\rangle = \delta_{ij}$

$$\boxed{\hat{P}_0 = \sum_{i=1}^{d_0} |\phi_i\rangle\langle\phi_i|}$$ is a projector to $\mathcal{H}_0$:

$$\hat{P}_0|\psi\rangle \begin{cases} = 0 & \text{for } |\psi\rangle\perp\mathcal{H}_0 \\ \in \mathcal{H}_0 & \text{otherwise} \end{cases}$$

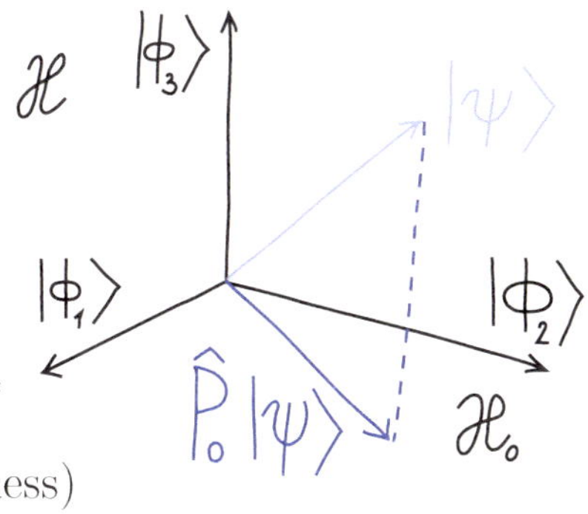

Projector to the whole $\mathcal{H}$ is $\hat{P}_\mathcal{H} = \sum_{i=1}^{d_\mathcal{H}} |\phi_i\rangle\langle\phi_i| = \hat{I}$ (completeness)

$\boxed{P_{\mathcal{H}_0}(\psi) \equiv \langle\psi|\hat{P}_0|\psi\rangle} = \sum_{i=1}^{d_0} |\langle\phi_i|\psi\rangle|^2 \equiv$ probability to identify the given state $|\psi\rangle$ with an arbitrary state $\in \mathcal{H}_0$ (cf. Sec. 2.1)

Matrix form: In an orthonormal basis such that $\{|\phi_i\rangle\}_{i=1}^{d_\mathcal{H}} \supset \{|\phi_i\rangle\}_{i=1}^{d_0}$ the projector expressed as a diagonal matrix:

$$\hat{P}_0 = \sum_i \begin{pmatrix}\vdots\\ \bullet_i\\ \vdots\end{pmatrix}(\cdots\ \bullet_i\ \cdots) = \begin{pmatrix}\bullet_1 & 0 & \cdots\\ 0 & \bullet_2 & \\ \vdots & & \ddots\end{pmatrix} \quad \text{with } \bullet_i = 0 \text{ or } 1$$

◀ **Historical remark**
1930: Paul Dirac writes the book *The Principles of Quantum Mechanics*, which provides a more intuitive (compared to von Neumann) path to quantum theory, using non-normalizable vectors and δ function (bra-kets in 3rd edition 1947)
1950-60's: I.M. Gelfand & N.Y. Vilenkin introduce rigged Hilbert spaces, putting Dirac's approach on more rigorous grounds. Systematic use in QM since 1966 (by A. Böhm et al.) but up to now rather scarce

■ Summing Hilbert spaces

One can combine one or more Hilbert spaces in the style of summation. The resulting space then contains the summed spaces as ordinary subspaces.

▶ Direct sum

Let $\{|\phi_{1i}\rangle\}_{i=1}^{d_1}$ be an orthonormal basis of $\mathcal{H}_1$ and $\{|\phi_{2j}\rangle\}_{j=1}^{d_2}$ one of $\mathcal{H}_2$

Direct sum $\boxed{\mathcal{H}=\mathcal{H}_1\oplus\mathcal{H}_2}$ is a space with the basis $\boxed{|\Phi_{ki}\rangle=\begin{cases}|\phi_{1i}\rangle \text{ for } k{=}1\\ |\phi_{2i}\rangle \text{ for } k{=}2\end{cases}}$

Dimension: $\boxed{d_{\mathcal{H}}=d_1+d_2}$ Orthonormality: $\langle\Phi_{ki}|\Phi_{k'i'}\rangle=\delta_{kk'}\delta_{ii'}$

$$\text{Any } |\Psi\rangle=\sum_{k,i}\alpha_{ki}|\Phi_{ki}\rangle\in\mathcal{H} \text{ is a sum } |\Psi\rangle=\underbrace{\sum_{i=1}^{d_1}\alpha_{1i}|\phi_{1i}\rangle}_{|\psi_1\rangle\equiv\hat{P}_1|\Psi\rangle\in\mathcal{H}_1}+\underbrace{\sum_{j=1}^{d_2}\alpha_{2j}|\phi_{2j}\rangle}_{|\psi_2\rangle\equiv\hat{P}_2|\Psi\rangle\in\mathcal{H}_2}$$

▶ Projector to the subspace $\mathcal{H}_k$ ($k=1,2$)

$$\hat{P}_k=\sum_i|\Phi_{ki}\rangle\langle\Phi_{ki}| \;\Rightarrow\; \begin{cases}\text{orthogonality}: & \hat{P}_1\hat{P}_2=\hat{P}_2\hat{P}_1=0\\ \text{completeness}: & \hat{P}_1+\hat{P}_2=\hat{I}_{\mathcal{H}}\end{cases}$$

▶ Scalar product: $\boxed{\langle\Psi|\Psi'\rangle_{\mathcal{H}}=\langle\psi_1|\psi_1'\rangle_{\mathcal{H}_1}+\langle\psi_2|\psi_2'\rangle_{\mathcal{H}_2}}$

▶ Finite-dim. representation: $|\psi_1\rangle=\begin{pmatrix}\alpha_{11}\\ \vdots\\ \alpha_{1d_1}\end{pmatrix}$, $|\psi_2\rangle=\begin{pmatrix}\alpha_{21}\\ \vdots\\ \alpha_{2d_2}\end{pmatrix}\Rightarrow|\Psi\rangle=\begin{pmatrix}\alpha_{11}\\ \vdots\\ \alpha_{1d_1}\\ \cdots\\ \alpha_{21}\\ \vdots\\ \alpha_{2d_2}\end{pmatrix}$

▶ Multiple sums: $\boxed{\mathcal{H}=\bigoplus_{k=1}^{n}\mathcal{H}_k}$

For instance, $\mathcal{H}_k$ = subspaces with different sharp values of a certain observable

■ Multiplying Hilbert spaces

Hilbert spaces can also be combined in the style of multiplication. This commonly happens in composite quantum systems which consist of distinct degrees of freedom. The multiplication is a rather interesting operation since it allows one to create physical states with no analogue in the classical world.

▶ Direct (tensor) product

Let $\{|\phi_{1i}\rangle\}_{i=1}^{d_1}$ be an orthonormal basis of $\mathcal{H}_1$ and $\{|\phi_{2j}\rangle\}_{j=1}^{d_2}$ one of $\mathcal{H}_2$

Tensor product $\boxed{\mathcal{H}=\mathcal{H}_1\otimes\mathcal{H}_2}$ has "dyadic product" basis $\boxed{|\Phi_{ij}\rangle\equiv|\phi_{1i}\rangle|\phi_{2j}\rangle}$

Note: non-product bases can also be constructed

Dimension: $\boxed{d_{\mathcal{H}} = d_1 \times d_2}$ Orthonormality: $\langle \Phi_{ij} | \Phi_{i'j'} \rangle = \delta_{ii'}\delta_{jj'}$

▶ **Factorized states**

$\forall$ pair $|\psi_1\rangle = \sum_i \alpha_i |\phi_{1i}\rangle \in \mathcal{H}_1$ and $|\psi_2\rangle = \sum_j \beta_j |\phi_{2j}\rangle \in \mathcal{H}_2$ there $\exists$ product state

$$|\Psi_\otimes\rangle \equiv \underbrace{|\psi_1\rangle \otimes |\psi_2\rangle}_{\equiv |\psi_1\rangle|\psi_2\rangle} = \sum_{i=1}^{d_1} \sum_{j=1}^{d_2} \underbrace{\alpha_i \beta_j}_{\gamma_{ij}} |\Phi_{ij}\rangle$$

For factorized states :
$\langle \Psi_\otimes | \Psi'_\otimes \rangle_{\mathcal{H}} = \langle \psi_1 | \psi'_1 \rangle_{\mathcal{H}_1} \times \langle \psi_2 | \psi'_2 \rangle_{\mathcal{H}_2}$

▶ **Entangled states**

Almost all states in $\mathcal{H}_1 \otimes \mathcal{H}_2$ are unfactorizable superpositions. Such states are called *entangled*.

$$|\Psi\rangle = \sum_{i=1}^{d_1} \sum_{j=1}^{d_2} \underbrace{\gamma_{ij}}_{\neq \alpha_i \beta_j} |\Phi_{ij}\rangle \neq |\psi_1\rangle|\psi_2\rangle$$

For entangled states :
$\langle \Psi | \Psi' \rangle_{\mathcal{H}} \neq \langle \psi_1 | \psi'_1 \rangle_{\mathcal{H}_1} \times \langle \psi_2 | \psi'_2 \rangle_{\mathcal{H}_2}$

▶ Multiple products of Hilbert spaces: $\boxed{\mathcal{H} = \bigotimes_{k=1}^{n} \mathcal{H}_k}$

▶ **The use in QM**

Hilbert space $\mathcal{H}$ of a composite system is the $\otimes$ product of partial spaces $\mathcal{H}_k$

$\mathcal{H}_k$ = spaces corresponding to different parts of the system (e.g. particles) or to different dynamical variables (e.g., spatial and spin degrees of freedom)
Entangled state vectors correspond to *non-classical* situations in which only the whole system and not its individual parts are attributed by a pure quantum-mechanical state (the subsystems are in mixed states, see Sec. 1.7). Entanglement represents a genuinely **quantum correlation** of the system's parts.

▶ More & less precise notations: $|\psi\rangle \in \mathcal{H}_k$ is denoted as $|\psi\rangle_k$

$$\bigotimes_{k=1}^{n} \mathcal{H}_k \ni |\psi_1\rangle_1 \otimes |\psi_2\rangle_2 \ldots \otimes |\psi_n\rangle_n \equiv |\psi_1\rangle_1 |\psi_2\rangle_2 \ldots |\psi_n\rangle_n \equiv |\psi_1\rangle|\psi_2\rangle \ldots |\psi_n\rangle$$

◀ **Historical remark**

1935: A. Einstein, B. Podolsky & N. Rosen use an entangled state to claim that QM is incomplete. E. Schrödinger analyzes such states and coins the term "entanglement"

2.1 Examples of quantum Hilbert spaces

In the following, we describe specific state spaces for particles with spin 0 and $\frac{1}{2}$, and the spaces assigned to collections of such particles. We will meet another essentially quantum phenomenon—indistinguishability of particles.

■ Single structureless and spinless particle

Particles with no internal degrees of freedom are described by ordinary scalar wavefunctions (cf. Introduction).

▶ Wavefunctions $|\psi\rangle \equiv \psi(\vec{x}) \in$ $\boxed{\mathcal{H} \equiv \mathcal{L}^2(\mathbb{R}^3)}$ $\quad \langle\psi| \equiv \psi^*(\vec{x})$

Scalar product: $$\boxed{\langle\psi_1|\psi_2\rangle \equiv \underbrace{\int \psi_1^*(\vec{x})\psi_2(\vec{x})\,d\vec{x}}_{\text{Cartesian}} \equiv \underbrace{\int \psi_1^*(\vec{y})\psi_2(\vec{y}) \left|\text{Det}\frac{\partial(x_1\ldots x_3)}{\partial(y_1\ldots y_3)}\right| d\vec{y}}_{\text{curvilinear coordinates}}}$$

Expansion of $\psi(\vec{x})$ in any discrete basis of orthonormal functions $\phi_i(\vec{x})$
$\Rightarrow$ isomorphism with ℓ^2

▶ Rigged Hilbert space of wavefunctions

Localized states $\delta(\vec{x}-\vec{x}\,') \notin \mathcal{L}^2(\mathbb{R}^3)$ and plane waves $e^{i\vec{k}\cdot\vec{x}} \notin \mathcal{L}^2(\mathbb{R}^3)$

Define a triple $\underline{\mathcal{H}} \subset \mathcal{H} \subset \overline{\mathcal{H}}$ with (in 1D case)

$\underline{\mathcal{H}} \equiv$ dense subset of functions: $\int_{-\infty}^{+\infty} |\psi(x)|^2(1+|x|)^m dx < \infty$ for $m = 0, 1, 2, \ldots$

$\overline{\mathcal{H}} \equiv$ functions satisfying $\int_{-\infty}^{+\infty} \psi'^*\psi\,dx < \infty\ \forall\psi' \in \underline{\mathcal{H}}$ (includes also polynomially diverging functions, plane waves and δ-functions)

◀ **Historical remark**

1926: Erwin Schrödiger formulates QM in terms of wavefunction and Max Born develops its probabilistic interpretation

■ Single structureless particle with spin $\frac{1}{2}$

Electrons, e.g., are particles with spin $\frac{1}{2}$. Their state space is formed by spinors, which represent the simplest generalization of scalar wavefunctions.

▶ **Spin = internal angular momentum** of a particle. For elementary (point-like) particles, it is a genuinely quantum property (general description of angular momentum in QM will be developed in Secs. 2.2, 2.3, and 4)

▶ The lowest nonzero spin is denoted as $\frac{1}{2}$ and has only 2 possible projections (spin states) in any spatial direction (conventionally direction z):

spin up $\quad s_z = +\frac{1}{2}\hbar \;\Rightarrow\; |\uparrow\rangle \equiv \binom{1}{0}$
spin down $\quad s_z = -\frac{1}{2}\hbar \;\Rightarrow\; |\downarrow\rangle \equiv \binom{0}{1}$
$\Rightarrow$ **general state** $\boxed{|\psi\rangle = \alpha|\uparrow\rangle + \beta|\downarrow\rangle \equiv \binom{\alpha}{\beta}}$

Spin Hilbert space $\boxed{\mathcal{H} \equiv \mathbb{C}^2}$ with $\langle\psi_1|\psi_2\rangle \equiv (\alpha_1^*, \beta_1^*)\binom{\alpha_2}{\beta_2} = \alpha_1^*\alpha_2 + \beta_1^*\beta_2$

▶ Combining spin with the spatial degrees of freedom:
direct product of "spatial" and "spin" Hilbert spaces: $\boxed{\mathcal{H} \equiv \mathcal{L}^2(\mathbb{R}^3) \otimes \mathbb{C}^2}$

Expansion of a general state: $|\psi\rangle = \sum_i \left[\alpha_i\phi_i(\vec{x})|\uparrow\rangle + \beta_i\phi_i(\vec{x})|\downarrow\rangle\right]$

$$= \sum_i \binom{\alpha_i}{\beta_i}\phi_i(\vec{x}) = \begin{pmatrix} \sum_i \alpha_i\phi_i(\vec{x}) \\ \sum_i \beta_i\phi_i(\vec{x}) \end{pmatrix} = \boxed{\begin{pmatrix} \psi_\uparrow(\vec{x}) \\ \psi_\downarrow(\vec{x}) \end{pmatrix} \equiv \boldsymbol{\Psi}(\vec{x}) \equiv \psi(\vec{x}, \underbrace{m_s}_{\pm\frac{1}{2}})} \quad \textbf{spinor}$$

Spinor ≡ two-component wavefunction ≡ wavefunction with continuous + discrete 2-valued variables (transformation properties of spinors under spatial rotations are different from ordinary vectors and will be derived later, see p. 60)

► Scalar product: $$\langle\psi|\psi'\rangle \equiv \int (\psi^*_\uparrow(\vec{x}),\psi^*_\downarrow(\vec{x})) \begin{pmatrix}\psi'_\uparrow(\vec{x})\\ \psi'_\downarrow(\vec{x})\end{pmatrix} d\vec{x} = \sum_{m_s}\int \Psi^*(\vec{x},m_s)\Psi'(\vec{x},m_s)d\vec{x}$$

◄ **Historical remark**

1922: O. Stern & W. Gerlach observe the first indication of spin

1924: Wolfgang Pauli introduces "two-valued quantum degree of freedom" and formulates the exclusion principle (see below), in 1927 he introduces spinors

1925: R. Kronig and G. Uhlenbeck & S. Goudsmit provide an interpretation of spin in terms of intrinsic rotation (refused at that time)

■ Two or more distinguishable structureless particles with spin $\frac{1}{2}$

We are ready to construct state spaces for collections of particles. At first we assume that the particles are of different types—*distinguishable*. We assume particles with spin $\frac{1}{2}$, but the same procedure can be applied regardless of spin.

► $\mathcal{H}_1, \mathcal{H}_2, \dots \mathcal{H}_N$= Hilbert spaces of individual particles: $\mathcal{H}_i = \mathcal{L}^2(\mathbb{R}^3)\otimes\mathbb{C}^2$

$$\mathcal{H}^{(N)} \equiv \mathcal{H}_1 \otimes \mathcal{H}_2 \otimes \dots \otimes \mathcal{H}_N$$

Wavefunction $\Psi(\underbrace{\vec{x}_1,m_1}_{\xi_1},\underbrace{\vec{x}_2,m_2}_{\xi_2},\dots\underbrace{\vec{x}_N,m_N}_{\xi_N})$

Scalar product

$$\langle\psi|\psi'\rangle \equiv \sum_{m_1}\cdots\sum_{m_N}\int d\vec{x}_1\dots\int d\vec{x}_N\ \Psi^*(\vec{x}_1,m_1,\dots\vec{x}_N,m_N)\Psi'(\vec{x}_1,m_1,\dots\vec{x}_N,m_N)$$

► **Probability expressions**

The wavefunction $\Psi(\boldsymbol{\xi}_1\dots\boldsymbol{\xi}_N)$ lives in the multidimensional configuration space containing all generalized coordinates $\boldsymbol{\xi}_i \equiv (\vec{x}_i, m_i)$ of individual particles. It contains all mutual correlations between the particles and allows one to extract two extremal types of probability distributions:

(a) **Joint probability** density to find particle #1 at $\boldsymbol{\xi}_1$ ... particle #N at $\boldsymbol{\xi}_N$

$$\rho(\boldsymbol{\xi}_1\dots\boldsymbol{\xi}_N) \equiv |\Psi(\boldsymbol{\xi}_1\dots\boldsymbol{\xi}_N)|^2$$ (contains all particle correlations)

Normalization: $\int\dots\int \rho(\boldsymbol{\xi}_1\dots\boldsymbol{\xi}_N)\,d\boldsymbol{\xi}_1\dots d\boldsymbol{\xi}_N = 1$

(b) **Integrated probability** density to find *any* of particles at $\boldsymbol{\xi}$

$$\rho(\boldsymbol{\xi}) = \frac{1}{N}\sum_{i=1}^{N}\underbrace{\int\dots\int}_{N-1}|\Psi(\boldsymbol{\xi}_1\dots\boldsymbol{\xi}_{i-1}\underbrace{\boldsymbol{\xi}}_{i}\boldsymbol{\xi}_{i+1}\dots\boldsymbol{\xi}_N)|^2\,d\boldsymbol{\xi}_1\dots d\boldsymbol{\xi}_{i-1}d\boldsymbol{\xi}_{i+1}\dots d\boldsymbol{\xi}_N$$

Normalization $\int\rho(\boldsymbol{\xi})\,d\boldsymbol{\xi} = 1$

Two indistinguishable particles

We are coming to the problem of indistinguishable particles. In quantum mechanics, if two particles are the same, there exists really *no way* to distinguish them. One cannot, for instance, think on virtual numbers associated with them. We start the analysis with the case of just two indistinguishable particles.

▶ Two distinguishable particles: $\Psi(\boldsymbol{\xi}_1,\boldsymbol{\xi}_2)\equiv|\Psi\rangle\in\mathcal{H}^{(2)}\equiv\mathcal{H}_1\otimes\mathcal{H}_2$

Introduce particle exchange operator: $\boxed{\hat{E}_{1\rightleftharpoons 2}\Psi(\boldsymbol{\xi}_1,\boldsymbol{\xi}_2)=\Psi(\boldsymbol{\xi}_2,\boldsymbol{\xi}_1)}$ with $\hat{E}^2_{1\rightleftharpoons 2}=\hat{I}$, i.e., in Dirac notation: $|\Psi\rangle=\sum_{ij}\alpha_{ij}|\phi_i\rangle_1|\phi_j\rangle_2\Rightarrow\hat{E}_{1\rightleftharpoons 2}|\Psi\rangle=\sum_{ij}\alpha_{ij}|\phi_j\rangle_1|\phi_i\rangle_2$

▶ For indistinguishable particles we require that the exchange only affects the phase: $\boxed{\hat{E}_{1\rightleftharpoons 2}|\Psi\rangle=e^{i\varphi}|\Psi\rangle}$ and that two subsequent exchanges yield the original state: $e^{2i\varphi}=1$

$$\Rightarrow\begin{cases}\textbf{bosons} & \varphi=0 \\ \textbf{fermions} & \varphi=\pi\end{cases}\quad\boxed{\begin{array}{rcl}\Psi(\boldsymbol{\xi}_1,\boldsymbol{\xi}_2)&=&+\Psi(\boldsymbol{\xi}_2,\boldsymbol{\xi}_1)\\ \Psi(\boldsymbol{\xi}_1,\boldsymbol{\xi}_2)&=&-\Psi(\boldsymbol{\xi}_2,\boldsymbol{\xi}_1)\end{array}}\quad\begin{array}{l}\textbf{symmetric}\\ \textbf{antisymmetric}\end{array}$$

▶ Any 2-body wavefunction decomposed into symmetric & antisymmetric parts

$$\Psi(\boldsymbol{\xi}_1,\boldsymbol{\xi}_2)=\underbrace{\tfrac{1}{2}\left[\Psi(\boldsymbol{\xi}_1,\boldsymbol{\xi}_2)+\Psi(\boldsymbol{\xi}_2,\boldsymbol{\xi}_1)\right]}_{\hat{P}_+\Psi(\boldsymbol{\xi}_1,\boldsymbol{\xi}_2)}+\underbrace{\tfrac{1}{2}\left[\Psi(\boldsymbol{\xi}_1,\boldsymbol{\xi}_2)-\Psi(\boldsymbol{\xi}_2,\boldsymbol{\xi}_1)\right]}_{\hat{P}_-\Psi(\boldsymbol{\xi}_1,\boldsymbol{\xi}_2)}\qquad \hat{P}_\pm=\tfrac{1}{2}[\hat{I}\pm\hat{E}_{1\rightleftharpoons 2}]$$

$\hat{P}_+$ and $\hat{P}_-$ = projectors to the symmetric and antisymmetric subspaces

$$\hat{P}_++\hat{P}_-=\hat{I}\quad\Rightarrow\quad\boxed{\mathcal{H}^{(2)}=\mathcal{H}^{(2)}_+\oplus\mathcal{H}^{(2)}_-}$$

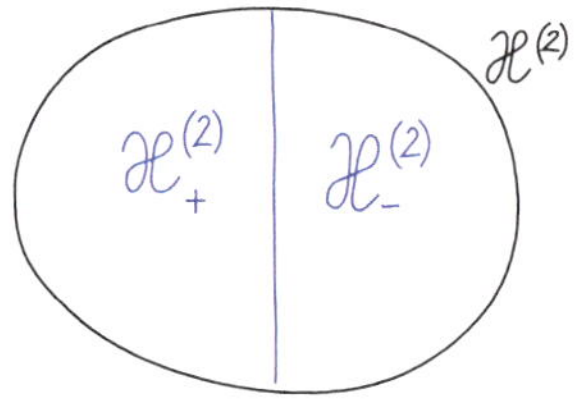

Dirac notation: $\hat{P}_\pm|\Psi\rangle=\sum_{ij}\alpha_{ij}\frac{1}{2}\left[|\phi_i\rangle_1|\phi_j\rangle_2\pm|\phi_j\rangle_1|\phi_i\rangle_2\right]$

▶ **Pauli principle**: $\boxed{\hat{P}_-|\psi\rangle_1|\psi\rangle_2=0}$

⇒ Two/more fermions cannot occur in the same single-particle state. Each such a state can be occupied at most by one fermion. This has tremendous consequences for the structure of matter! "Without Pauli principle, the world would be a boring place" (probably with no bored creature present).

▶ **Interference effects caused by indistinguishability**

Two distinguishable particles in a separable state: $\Psi(\boldsymbol{\xi}_1,\boldsymbol{\xi}_2)=\psi_1(\boldsymbol{\xi}_1)\psi_2(\boldsymbol{\xi}_2)$
Joint prob.density: $\rho(\boldsymbol{\xi}_1,\boldsymbol{\xi}_2)=\rho_1(\boldsymbol{\xi}_1)\rho_2(\boldsymbol{\xi}_2)$
Integrated prob.density: $\rho(\boldsymbol{\xi})=\frac{1}{2}\left[\rho_1(\boldsymbol{\xi})+\rho_2(\boldsymbol{\xi})\right]$ ⇒ no interference

Indistinguishable particles: $\hat{P}_\pm\Psi(\boldsymbol{\xi}_1,\boldsymbol{\xi}_2)\propto[\psi_1(\boldsymbol{\xi}_1)\psi_2(\boldsymbol{\xi}_2)\pm\psi_1(\boldsymbol{\xi}_2)\psi_2(\boldsymbol{\xi}_1)]$
Joint: $\rho(\boldsymbol{\xi}_1,\boldsymbol{\xi}_2)\propto\rho_1(\boldsymbol{\xi}_1)\rho_2(\boldsymbol{\xi}_2)+\rho_1(\boldsymbol{\xi}_2)\rho_2(\boldsymbol{\xi}_1)\pm 2\mathrm{Re}[\psi_1(\boldsymbol{\xi}_1)\psi_2^*(\boldsymbol{\xi}_1)\psi_1^*(\boldsymbol{\xi}_2)\psi_2(\boldsymbol{\xi}_2)]$
Integrated prob.dens.: $\rho(\boldsymbol{\xi})\propto\rho_1(\boldsymbol{\xi})+\rho_2(\boldsymbol{\xi})\pm 2\mathrm{Re}\left[\langle\psi_1|\psi_2\rangle\psi_1^*(\boldsymbol{\xi})\psi_2(\boldsymbol{\xi})\right]$

The indistinguishability of particles makes the state $\hat{P}_{\pm}\Psi(\boldsymbol{\xi}_1,\boldsymbol{\xi}_2)$ entangled! It may create interference terms in both probability densities $\rho(\boldsymbol{\xi}_1,\boldsymbol{\xi}_2)$ and $\rho(\boldsymbol{\xi})$, but only if the states $\psi_1(\boldsymbol{\xi})$ and $\psi_2(\boldsymbol{\xi})$ have a nonzero overlap ($\Rightarrow$ no interference effects e.g. for very distant particles or for particles with opposite spins).

■ Many indistinguishable particles

It is straightforward (but a bit more laborious) to generalize the above results to $N>2$ indistinguishable particles. Particle permutations are decomposed into pairwise exchanges, the states of identical bosons (fermions) being identified with symmetric (antisymmetric) subspaces with respect to these exchanges. A general theory of bosonic & fermionic systems will be elaborated in Chapter 7.

► N distinguishable particles: $\Psi(\boldsymbol{\xi}_1,\ldots\boldsymbol{\xi}_N)\equiv|\Psi\rangle\in\mathcal{H}^{(N)}\equiv\otimes_{k=1}^{N}\mathcal{H}_k$

Basis: $|\phi_{i_1}\rangle_1|\phi_{i_2}\rangle_2\ldots|\phi_{i_N}\rangle_N\equiv|\Phi_{i_1i_2\ldots i_N}\rangle$ with $i_k=1,2,3,\ldots$

► **Particle exchange operators**: $\boxed{\hat{E}_{k\rightleftharpoons l}|\Phi_{i_1\ldots i_k\ldots i_l\ldots i_N}\rangle=|\Phi_{i_1\ldots i_l\ldots i_k\ldots i_N}\rangle}$

Particle permutations: $(1,2,\ldots N)\mapsto(k_1^{\pi},k_2^{\pi},\ldots k_N^{\pi})$ with $\pi=1,\ldots N!$

Permutation operators: $\boxed{\hat{\mathcal{E}}_{\pi}|\Phi_{i_1i_2\ldots i_N}\rangle=|\Phi_{i_{k_1^{\pi}}i_{k_2^{\pi}}\ldots i_{k_N^{\pi}}}\rangle}$

$\hat{\mathcal{E}}_{\pi}=$ products of $\hat{E}_{k\rightleftharpoons l}\Rightarrow$ odd/even number of factors $\Rightarrow$ odd/even permutation

Permutation sign $\sigma_{\pi}=\begin{cases}+ & \text{for even permutation}\\ - & \text{for odd permutation}\end{cases}$

► **Hilbert space decomposition**

$$\boxed{\mathcal{H}^{(N)}=\mathcal{H}_+^{(N)}\oplus\cdots\oplus\mathcal{H}_-^{(N)}}$$

$\mathcal{H}^{(N)}$ $\mathcal{H}_+^{(N)}$ $\mathcal{H}_-^{(N)}$

where the fully symmetric (+) and fully antisymmetric (−) subspaces satisfy:

$\boxed{\hat{\mathcal{E}}_{\pi}|\Psi\rangle=+|\Psi\rangle}\ \forall|\Psi\rangle\in\mathcal{H}_+^{(N)}$ and $\boxed{\hat{\mathcal{E}}_{\pi}|\Psi\rangle=\sigma_{\pi}|\Psi\rangle}\ \forall|\Psi\rangle\in\mathcal{H}_-^{(N)}$

Postulate: Hilbert space for N identical particles is either $\mathcal{H}_+^{(N)}$ (for **bosons**) or $\mathcal{H}_-^{(N)}$ (for **fermions**)

$$\hat{P}_+=\frac{1}{N!}\sum_{\pi=1}^{N!}\hat{\mathcal{E}}_{\pi}\qquad\hat{P}_-=\frac{1}{N!}\sum_{\pi=1}^{N!}\sigma_{\pi}\hat{\mathcal{E}}_{\pi}$$

projectors $\hat{P}_{\pm}^2=\hat{P}_{\pm}$

$\to\mathcal{H}_+^{(N)}$ $\hat{\mathcal{E}}_{\pi}\hat{P}_+|\Psi\rangle=\hat{P}_+|\Psi\rangle$

$\to\mathcal{H}_-^{(N)}$ $\hat{\mathcal{E}}_{\pi}\hat{P}_-|\Psi\rangle=\sigma_{\pi}\hat{P}_-|\Psi\rangle$

$\hat{P}_++\hat{P}_-\neq\hat{I}$: the rest of the space, $(\hat{I}-\hat{P}_+-\hat{P}_-)\mathcal{H}$, contains mixed symmetry subspaces (corresponding e.g. to mixtures of several types of identical particles)

► Expression of a basis in the **fermionic** space through **Slater determinant**:

$$\hat{P}_-\Big[\underbrace{|\phi_1\rangle_1|\phi_2\rangle_2\ldots|\phi_N\rangle_N}_{|\Phi_{12\ldots N}\rangle}\Big]=\frac{1}{N!}\,\mathrm{Det}\begin{pmatrix}|\phi_1\rangle_1 & |\phi_1\rangle_2 & \ldots & |\phi_1\rangle_N\\ |\phi_2\rangle_1 & |\phi_2\rangle_2 & \ldots & |\phi_2\rangle_N\\ \vdots & & \ldots & \vdots\\ |\phi_N\rangle_1 & |\phi_N\rangle_2 & \ldots & |\phi_N\rangle_N\end{pmatrix}$$

Analogous expression (but symmetrized, $\neq$Det) can be given for bosons (Sec. 7.1)

Slater-determinant (or symmetrized) states originate from *separable* states in the space of distinguishable particles, therefore they carry just a minimal unavoidable entanglement caused by indistinguishability of particles. Slater-det. (or symmetrized) states form a **basis** in $\mathcal{H}_-^{(N)}$ (or $\mathcal{H}_+^{(N)}$) $\Rightarrow$ general N-body fermionic (or bosonic) state can be expressed as a *superposition of such states.*

◀ **Historical remark**

1924: S.N. Bose derives Planck blackbody law from indistinguishability of photons
1924: Wolfgang Pauli formulates the exclusion principle to explain periodic table
1926: Werner Heisenberg and Paul Dirac relate Pauli principle to antisymmetric wavefunctions and Bose-Einstein statistics to symmetric wavefunctions. Dirac and Enrico Fermi derive statistical law for "fermions"
1927: D. Hartree & Vladimir Fock derive approximation for atomic N-electron wavefunctions, in 1929 J. Slater facilitates the description by using the determinant
1939-50: M. Fierz, W. Pauli, J. Schwinger provide proofs (within the relativistic quantum theory) of the general theorem relating the "type of statistics" to spin

■ Systems with unbounded number of particles

At last, we come to the case in which the particle number is not fixed. Indeed, if special relativity is taken into account, particles can be repeatedly created and annihilated, conserving the total energy $\Leftrightarrow$ mass of the system. It turns out that with no upper bound on the particle number we leave the safe harbor of separable Hilbert spaces and face the limitless ocean of continuum. This is a transition to the field theory. Work with the Fock space within the nonrelativistic QM will be practiced in Sec. 6.

▶ Fock space

Sum of spaces for all particle numbers $N = 0, 1, 2, 3, \dots$

$$\mathcal{H} \equiv \underbrace{\mathcal{H}^{(0)}}_{\text{vacuum }|0\rangle} \oplus \underbrace{\mathcal{H}^{(1)}}_{\text{1 particle}} \oplus \underbrace{\mathcal{H}_\bullet^{(2)}}_{\text{2 particles}} \cdots \oplus \underbrace{\mathcal{H}_\bullet^{(N)}}_{N\text{ particles}} \oplus \cdots\cdots$$

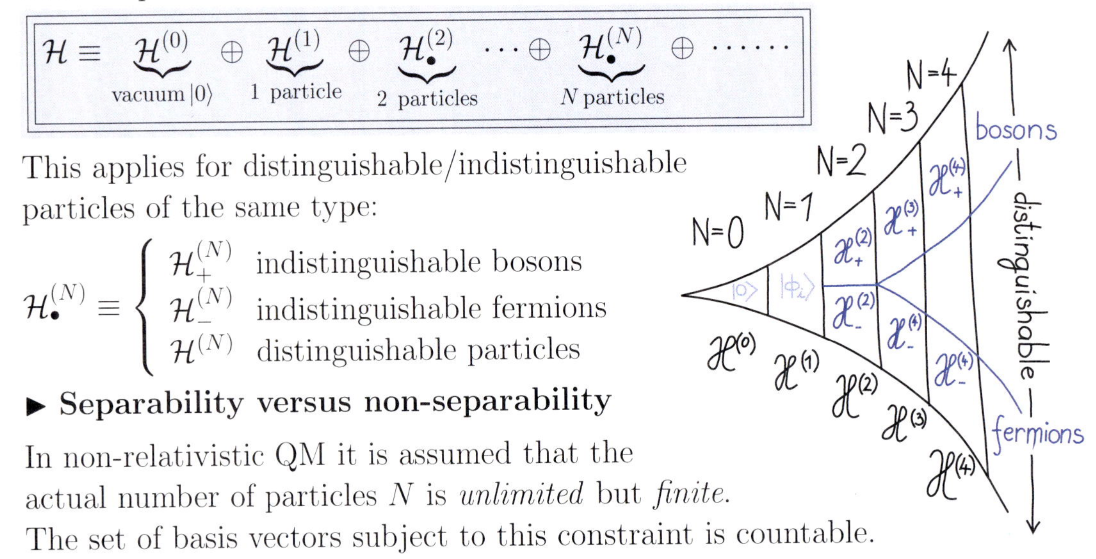

This applies for distinguishable/indistinguishable particles of the same type:

$$\mathcal{H}_\bullet^{(N)} \equiv \begin{cases} \mathcal{H}_+^{(N)} & \text{indistinguishable bosons} \\ \mathcal{H}_-^{(N)} & \text{indistinguishable fermions} \\ \mathcal{H}^{(N)} & \text{distinguishable particles} \end{cases}$$

▶ Separability versus non-separability

In non-relativistic QM it is assumed that the actual number of particles N is *unlimited* but *finite*.
The set of basis vectors subject to this constraint is countable.

Under this assumption, the Fock space is **separable**. However, the *closure* of the Fock space including $\mathcal{H}_{\bullet}^{(\infty)}$ is **non-separable**. Rasoning: basis states $|\Phi_{i_1 i_2 \ldots}\rangle \equiv |\phi_{i_1}\rangle_1 |\phi_{i_2}\rangle_2 \ldots$ for $N=\infty$ are specified by an infinite number of integer indices $i_1, i_2, \ldots$ identifying basis states of individual particles. This set is uncountable for the same reason why real numbers (infinite sequences of digits) are uncountable (see Cantor's "diagonal slash" argument).

◀ **Historical remark**

1932: V. Fock introduced the space for indefinite particle number

1958: Paul Dirac relates the Fock space to field quantization & continuum problems

1.2 Representation of observables

Our next task is to let observables into the Hilbert space and to predict results of actual measurements. In classical mechanics, observables were just ordinary functions on the phase space. In quantum mechanics, the thing is more complicated since—as we know from experiments—many observables yield discrete values and results of measurements are generally *indeterministic*. We need a mathematical tool capable to cope with these unusual properties.

■ Operators associated with observables

Associated with each state vector $|\psi\rangle \in \mathcal{H}$ there must be a *probability distribution* $p_\psi(a)$ characterizing all possible measurement outputs $\{a\}$ of any physical quantity A. A suitable path to obtain such distributions proceeds via the association of each quantity A with an operator $\hat{A}$, which represents a specific mapping $\mathcal{H} \mapsto \mathcal{H}$. We first present a plausible (but not unique) motivation for launching out in this direction and then briefly outline some subtleties of the operator theory that will be needed for mastering the QM formalism.

▶ Moments of statistical distribution

Observable $A \to$ values a (potential measurement outcomes)

Measurement of A on system in state $|\psi\rangle \to$ **probability distribution** $p_\psi(a)$ of outcomes, which is uniquely associated with **statistical moments**

$$\langle A^1\rangle_\psi,\ \langle A^2\rangle_\psi,\ \langle A^3\rangle_\psi,\ \ldots \qquad \boxed{\langle A^n\rangle_\psi \equiv \int a^n p_\psi(a)\, da}$$

▶ Calculation of moments via operators

Postulate: Observable A is associated with an operator $\hat{A}$ acting on $\mathcal{H}$

$$\boxed{\hat{A}|\psi\rangle \equiv |\hat{A}\psi\rangle} \equiv |\psi'\rangle \in \mathcal{H} \qquad \text{Powers of operator: } \hat{A}^n|\psi\rangle \equiv \underbrace{\hat{A}\hat{A}\ldots\hat{A}}_{n\,\text{times}}|\psi\rangle \equiv |\hat{A}^n\psi\rangle \in \mathcal{H}$$

Moments of $p_\psi(a)$ calculated as $\boxed{\langle A^n\rangle_\psi = \langle\psi|\hat{A}^n\psi\rangle}$

▸ **Requirements** upon $\hat{A}$ (should be considered as a part of the postulate)

(a) **Linearity** $\boxed{\hat{A}(\alpha|\psi_1\rangle+\beta|\psi_2\rangle) = \alpha\hat{A}|\psi_1\rangle+\beta\hat{A}|\psi_2\rangle}$

$\Rightarrow$ representation by matrices: $\hat{A} = \begin{pmatrix} A_{11} & A_{12} & \cdots \\ A_{21} & A_{22} & \\ \vdots & & \ddots \end{pmatrix}$

(b) **Hermiticity** $\boxed{\langle\psi_1|\hat{A}\psi_2\rangle = \langle\hat{A}\psi_1|\psi_2\rangle} = \langle\psi_2|\hat{A}\psi_1\rangle^* \quad \Rightarrow \quad \boxed{\langle\psi|\hat{A}^n\psi\rangle \in \mathbb{R}}$

$\Rightarrow \quad A_{ij} = A^*_{ji}$ for $i{\neq}j$ and $A_{ii} \in \mathbb{R}$

Hermiticity is sufficient (not necessary) condition for $\langle A^n\rangle_\psi$ being **real**

▸ **Definition domain**: Operator $\hat{A}$ defined on $\mathrm{Def}(\hat{A}) \subseteq \mathcal{H}$

For physics purposes it often suffices if $\boxed{\mathrm{Def}(\hat{A}) \equiv \text{a } \textbf{dense subset } \underline{\mathcal{H}} \subset \mathcal{H}}$

(cf. rigged Hilbert space)

▸ **Operator norm**: $||\hat{A}||^2 \equiv \mathrm{Sup}\left\{\frac{\langle\hat{A}\psi|\hat{A}\psi\rangle}{\langle\psi|\psi\rangle}\right\}_{|\psi\rangle\in\mathrm{Def}(\hat{A})}$

$||\hat{A}|| < \infty$ for **bounded** operators, $||\hat{A}|| = \infty$ for **unbounded** operators

▸ **Hermitian adjoint operator**

$\hat{A}^\dagger$ such that: $\boxed{\langle\psi_1|\hat{A}\psi_2\rangle{=}\langle\hat{A}^\dagger\psi_1|\psi_2\rangle} = \langle\psi_2|\hat{A}^\dagger\psi_1\rangle^* \quad \begin{cases} \forall|\psi_2\rangle\in\mathrm{Def}(\hat{A}) \\ \forall|\psi_1\rangle\in\mathrm{Def}(\hat{A}^\dagger)\supseteq\mathrm{Def}(\hat{A}) \end{cases}$

Matrix representation: $\hat{A}^\dagger = \begin{pmatrix} A^*_{11} & A^*_{21} & \cdots \\ A^*_{12} & A^*_{22} & \\ \vdots & & \ddots \end{pmatrix} \equiv \hat{A}^{\mathrm{T}*}$

▸ **Hermitian vs. selfadjoint operators**

Hermitian (symmetric) operator: $\boxed{\hat{A}|\psi\rangle = \hat{A}^\dagger|\psi\rangle} \quad \forall|\psi\rangle \in \mathrm{Def}(\hat{A})$

Note: The term "Hermitian operator" in an ∞-dim. space is often used in physics literature instead of "symmetric operator" used in mathematics.

Selfadjoint operator: $\hat{A} = \hat{A}^\dagger$ with a stronger condition $\mathrm{Def}(\hat{A}){=}\mathrm{Def}(\hat{A}^\dagger) \subseteq \mathcal{H}$

▸ **Function of operator**

Only functions expressible as Taylor series: $f(x) = \sum_k f_k\, x^k$

(more general definition will be given below)

$$\boxed{f(\hat{A}) \equiv \sum_k f_k\, \hat{A}^k} \quad f_k \in \mathbb{R} \Rightarrow f(\hat{A}) \text{ Hermitian}$$

▸ **Tensor products of operators**

Let us have $\hat{A}_1$ acting on $\mathcal{H}_1$ ($\hat{A}_1|\phi_{1i}\rangle{\equiv}|\phi'_{1i}\rangle$) and $\hat{A}_2$ acting on $\mathcal{H}_2$ ($\hat{A}_2|\phi_{2j}\rangle{\equiv}|\phi'_{2j}\rangle$)

We define $\boxed{\hat{A} \equiv \hat{A}_1 \otimes \hat{A}_2}$ acting on $\mathcal{H} = \mathcal{H}_1 \otimes \mathcal{H}_2$:

$$\hat{A}|\psi\rangle \equiv \hat{A}\Big[\sum_{i,j}\gamma_{ij}\overbrace{|\phi_{1i}\rangle|\phi_{2j}\rangle}^{|\Phi_{ij}\rangle}\Big] = \sum_{i,j}\gamma_{ij}\overbrace{|\phi'_{1i}\rangle|\phi'_{2j}\rangle}^{|\Phi'_{ij}\rangle}$$

Possible extension of $\hat{A}_1$ and $\hat{A}_2$ to $\mathcal{H} = \mathcal{H}_1 \otimes \mathcal{H}_2$:

$$\boxed{\begin{aligned} \hat{A}_1^{(\mathrm{ext})} &\equiv \hat{A}_1 \otimes \hat{I}_2 \\ \hat{A}_2^{(\mathrm{ext})} &\equiv \hat{I}_1 \otimes \hat{A}_2 \end{aligned}} \quad \begin{aligned} \hat{I}_1 &\equiv \text{unit op. in } \mathcal{H}_1 \\ \hat{I}_2 &\equiv \text{unit op. in } \mathcal{H}_2 \end{aligned}$$

Eigenvalues and eigenvectors of Hermitian operators

The key characteristic of any operator in the Hilbert space is its spectrum of eigenvalues and the set of the corresponding eigenvectors. Not only these eigensolutions constitute a subject of an involved mathematical theory, they also play the most essential role in the formulation of quantum mechanics.

▶ "Dispersion-free" states

Consider a state $|\psi\rangle$ in which the observable A yields a sharp value:

$$\boxed{\langle\langle A^2\rangle\rangle_\psi \equiv \langle A^2\rangle_\psi - \langle A\rangle_\psi^2 \stackrel{!}{=} 0}$$

$$\langle\psi|\hat{A}^2\psi\rangle - \langle\psi|\hat{A}\psi\rangle^2 = 0 \quad \Rightarrow \quad \langle\psi|[\hat{A} - \langle A\rangle_\psi \hat{I}\,]^2\psi\rangle = 0$$

$$\Rightarrow [\hat{A} - \langle A\rangle_\psi \hat{I}\,]|\psi\rangle = 0 \quad \Rightarrow \quad \boxed{\hat{A}|\psi\rangle = a|\psi\rangle}$$

$\Rightarrow |\psi\rangle$ **eigenvector** and $a = \langle A\rangle_\psi$ **eigenvalue** of operator $\hat{A}$

For $\hat{A} = \hat{A}^\dagger$ the eigenvalues $a \in \mathbb{R}$

Postulate: { possible measurement outcomes of A } $\equiv$ { eigenvalues of $\hat{A}$ }

i.e., each possible value a has its associated dispersion-free state $|a\rangle$

We will use a "stammering" notation: $\boxed{\hat{A}|a\rangle = a|a\rangle}$

▶ Orthogonality of eigenvectors with different eigenvalues (for Hermitian op.)

$$\left.\begin{array}{ll} \hat{A}|a\rangle = a|a\rangle & |\langle a'| \\ \hat{A}|a'\rangle = a'|a'\rangle & |\langle a| \end{array}\right\} \Rightarrow \underbrace{(a'-a)}_{\neq 0}\langle a|a'\rangle = 0 \Rightarrow \boxed{\langle a|a'\rangle = 0}$$

$\Rightarrow$ Eigenstates with different eigenvalues are perfectly distinguishable

▶ Degeneracy

It can happen that a single eigenvalue a has $n \geq 2$ linearly independent eigenvectors $\{|a;k\rangle\}_{k=1}^{n}$. Then all linear combinations are eigenvectors with the same eigenvalue a:

$$\hat{A}\Big(\sum_{k=1}^{n} \alpha_k |a;k\rangle\Big) = a \sum_{k=1}^{n} \alpha_k |a;k\rangle$$

$\Rightarrow \quad \mathcal{H}_a \equiv \mathrm{Span}\{|a;1\rangle, \ldots |a;n\rangle\} \equiv$ **degeneracy subspace**

with $d_{\mathcal{H}_a} \equiv \boxed{d_a = n_{\max}}$ maximal number of linearly independent eigenvectors

$\Rightarrow \exists$ an orthonormal basis $\{|a^{(k)}\rangle\}_{k=1}^{d_a}$ of $\mathcal{H}_a$ $\quad \boxed{\langle a^{(k)}|a^{(l)}\rangle = \delta_{kl}}$

▶ Eigensolutions for finite dimension

$\hat{A}|a\rangle = a|a\rangle \quad \Rightarrow \quad (\hat{A} - a\hat{I})|a\rangle = 0$

For dimension $d_{\mathcal{H}} < \infty$ the last relation represents a finite set (number $= d_{\mathcal{H}}$) of linear equations with null right-hand side. The solution exists *iff*:

$\boxed{\mathrm{Det}(\hat{A} - a\hat{I}) = 0}$ **polynomial equation** of order $d_{\mathcal{H}}$ in the variable a
$\Rightarrow \exists\, d_{\mathcal{H}}$ solutions $a = a_i$ (with i=1,...$d_{\mathcal{H}}$)

For $\hat{A}=\hat{A}^\dagger$ there $\exists\, d_{\mathcal{H}}$ lin. independent vectors solving $\begin{pmatrix} A_{11}-a & A_{12} & \cdots \\ A_{21} & A_{22}-a & \\ \vdots & & \ddots \end{pmatrix}\begin{pmatrix} \alpha_1 \\ \alpha_2 \\ \vdots \end{pmatrix}=0$

Degeneracy $\equiv$ equality of $n\geq 2$ eigenvalues: $a_{i_1}=\cdots=a_{i_n}\equiv a$ with lin. independent eigenvectors $\{|a_{i_1}\rangle,\ldots|a_{i_n}\rangle\}$ forming the degeneracy subspace $\mathcal{H}_a$ with $d_a=n$

▶ Completeness for finite dimension

Eigenvectors $\{|a_i^{(k)}\rangle\}_{i,k}$ of any *Hermitian* operator $\hat{A}$ form an **orthonormal basis** of $\mathcal{H}$,

where $\begin{cases} i & \text{enumerates } \textit{different} \text{ eigenvalues } a_i \\ k=1\cdots d_i & \text{counts basis vectors} \in \text{degeneracy subspace } \mathcal{H}_{a_i} \end{cases}$

For nondegenerate eigenvalues ($d_i=1$) we use the notation $|a_i\rangle \equiv |a_i^{(1)}\rangle$

Completeness relation then reads as:

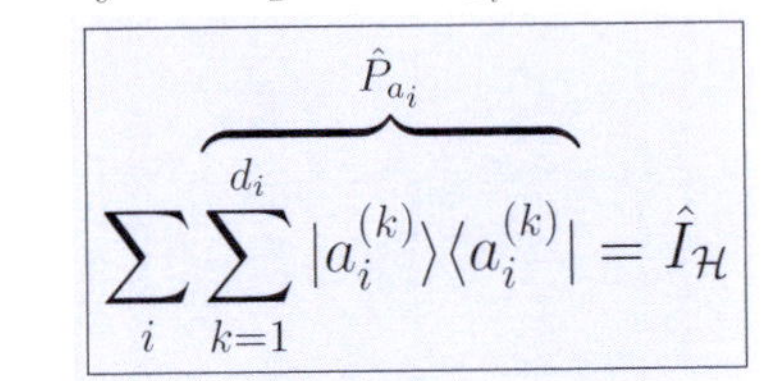

$$\sum_i \overbrace{\sum_{k=1}^{d_i} |a_i^{(k)}\rangle\langle a_i^{(k)}|}^{\hat{P}_{a_i}} = \hat{I}_{\mathcal{H}}$$

$\hat{A} \equiv \begin{pmatrix} a_1 & 0 & \cdots \\ 0 & a_2 & \\ \vdots & & \ddots \end{pmatrix}$ matrix representation of $\hat{A}$ in its own discrete eigenbasis

▶ Eigensolutions for infinite dimension

For $d_{\mathcal{H}}=\infty$, the expression $\mathrm{Det}(\hat{A}-a\hat{I})$ has no sense. To find solutions of $\hat{A}|a\rangle=a|a\rangle$ and to determine their properties is much more difficult in this case. In general, an ∞-dimensional operator $\hat{A}$ may have both discrete and **continuous spectrum** of eigenvalues. Moreover, eigenvalues from the continuous spectrum have no eigenvectors $\in \mathcal{H}$. Note that a rigorous analysis of these issues goes beyond our present level of advancement. We will just indicate two alternative mathematical treatments: one by von Neumann, who considered finite intervals of continuous eigenvalues and used a standard Hilbert space, and one initiated by Dirac, who stepped out towards the rigged Hilbert space.

Example:

$$\underbrace{\begin{pmatrix} 0&1&0&0&0&\cdots \\ 1&0&1&0&0& \\ 0&1&0&1&0& \\ 0&0&1&0&1& \\ \vdots&&&&&\ddots \end{pmatrix}}_{\hat{A}} \underbrace{\begin{pmatrix} \alpha_1 \\ \alpha_2 \\ \alpha_3 \\ \alpha_4 \\ \vdots \end{pmatrix}}_{|a\rangle} = a \underbrace{\begin{pmatrix} \alpha_1 \\ \alpha_2 \\ \alpha_3 \\ \alpha_4 \\ \vdots \end{pmatrix}}_{|a\rangle} \Rightarrow \left\{ \begin{matrix} \alpha_2=a\alpha_1 \\ \alpha_1+\alpha_3=a\alpha_2 \\ \alpha_2+\alpha_4=a\alpha_3 \\ \alpha_3+\alpha_5=a\alpha_4 \\ \vdots \\ \alpha_{d-2}+\alpha_d=a\alpha_{d-1} \\ \alpha_{d-1}=a\alpha_d \end{matrix} \right\} \Rightarrow \left\{ \begin{matrix} \text{Recurs. conditions} \\ (i=0,1,\ldots,d+1) \\ \alpha_{i-1}+\alpha_{i+1}=a\alpha_i \\ \alpha_0=0 \qquad \alpha_{d+1}=0 \end{matrix} \right.$$

For $d<\infty$ the solution $\exists$ only for some discrete values $a_j \in [-2,+2]$, with eigenvectors $|a_j\rangle$ being trivially normalizable. For $d=\infty$ the solution $\exists\ \forall\ a \in (-\infty,+\infty)$, but the eigenvectors are not normalizable: $|a\rangle \notin \ell^2$. It can be shown that for $|a|>2$ the components α_i diverge exponentially with i, thus $|a\rangle \notin \overline{\ell^2}$ ($\equiv \overline{\mathcal{H}}$ of the Gelfand triple), whereas for $|a|<2$ the components α_i are bounded and oscillate with i, thus $|a\rangle \in \overline{\ell^2}$. So the matrix $\hat{A}$ for $d=\infty$ has a continuous spectrum $a\in[-2,+2]$ with eigenvectors $|a\rangle \in \overline{\mathcal{H}} \supset \mathcal{H}$.

In general: $\underbrace{\mathcal{S}(\hat{A})}_{\text{spectrum}} = \underbrace{\mathcal{D}(\hat{A})}_{\text{discrete part}} \cup \underbrace{\mathcal{C}(\hat{A})}_{\text{continuous part}}$

Eigenvalues $a_i \in \mathcal{D}(\hat{A})$ have eigenvectors $|a_i^{(k)}\rangle \in \mathcal{H}$

Projectors: $\hat{P}_{a_i}^{(k)} = |a_i^{(k)}\rangle\langle a_i^{(k)}|$, $\hat{P}_{a_i} = \sum_{k=1}^{d_i} \hat{P}_{a_i}^{(k)}$

Orthonormality: $\langle a_i^{(k)}|a_j^{(l)}\rangle = \delta_{ij}\delta_{kl}$

In contrast, eigenvalues $a \in \mathcal{C}(\hat{A})$ have eigenvectors $|a^{(k)}\rangle \in \overline{\mathcal{H}}$ $\supset \mathcal{H} \supset \underline{\mathcal{H}}$
with $k \in \mathcal{D}_a \left\{ \begin{smallmatrix} \text{discrete } (k=1,\dots,d_a) \\ \text{continuous} \end{smallmatrix} \right\}$ degeneracy index

▶ Alternative approaches to continuous spectrum

(a) Dirac works in the **extended space** $\overline{\mathcal{H}} \supset \mathcal{H}$ which accommodates not only discrete eigenvectors $|a_i^{(k)}\rangle$, but also continuous eigenvectors $|a^{(k)}\rangle$ for $a \in \mathcal{C}(\hat{A})$

$\hat{\Pi}_a^{(k)} \equiv |a^{(k)}\rangle\langle a^{(k)}| \equiv$ projector to $|a^{(k)}\rangle \in \overline{\mathcal{H}}$

$\hat{\Pi}_a \equiv \sum_{k\in\mathcal{D}_a} \hat{\Pi}_a^{(k)} \equiv$ projector to deg.subspace, where $\sum_{k\in\mathcal{D}_a} \equiv \begin{cases} \sum_{k=1}^{d_a} & \text{(discrete deg.index)} \\ \int_{\mathcal{D}_a} dk & \text{(continuous deg.index)} \end{cases}$

The following "orthonormality" conditions are formally required:

$$\begin{array}{l} \langle a^{(k)}|a'^{(l)}\rangle = \delta(a-a')\delta_{kl} \quad (k,l \text{ discrete}) \\ \langle a^{(k)}|a'^{(l)}\rangle = \delta(a-a')\delta(k-l) \quad (k,l \text{ continuous}) \end{array} \quad \langle a^{(k)}|a_i^{(l)}\rangle = 0$$

Normalization to δ-function (that is not a well-defined scalar product!)

(b) Von Neumann works in the **standard Hilbert space** $\mathcal{H}$ in which $\nexists$ eigenvectors for continuous eigenvalues, but $\exists$ subspaces $\mathcal{H}_{(a',a'')} \subset \mathcal{H}$ corresponding to any interval (a', a'') of these eigenvalues in the sense that probability distributions of observable $\hat{A}$ vanish outside (a', a'') for any state $\in \mathcal{H}_{(a',a'')}$

$\hat{\mathbf{\Pi}}_{(a',a'')} \equiv$ projector to $\mathcal{H}_{(a',a'')}$ $\Rightarrow$ in Dirac's language: $\hat{\mathbf{\Pi}}_{(a',a'')} \sim \int_{a'}^{a''} \hat{\Pi}_a \, da$

$\hat{\mathbf{\Pi}}_{(a',a''')} = \hat{\mathbf{\Pi}}_{(a',a'')} + \hat{\mathbf{\Pi}}_{(a'',a''')}$ for $a' \leq a'' \leq a'''$

$\hat{\mathbf{\Pi}}_{(-\infty,a')} \equiv \hat{\mathbf{\Pi}}(a')$ $\equiv$ "cummulative" projector to a subspace with $a \leq a'$

Projector to an infinitesimal eigenvalue interval is related to $\hat{\Pi}_{a'}$:

$$\hat{\mathbf{\Pi}}_{(a',a'+da)} = \hat{\mathbf{\Pi}}(a'+da) - \hat{\mathbf{\Pi}}(a') \equiv \tfrac{d}{da}\hat{\mathbf{\Pi}}(a)|_{a=a'} da$$

$$\Rightarrow \quad \hat{\Pi}_{a'} \sim \tfrac{d}{da}\hat{\mathbf{\Pi}}(a)|_{a=a'}$$

Schematic illustration:

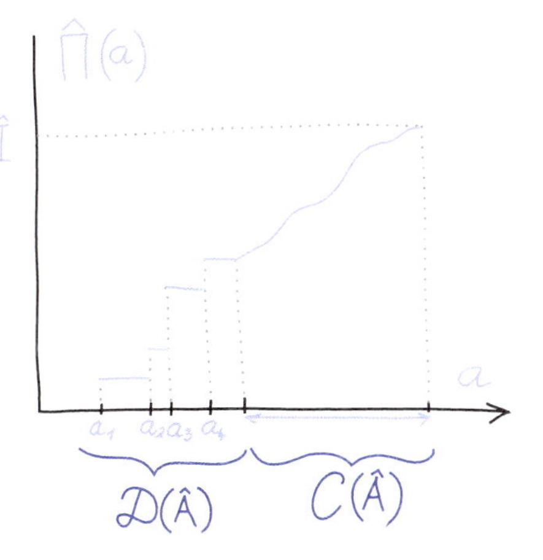

▸ Completeness for infinite dimension (Dirac's approach)

$$\sum_{\substack{i\\a_i\in\mathcal{D}(\hat{A})}} \underbrace{\sum_{k=1}^{d_i} |a_i^{(k)}\rangle\langle a_i^{(k)}|}_{\hat{P}_{a_i}} + \int_{\mathcal{C}(\hat{A})} \underbrace{\sum_{k\in\mathcal{D}_a} |a^{(k)}\rangle\langle a^{(k)}|}_{\hat{\Pi}_a}\, da = \hat{I}_{\mathcal{H}}$$

$$\hat{P}_{a_i}\hat{P}_{a_j} = \delta_{ij}\hat{P}_{a_i} \qquad \hat{\Pi}_a\hat{\Pi}_{a'} = \delta(a-a')\hat{\Pi}_a \qquad \hat{P}_{a_i}\hat{\Pi}_a = 0$$

Consistency: unique expansion of $|\psi\rangle$ in the "eigenbasis" of $\hat{A}$:

$|\psi\rangle = \sum_{i,k}\langle a_i^{(k)}\underbrace{|\psi\rangle}_{*}|a_i^{(k)}\rangle + \int\sum_{l}\langle a^{(l)}\underbrace{|\psi\rangle}_{*}|a^{(l)}\rangle\, da$, for * substitute the same expr.:

$$|\psi\rangle = \sum_{i,k,i',k'} \underbrace{\langle a_i^{(k)}|a_{i'}^{(k')}\rangle}_{\delta_{ii'}\delta_{kk'}}\langle a_{i'}^{(k')}|\psi\rangle|a_i^{(k)}\rangle + \sum_{i,k,l}\int \underbrace{\langle a^{(l)}|a_i^{(k)}\rangle}_{0}\, da\langle a_i^{(k)}|\psi\rangle|a_i^{(k)}\rangle +$$

$$+\int\sum_{i,k}\underbrace{\langle a_i^{(k)}|a^{(l)}\rangle}_{0}\langle a^{(l)}|\psi\rangle|a^{(l)}\rangle\, da + \iint\sum_{ll'}\underbrace{\langle a^{(l)}|a'^{(l')}\rangle}_{\delta(a-a')\delta_{ll'}}\langle a'^{(l')}|\psi\rangle|a^{(l)}\rangle\, da\, da' = \left\{\begin{array}{l}\text{previous}\\ \text{expression}\end{array}\right.$$

▸ Completeness for infinite dimension (von Neumann's approach)

$$\int_{\mathcal{S}(\hat{A})} d\hat{\mathbf{\Pi}}(a) = \hat{I}_{\mathcal{H}}$$

where use is made of Stieltjes method of integration:

$$\int f(x)\underbrace{d\hat{\sigma}(x)}_{\substack{\text{operator}\\ \text{measure}}} \equiv \lim_{n\to\infty}\sum_{k=1}^{n} f(x_k)[\hat{\sigma}(x_{k+1})-\hat{\sigma}(x_k)]$$

▸ Spectral decomposition of operator

The above completeness relations lead to the expression of an operator in terms of its eigenvalues and the projectors to the corresponding eigenspaces.

discrete spectrum: $\hat{A} = \sum_i a_i\hat{P}_{a_i} \qquad f(\hat{A}) = \sum_i f(a_i)\hat{P}_{a_i}$

$$|\psi\rangle = \sum_i\sum_{k=1}^{d_i}\langle a_i^{(k)}|\psi\rangle|a_i^{(k)}\rangle \Rightarrow \hat{A}^n|\psi\rangle = \sum_i a_i^n\underbrace{\sum_{k=1}^{d_i}|a_i^{(k)}\rangle\langle a_i^{(k)}|}_{\hat{P}_{a_i}}\psi\rangle \Rightarrow \hat{A}^n = \sum_i (a_i)^n\hat{P}_{a_i}$$

general (combined) spectrum:

$$\begin{aligned}\hat{A} &= \sum_{\mathcal{D}(\hat{A})} a_i\hat{P}_{a_i} + \int_{\mathcal{C}(\hat{A})} a\,\hat{\Pi}_a\, da &&\equiv \int_{\mathcal{S}(\hat{A})} a\, d\hat{\mathbf{\Pi}}(a)\\ f(\hat{A}) &= \sum_{\mathcal{D}(\hat{A})} f(a_i)\hat{P}_{a_i} + \int_{\mathcal{C}(\hat{A})} f(a)\,\hat{\Pi}_a\, da &&\equiv \int_{\mathcal{S}(\hat{A})} f(a)\, d\hat{\mathbf{\Pi}}(a)\end{aligned}$$

▸ Definition of **irregular operator functions** $f(\hat{A}) \neq \sum_k f_k\hat{A}^k$

$$f(\hat{A}) \equiv \int_{\mathcal{S}(\hat{A})\cap\mathrm{Def}[f(a)]} f(a)\, d\hat{\mathbf{\Pi}}(a)$$

with $\mathrm{Def}[f(\hat{A})] \equiv$ subspace of $\mathcal{H}$ spanned by eigenvectors with $|f(a)| < \infty$

▸ Eigenvalue expression of **operator norm**: $||\hat{A}||^2 = \mathrm{Sup}\left\{|a|^2\right\}_{a\in\mathcal{S}(\hat{A})}$

Bounded (unbounded) operator $\hat{A} \Leftrightarrow$ bounded (unbounded) spectrum $\mathcal{S}(\hat{A})$

■ Probability distribution for the outcomes of measurements

The spectral decomposition of an operator associated with observable A and the postulate on the statistical moments $\langle A^n \rangle_\psi$ enables us to finally deduce the desired probability distribution $p_\psi(a)$. Note that the resulting formula for $p_\psi(a)$ can be used as an alternative (equivalent) postulate instead of that for $\langle A^n \rangle_\psi$.

► Moments of the probability distribution for observable A in state $|\psi\rangle$:

$$\langle A^n \rangle_\psi \equiv \begin{cases} \sum\limits_{\mathcal{D}(\hat{A})} (a_i)^n p_\psi(a_i) + \int\limits_{\mathcal{C}(\hat{A})} a^n p_\psi(a)\, da & \text{defining formula} \\ \sum\limits_{\mathcal{D}(\hat{A})} (a_i)^n \langle \psi | \hat{P}_{a_i} \psi \rangle + \int\limits_{\mathcal{C}(\hat{A})} a^n \langle \psi | \hat{\Pi}_a \psi \rangle\, da \equiv \langle \psi | \hat{A}^n \psi \rangle & \text{expression from spectral decomp.} \end{cases}$$

► For $p_\psi(a) \equiv$ probability (density) of finding value a of A in $|\psi\rangle$ we then get:

Discrete case : $$p_\psi(a_i) = \langle \psi | \hat{P}_{a_i} \psi \rangle = \sum_{k=1}^{d_i} |\langle a_i^{(k)} | \psi \rangle|^2$$

Continuous case : $$p_\psi(a)\, da = \langle \psi | \hat{\Pi}_a \psi \rangle\, da = \sum_{k \in \mathcal{D}_a} |\langle a^{(k)} | \psi \rangle|^2\, da = \langle \psi | d\hat{\mathbf{\Pi}}(a) \psi \rangle$$

$$\left. \begin{array}{ll} \langle a | \psi \rangle & \equiv \text{amplitude} \\ |\langle a | \psi \rangle|^2 & \equiv \text{probability} \end{array} \right\} \text{ to measure } a \text{ on } |\psi\rangle \Leftrightarrow \text{to associate } |\psi\rangle \text{ with } |a\rangle$$

◄ **Historical remark**

1900-10: David Hilbert studies spectral properties of integral operators

1924: D. Hilbert and R. Courant publish the book *Methoden der mathematischen Physik* containing methods that later became relevant in QM

1925: Werner Heisenberg (and M. Born & P. Jordan) formulate "matrix mechanics"

1926: Erwin Schrödinger in his wave mechanics makes use of operators associated with observables, he shows the equivalence with matrix mechanics

1926-32: John von Neumann unifies Schrödinger's and Heisenberg's approaches using self-adjoint operators acting on a general Hilbert space, with M. Stone they work out the theory of such operators

1927-30: Paul Dirac develops "symbolic" formalism transcending ordinary Hilbert space, this is formalized in the 1950's in terms of rigged Hilbert spaces

2.2 Examples of quantum operators

The formalism developed in the previous section is now ready to bear fruit. We will introduce the operators associated with observables characterizing a single particle.

■ Spin-$\frac{1}{2}$ operators

Spin operators are the clearest examples of quantum observables since they work in the best of all possible Hilbert spaces—that with dimension 2.

▶ **Operators of spin components** along x, y, z axes in $\mathcal{H} \equiv \mathbb{C}^2$

$$\hat{S}_x = \frac{\hbar}{2}\underbrace{\begin{pmatrix} 0 & 1 \\ 1 & 0 \end{pmatrix}}_{\hat{\sigma}_x} \quad \hat{S}_y = \frac{\hbar}{2}\underbrace{\begin{pmatrix} 0 & -i \\ +i & 0 \end{pmatrix}}_{\hat{\sigma}_y} \quad \hat{S}_z = \frac{\hbar}{2}\underbrace{\begin{pmatrix} 1 & 0 \\ 0 & -1 \end{pmatrix}}_{\hat{\sigma}_z}$$

Pauli matrices

▶ **Projection to general direction** $\vec{n} = (\underbrace{\sin\vartheta\cos\varphi}_{n_x}, \underbrace{\sin\vartheta\sin\varphi}_{n_y}, \underbrace{\cos\vartheta}_{n_z})$
$|\vec{n}|^2=1$

$$\hat{S}_{\vec{n}} = \vec{n}\cdot\hat{\vec{S}} = \frac{\hbar}{2}(\vec{n}\cdot\hat{\vec{\sigma}}) = \frac{\hbar}{2}\begin{pmatrix} n_z & n_x - in_y \\ n_x + in_y & -n_z \end{pmatrix} = \frac{\hbar}{2}\begin{pmatrix} \cos\vartheta & e^{-i\varphi}\sin\vartheta \\ e^{+i\varphi}\sin\vartheta & -\cos\vartheta \end{pmatrix}$$

▶ **Eigenvalues of spin projection** $\hat{S}_{\vec{n}}$

$$\text{Det}\left[\frac{\hbar}{2}\begin{pmatrix} n_z - \lambda & n_x - in_y \\ n_x + in_y & -(n_z + \lambda) \end{pmatrix}\right] = 0 \quad \Rightarrow \quad \lambda^2 = 1 \quad \Rightarrow \quad s_{\vec{n}} = \begin{cases} +\frac{\hbar}{2} \\ -\frac{\hbar}{2} \end{cases}$$

▶ **Eigenvectors of spin projection** $\hat{S}_{\vec{n}}$

Eigenequation $\begin{pmatrix} n_z & n_x - in_y \\ n_x + in_y & -n_z \end{pmatrix}\begin{pmatrix} \alpha_\pm \\ \beta_\pm \end{pmatrix} = \pm\begin{pmatrix} \alpha_\pm \\ \beta_\pm \end{pmatrix}$ has ∞ solutions:

$n_z \neq \pm 1$ (otherwise solutions known) $\Rightarrow \alpha_\pm = -\frac{n_x - in_y}{n_z \mp 1}\beta_\pm$

Normalized solutions: $\left|s_{\vec{n}} = +\frac{\hbar}{2}\right\rangle = \begin{pmatrix} e^{-i\varphi}\cos\frac{\vartheta}{2} \\ \sin\frac{\vartheta}{2} \end{pmatrix}$ $\quad \left|s_{\vec{n}} = -\frac{\hbar}{2}\right\rangle = \begin{pmatrix} -e^{-i\varphi}\sin\frac{\vartheta}{2} \\ \cos\frac{\vartheta}{2} \end{pmatrix}$

Orthogonality: $\begin{pmatrix} \alpha_-^* & \beta_-^* \end{pmatrix}\begin{pmatrix} \alpha_+ \\ \beta_+ \end{pmatrix} = 0$

Projectors to eigenspaces:

$$\hat{P}_{\pm\vec{n}} = \begin{pmatrix} \alpha_\pm \\ \beta_\pm \end{pmatrix}\begin{pmatrix} \alpha_\pm^* & \beta_\pm^* \end{pmatrix} = \begin{cases} \begin{pmatrix} \cos^2\frac{\vartheta}{2} & \frac{e^{-i\varphi}}{2}\sin\vartheta \\ \frac{e^{+i\varphi}}{2}\sin\vartheta & \sin^2\frac{\vartheta}{2} \end{pmatrix} & \text{for } s_{\vec{n}} = +\frac{\hbar}{2} \\ \begin{pmatrix} \sin^2\frac{\vartheta}{2} & -\frac{e^{-i\varphi}}{2}\sin\vartheta \\ -\frac{e^{+i\varphi}}{2}\sin\vartheta & \cos^2\frac{\vartheta}{2} \end{pmatrix} & \text{for } s_{\vec{n}} = -\frac{\hbar}{2} \end{cases}$$

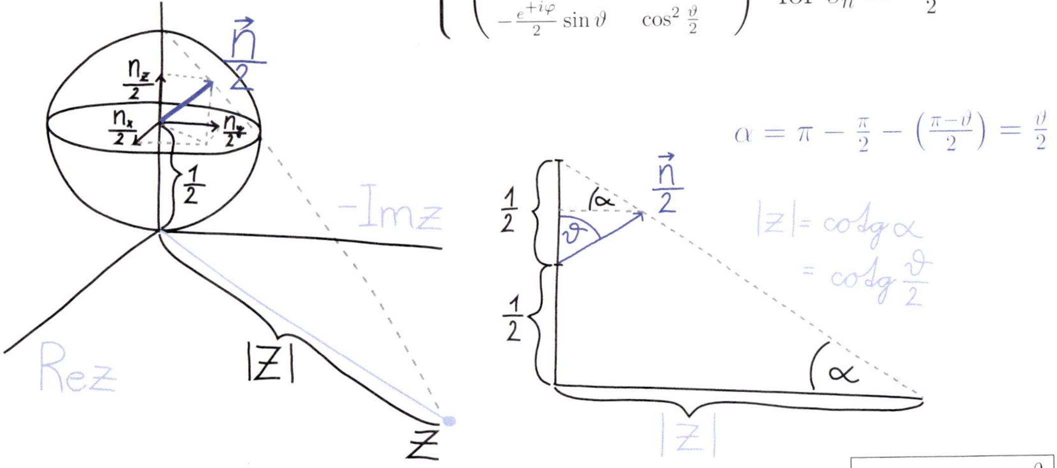

Unnormalized eigenvector: $\left|s_{\vec{n}} = +\frac{\hbar}{2}\right\rangle = \underbrace{-\frac{n_x - in_y}{n_z - 1}}_{z}|\uparrow\rangle + |\downarrow\rangle$ with $z = e^{-i\varphi}\cot\frac{\vartheta}{2}$

$\Rightarrow z \equiv$ stereographic projection of vector $\frac{\vec{n}}{2}$ onto $\mathbb{C}$

Any superposition $|\psi\rangle = \alpha|\uparrow\rangle + \beta|\downarrow\rangle$ represents a state of spin pointing in a fixed direction $\vec{n}$, which is obtained from $z = \alpha/\beta$ by the stereographic projection.

■ Coordinate & momentum

The most important observables in classical mechanics (such that all the other observables are made of them) are the coordinates and momenta. Unfortunately, these are precisely the observables whose QM operators make troubles.

► Coordinate & momentum eigenfunctions

Hilbert space $\mathcal{H}=\mathcal{L}^2(\mathbb{R}^3)$ & rigged Hilbert space $\underline{\mathcal{H}}\subset\mathcal{H}\subset\overline{\mathcal{H}}$
with $\underline{\mathcal{H}}\equiv$ *differentiable* functions satisfying $|\psi(\vec{x})|_{|\vec{x}|\to\infty}\lesssim|\vec{x}|^{-m}$ for any $m>0$

Postulate: δ function & plane wave $\equiv$ eigenstates of position & momentum

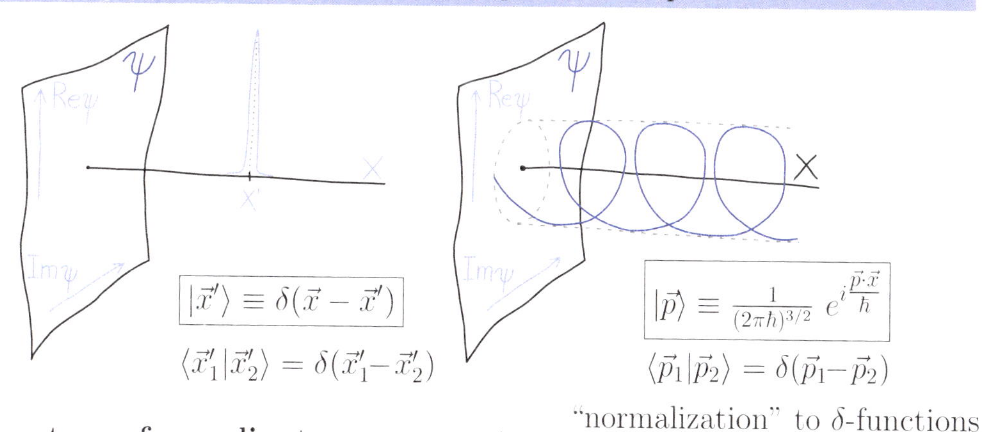

$$|\vec{x}'\rangle\equiv\delta(\vec{x}-\vec{x}')$$
$$\langle\vec{x}'_1|\vec{x}'_2\rangle=\delta(\vec{x}'_1-\vec{x}'_2)$$

$$|\vec{p}\rangle\equiv\frac{1}{(2\pi\hbar)^{3/2}}\,e^{i\frac{\vec{p}\cdot\vec{x}}{\hbar}}$$
$$\langle\vec{p}_1|\vec{p}_2\rangle=\delta(\vec{p}_1-\vec{p}_2)$$

"normalization" to δ-functions

► Operators of coordinate components

$\vec{x}=(x,y,z)\equiv(x_1,x_2,x_3)$

Action of operator $\hat{x}_i\equiv$ multiplication by variable x_i:
$$\underbrace{\hat{x}_i\psi(\vec{x})}_{[\hat{x}_i\psi](\vec{x})}\equiv\underbrace{x_i\psi(\vec{x})}_{\psi'(\vec{x})}$$

$\mathrm{Def}(\hat{x}_i):=\underline{\mathcal{H}}$

Hermiticity: $\int\psi_1(\vec{x})^*[x_i\psi_2(\vec{x})]d\vec{x}=\int[x_i\psi_1(\vec{x})]^*\psi_2(\vec{x})d\vec{x}$

Eigenstates: $x_i\delta(\vec{x}-\vec{x}')=x'_i\delta(\vec{x}-\vec{x}')$

$\Rightarrow$ continuous spectrum $x'_i\in(-\infty,+\infty)$ with $\delta(\vec{x}-\vec{x}')\in\overline{\mathcal{H}}$

► Operators of momentum components

$\vec{p}=(p_x,p_y,p_z)\equiv(p_1,p_2,p_3)$

Action of $\hat{p}_i\propto$ derivative by x_i:
$$\underbrace{\hat{p}_i\psi(\vec{x})}_{[\hat{p}_i\psi](\vec{x})}\equiv\underbrace{-i\hbar\frac{\partial}{\partial x_i}\psi(\vec{x})}_{\psi'(\vec{x})}\quad\Leftrightarrow\quad\hat{\vec{p}}=-i\hbar\vec{\nabla}$$

$\mathrm{Def}(\hat{p}_i):=\underline{\mathcal{H}}$

Hermiticity:
$$\int\psi_1(\vec{x})^*[-i\hbar\tfrac{\partial}{\partial x_i}\psi_2(\vec{x})]d\vec{x}=\int[-i\hbar\tfrac{\partial}{\partial x_i}\psi_1(\vec{x})]^*\psi_2(\vec{x})d\vec{x}+\overbrace{[\psi_1(\vec{x})^*\psi_2(\vec{x})]_{-\infty}^{+\infty}}^{0}$$

Eigenstates: $-i\hbar\frac{\partial}{\partial x_i}e^{i\vec{p}\cdot\vec{x}/\hbar}=p_ie^{i\vec{p}\cdot\vec{x}/\hbar}$

$\Rightarrow$ continuous spectrum $p_i\in(-\infty,+\infty)$ with $e^{i\vec{p}\cdot\vec{x}/\hbar}\in\overline{\mathcal{H}}$

$\boxed{\vec{p}=\hbar\vec{k}=\frac{2\pi\hbar}{\lambda}\vec{n}}$ with $\vec{k}\equiv$ **wave vector** pointing along unit vector $\vec{n}$

We obtain **de Broglie relation** for the wavelength: $\boxed{\lambda=\frac{2\pi\hbar}{p}\equiv\frac{h}{p}}$

◀ **Historical remark**
1924: Louis de Broglie associates plane waves with moving particles
1926: Erwin Schrödinger applies operators within the wave mechanics
1927: Wolfgang Pauli introduces spin matrices
1930: Paul Dirac introduces explicit momentum and position operators
1940's-60's: Rigorous mathematical treatment in terms of the distribution theory (L. Schwarz *et al.*) and rigged Hilbert spaces (I. Gelfand *et al.*)

■ Hamiltonian of a structureless particle

The incorrigible *enfants terribles* coordinate and momentum give birth to a respected (although not always well-behaved) operator named Hamiltonian. In the nonrelativistic QM, the Hamiltonian is of central importance as it represents energy and generates evolution (as we will see in Sec. 1.5).

▶ Hamiltonian $\hat{H} \equiv$ **operator of energy**

Eigenequation $\boxed{\hat{H}|E\rangle = E|E\rangle}$ **stationary Schrödinger equation**

▶ **Free particle** of mass M

$$\boxed{\hat{H} = \frac{1}{2M}(\hat{\vec{p}}\cdot\hat{\vec{p}}) = -\frac{\hbar^2}{2M}\underbrace{\left(\vec{\nabla}\cdot\vec{\nabla}\right)}_{\Delta}}$$ operator of **kinetic energy**

Eigenequation $\Big(\Delta + \underbrace{\tfrac{2ME}{\hbar^2}}_{\pm k^2=\pm(k_1^2+k_2^2+k_3^2)}\Big)\psi(\vec{x}) = 0$

Solutions for $E \geq 0$ physical: $\psi \propto e^{\pm i\vec{k}\cdot\vec{x}} \in \overline{\mathcal{H}}$
Solutions for $E < 0$ nonphysical: $\psi \propto e^{\pm\vec{k}\cdot\vec{x}} \notin \overline{\mathcal{H}}$

Continuous spectrum $E \in [0, +\infty)$ infinitely degenerate (except $E{=}0$)
Eigenstates: $|E_{\vec{k}}\rangle = e^{i\vec{k}\cdot\vec{x}} \equiv |\vec{p} = \hbar\vec{k}\rangle$ with eigenvalues $E \equiv E_{\vec{k}} = \frac{(\hbar k)^2}{2M}$

▶ **Particle in scalar potential**

$V(\vec{x}) \equiv$ potential energy in an external field

$$\boxed{\hat{H} = \underbrace{\frac{1}{2M}(\hat{\vec{p}}\cdot\hat{\vec{p}})}_{\hat{T}\ \text{kinetic}} + \underbrace{V(\hat{\vec{x}})}_{\hat{V}\ \text{potential energy}} \equiv -\frac{\hbar^2}{2M}\Delta + V(\vec{x})}$$

Stationary Schrödinger eq.
$\left[-\frac{\hbar^2}{2M}\Delta + V(\vec{x}) - E\right]\psi(\vec{x}) = 0$

▶ **Bound vs. unbound states**

Eigenstates of the above Schrödinger equation are of two types:
(1) bound states (correspond to finite motion) ⇒ discrete spectrum, normalizable wavefunction $\psi(\vec{x}) \in \mathcal{H} = \mathcal{L}^2(\mathbb{R}^3)$
(2) unbound states (correspond to infinite motion) ⇒ continuous spectrum, non-normalizable wavefunction $\psi(\vec{x}) \in \overline{\mathcal{H}}$

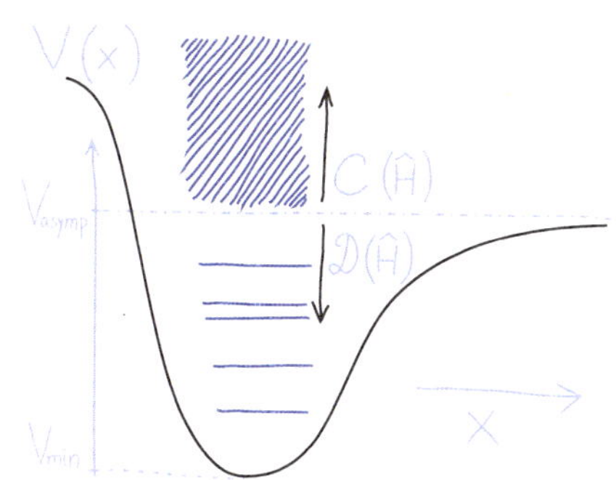

Consider two types of potential $V(\vec{x})$ (derivations of the following statements are not presented here):

(a) Potential wells of a general shape:

Define values:

$$V_{\rm asymp} = \lim_{r\to\infty} {\rm Min}\{V(\overbrace{r,\vartheta,\varphi}^{\text{sph.coord.}})\}_{\vartheta,\varphi}$$

$$V_{\rm min} \equiv {\rm Min}\{V(\vec{x})\}_{\vec{x}}$$

Eigenvalues of $\hat{H} \to \begin{cases} E_i \in (V_{\rm min}, V_{\rm asymp}) & \textbf{discrete spectrum} \text{ (bound states)} \\ E \in (V_{\rm asymp}, +\infty) & \textbf{continuous spect.} \text{ (unbound states)} \end{cases}$

(b) Periodic potentials of a general shape (solids, crystals):
$V_{\rm asymp}$ does not exist; the spectrum is continuous and has a **band structure**; eigenfunctions are not normalizable (unbound states)

▶ Nonanalytic potentials: conditions upon eigenfunctions

From the stationary Schrödinger equation it follows that:

$$V, \frac{\partial V}{\partial x_i}, \ldots \frac{\partial^n V}{\partial x_i^n}\bigg|_{\vec{x}=\vec{a}} \text{ continuous} \quad \Leftrightarrow \quad \psi, \frac{\partial \psi}{\partial x_i}, \ldots \frac{\partial^n \psi}{\partial x_i^n}, \frac{\partial^{n+1} \psi}{\partial x_i^{n+1}}, \frac{\partial^{n+2} \psi}{\partial x_i^{n+2}}\bigg|_{\vec{x}=\vec{a}} \text{ continuous}$$

$V(\vec{x})|_{\vec{x}=\vec{a}}$ discontinuous (**finite jump** of the potential)

$$\Rightarrow \quad \psi, \frac{\partial \psi}{\partial x_i}\bigg|_{\vec{x}=\vec{a}} \text{ continuous}$$

$$\Rightarrow \boxed{\beta_i(\vec{x})|_{\vec{x}=\vec{a}} \equiv \frac{\frac{\partial\psi}{\partial x_i}(\vec{x})}{\psi(\vec{x})}\Bigg|_{\vec{x}=\vec{a}} = \underbrace{\frac{\partial}{\partial x_i}\ln\psi(\vec{x})}_{\substack{\text{logarithmic}\\ \text{derivative}}}\Bigg|_{\vec{x}=\vec{a}}} \text{ continuous}$$

Example: 1D potential well of a finite range $x \in [x_1, x_2]$

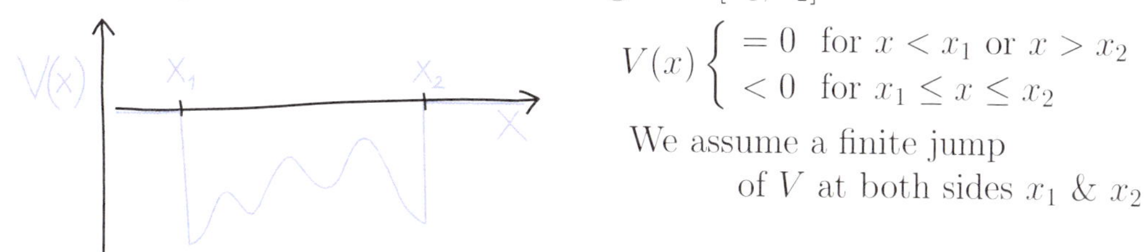

$$V(x)\begin{cases} = 0 & \text{for } x < x_1 \text{ or } x > x_2 \\ < 0 & \text{for } x_1 \le x \le x_2 \end{cases}$$

We assume a finite jump of V at both sides x_1 & x_2

Eigenfunctions for bound ($E < 0$) and unbound ($E \ge 0$) states:

	$x < x_1$	$x_1 \le x \le x_2$	$x_2 < x$
$E < 0$	$A_1 e^{+kx} + \cancel{A_2 e^{-kx}}$	$B_1\psi_1(x)_E + B_2\psi_2(x)_E$	$\cancel{C_1 e^{+kx}} + C_2 e^{-kx}$
$E \ge 0$	$A_1\cos(kx) + A_2\sin(kx)$	$B_1\psi_1(x)_E + B_2\psi_2(x)_E$	$C_1\cos(kx) + C_2\sin(kx)$

$\{\psi_1(x)_E, \psi_2(x)_E\} \equiv 2$ independent eigensolutions inside the well,

$k = \frac{\sqrt{2ME}}{\hbar}$, $\{A_1, A_2, B_1, B_2, C_1, C_2\} \equiv$ free parameters

$E < 0$: (2 matching conditions at x_1)+(2 match.conds.at x_2)+(1 norm.condition)
$\Rightarrow$ cannot be solved with 4 free parameters $\forall E \Rightarrow$ discrete E spectrum

$E \geq 0$: (2 matching conditions at x_1)+(2 matching conditions at x_2)
$\Rightarrow$ can be solved with 6 free parameters $\forall E \Rightarrow$ continuous E spectrum

For **infinite jump** of $V(\vec{x})|_{\vec{x}=\vec{a}}$ only $\psi(\vec{x})|_{\vec{x}=\vec{a}}$ must be continuous

■ Hamiltonian with a separable potential

We look now at the special case of a separable potential, i.e., potential of the form $\boxed{V(\vec{x}) = V_1(x_1) + V_2(x_2) + V_3(x_3)}$ with $V_k(x_k) \equiv$ arbitrary 1D potential. There are just a few (two?) practical examples of such potentials, but the analysis will help us to understand a rather important technique: separation of variables in differential equations.

▶ Let us solve 3 × 1D equation $\overbrace{\left[-\frac{\hbar^2}{2M}\frac{\partial^2}{\partial x_k^2} + V_k(x_k)\right]}^{\hat{H}_k} \psi_{i_k}(x_k) = E_{i_k}\psi_{i_k}(x_i)$

$\Rightarrow$ solution of the 3D problem can be written as:

$$\boxed{\underbrace{[\hat{H}_1 + \hat{H}_2 + \hat{H}_3]}_{\hat{H}} \underbrace{\psi_{i_1}(x_1)\psi_{i_2}(x_2)\psi_{i_3}(x_3)}_{\psi_{i_1i_2i_3}(\vec{x})} = \underbrace{(E_{i_1} + E_{i_2} + E_{i_3})}_{E_{i_1i_2i_3}} \underbrace{\psi_{i_1}(x_1)\psi_{i_2}(x_2)\psi_{i_3}(x_3)}_{\psi_{i_1i_2i_3}(\vec{x})}}$$

▶ 1D eigenfunctions $\left\{\psi_{i_k}(x_k) \equiv |\psi_{i_k}\rangle\right\}_{i_k=1,2,\dots} \equiv$ basis in Hilbert space $\mathcal{H}_k$

$\left\{\psi_{i_1}(x_1)\psi_{i_2}(x_2)\psi_{i_3}(x_3) \equiv |\psi_{i_1}\rangle|\psi_{i_2}\rangle|\psi_{i_3}\rangle\right\}_{i_k=1,2,\dots} \equiv$ basis in $\mathcal{H} = \mathcal{H}_1 \otimes \mathcal{H}_2 \otimes \mathcal{H}_3$

▶ **Examples** of separable potentials

(a) Particle in a box

$$\boxed{V(\vec{x}) = \left\{\begin{smallmatrix} 0 \text{ for } x_k \in (a_k, b_k), k=1,2,3 \\ \infty \text{ otherwise} \end{smallmatrix}\right\}}$$
$$= V_{(a_1,b_1)}(x) + V_{(a_2,b_2)}(y) + V_{(a_3,b_3)}(z)$$

$V_{(a_k,b_k)}(x_k) \equiv$ **1D** infinite square well which has the following solution:

$$\underbrace{\left(\tfrac{1}{2M}\hat{p}_k^2 + V_{(a_k,b_k)}\right)}_{\hat{H}_k}|\psi_{n_k}\rangle = \underbrace{\left(\tfrac{\pi\hbar}{\sqrt{2M}L_k}n_k\right)^2}_{E_{n_k}}|\psi_{n_k}\rangle \quad \text{with } n_k = 1, 2, 3, \dots$$

$|\psi_{n_k}\rangle \equiv \psi_{n_k}(x_k) \propto \sin\left[\frac{n_k\pi}{L_k}(x - a_k)\right]$ where $L_k = b_k - a_k$

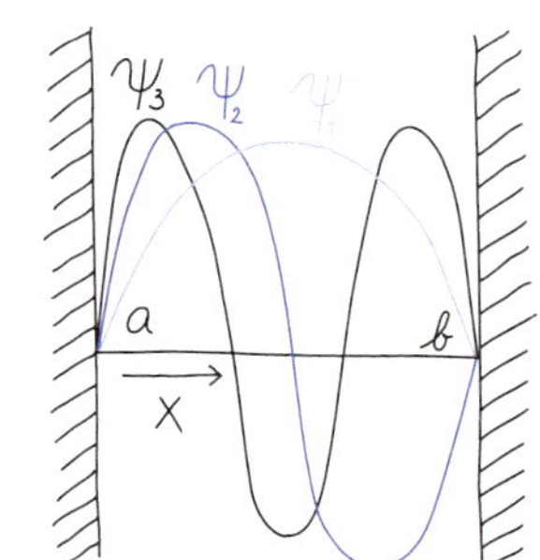

Solution of the **3D** problem:

$$\underbrace{[\hat{H}_1 + \hat{H}_2 + \hat{H}_3]}_{\hat{H}} \underbrace{|\psi_{n_1}\rangle|\psi_{n_2}\rangle|\psi_{n_3}\rangle}_{|\psi_{n_1n_2n_3}\rangle} = \underbrace{\tfrac{(\pi\hbar)^2}{2M}\left[\left(\tfrac{n_1}{L_1}\right)^2 + \left(\tfrac{n_2}{L_2}\right)^2 + \left(\tfrac{n_3}{L_3}\right)^2\right]}_{E_{n_1n_2n_3}} \underbrace{|\psi_{n_1}\rangle|\psi_{n_2}\rangle|\psi_{n_3}\rangle}_{|\psi_{n_1n_2n_3}\rangle}$$

Equilateral case: $L_k = L \Rightarrow E_{n_1n_2n_3} \mapsto E_N = \frac{(\pi\hbar)^2}{2ML^2}(\underbrace{n_1^2 + n_2^2 + n_3^2}_{N=3,6,9,11,12,14,\dots})$

$\Rightarrow$ degeneracy $d_N = 1, 3, 6, \dots$

Consequence: the ground state energy $E_{\rm gs} \propto \frac{1}{V^{2/3}}$ grows with volume $V \Rightarrow$ "Schrödinger pressure" against any increase of the particle containment

(b) Harmonic oscillator

$$V(\vec{x}) = \frac{M}{2}(\omega_1^2 x_1^2 + \omega_2^2 x_2^2 + \omega_3^2 x_3^2)$$

1D:

$$\underbrace{\left(\frac{1}{2M}\hat{p}_k^2 + \frac{M\omega_k^2}{2}\hat{x}_k^2\right)}_{\hat{H}_k} |\psi_{n_k}\rangle = \underbrace{\hbar\omega_k \left(n_k + \tfrac{1}{2}\right)}_{E_{n_k}} |\psi_{n_k}\rangle$$

with $n_k = 0, 1, 2, 3, \ldots$

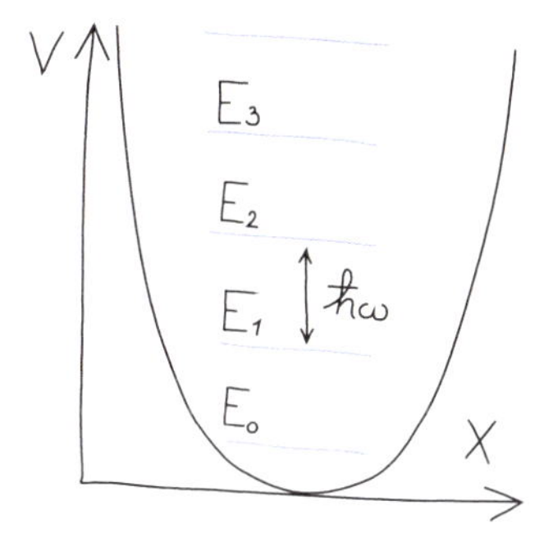

The solution obtained from the diff. form of Sch.eq.:

$$\left[\frac{d^2}{d\xi_k^2} + (\lambda - \xi_k^2)\right] \psi(\xi_k) = 0$$

where $\xi_k = \sqrt{\frac{M\omega_k}{\hbar}} x_k$ and $\lambda = \frac{2E}{M\omega}$

Eigenstates: $|\psi_{n_k}\rangle \equiv \psi_{n_k}(x_k) \propto e^{-\xi_k^2/2} H_{n_k}(\xi_k)$

$$H_n(\xi) \equiv \frac{d^n}{d\eta^n} \underbrace{e^{\xi^2-(\xi-\eta)^2}}_{\text{generating function}} \Big|_{\eta=0} \quad \equiv \textit{Hermite polynomials}$$

3D:

$$\underbrace{[\hat{H}_1 + \hat{H}_2 + \hat{H}_3]}_{\hat{H}} \underbrace{|\psi_{n_1}\rangle|\psi_{n_2}\rangle|\psi_{n_3}\rangle}_{|\psi_{n_1n_2n_3}\rangle} = \underbrace{\hbar\left(\omega_1 n_1 + \omega_2 n_2 + \omega_3 n_3 + \tfrac{3}{2}\right)}_{E_{n_1n_2n_3}} \underbrace{|\psi_{n_1}\rangle|\psi_{n_2}\rangle|\psi_{n_3}\rangle}_{|\psi_{n_1n_2n_3}\rangle}$$

Isotropic case: $\omega_k = \omega \Rightarrow E_{n_1n_2n_3} \mapsto E_N = \hbar\omega(N + \frac{3}{2})$

$$N = n_1 + n_2 + n_3 \Rightarrow \text{degeneracy} \left\{ \begin{matrix} N=0 & d_0=1 \\ N=1 & d_1=3 \\ N=2 & d_2=6 \\ \vdots & \end{matrix} \right\} d_N = \frac{(N+1)(N+2)}{2}$$

■ Orbital angular momentum

Before we continue with Hamiltonians and potentials, it is useful to construct angular momentum operators associated with orbital motions of a particle.

▶ In analogy with classical physics we introduce angular momentum of orbital motion:

components $\hat{L}_i = \varepsilon_{ijk}\hat{x}_j\hat{p}_k$ $\Leftrightarrow$ **vector** $\hat{\vec{L}} = \hat{\vec{x}} \times \hat{\vec{p}} = -i\hbar\left[\vec{x} \times \vec{\nabla}\right]$

Hermiticity: $\hat{L}_i^\dagger = \varepsilon_{ijk}\hat{p}_k^\dagger\hat{x}_j^\dagger = \varepsilon_{ijk}\hat{p}_k\hat{x}_j = \varepsilon_{ijk}\hat{x}_j\hat{p}_k = \hat{L}_i$ (since $j \neq k$)

▶ **Expression in spherical coordinates**

Transformation of wavefunctions: $\psi(x, y, z) \mapsto \psi(r, \vartheta, \varphi)$

Unit vectors

$$\begin{pmatrix} \vec{n}_r \\ \vec{n}_\vartheta \\ \vec{n}_\varphi \end{pmatrix} = \begin{pmatrix} \sin\vartheta\cos\varphi & \sin\vartheta\sin\varphi & \cos\vartheta \\ \cos\vartheta\cos\varphi & \cos\vartheta\sin\varphi & -\sin\vartheta \\ -\sin\varphi & \cos\varphi & 0 \end{pmatrix} \begin{pmatrix} \vec{n}_x \\ \vec{n}_y \\ \vec{n}_z \end{pmatrix}$$

$$\begin{pmatrix} \vec{n}_x \\ \vec{n}_y \\ \vec{n}_z \end{pmatrix} = \begin{pmatrix} \sin\vartheta\cos\varphi & \cos\vartheta\cos\varphi & -\sin\varphi \\ \sin\vartheta\sin\varphi & \cos\vartheta\sin\varphi & \cos\varphi \\ \cos\vartheta & -\sin\vartheta & 0 \end{pmatrix} \begin{pmatrix} \vec{n}_r \\ \vec{n}_\vartheta \\ \vec{n}_\varphi \end{pmatrix}$$

Orthogonal matrix $\Rightarrow$ [inverse=transpose]

$$\hat{\vec{L}} = -i\hbar\left[\underbrace{r\vec{n}_r}_{\vec{x}} \times \underbrace{\left(\vec{n}_r\frac{\partial}{\partial r} + \vec{n}_\vartheta\frac{1}{r}\frac{\partial}{\partial\vartheta} + \vec{n}_\varphi\frac{1}{r\sin\vartheta}\frac{\partial}{\partial\varphi}\right)}_{\vec{\nabla}}\right]$$

$\vec{n}_r \times \vec{n}_r = 0$
$\vec{n}_r \times \vec{n}_\vartheta = \vec{n}_\varphi$
$\vec{n}_r \times \vec{n}_\varphi = -\vec{n}_\vartheta$

$\boxed{\hat{\vec{L}} = -i\hbar\left[\vec{n}_\varphi \frac{\partial}{\partial\vartheta} - \vec{n}_\vartheta \frac{1}{\sin\vartheta}\frac{\partial}{\partial\varphi}\right]}$ acts only on the angular part of $\psi(r,\vartheta,\varphi)$

$\Rightarrow$ we consider factorized wavefunctions $\boxed{\psi(r,\vartheta,\varphi) \equiv R(r)\,\Omega(\vartheta,\varphi)}$

▶ **Angular-momentum component** along the z-axis

$$\vec{n}_z = \cos\vartheta\,\vec{n}_r - \sin\vartheta\,\vec{n}_\vartheta \quad\Rightarrow\quad \vec{n}_z\cdot\hat{\vec{L}} \equiv \boxed{\hat{L}_z = -i\hbar\frac{\partial}{\partial\varphi}}$$

Eigenvalue equation: $\hat{L}_z \underbrace{\Omega(\vartheta,\varphi)}_{f(\vartheta)g(\varphi)} = l_z\Omega(\vartheta,\varphi)$

$$-i\hbar\frac{\partial}{\partial\varphi}g(\varphi) = l_z g(\varphi) \quad \text{with condition } g(\varphi{+}2\pi) = g(\varphi)$$

$\Rightarrow \quad \boxed{l_z = m\hbar}$ with $m = 0, \pm1, \pm2, \pm3, \ldots$ and $\boxed{g_m(\varphi) = e^{im\varphi}}$

Additional condition $l_z^2 \leq L^2 \Rightarrow \boxed{|m| \leq m_{\max}}$ (see below and on p. 124)

From the symmetry argument, the same must be true for *any component.*

▶ **Squared orbital momentum**

The size of the angular-momentum vector is determined by the square:

$$\boxed{\hat{L}^2 = \hat{\vec{L}}\cdot\hat{\vec{L}}} = -\hbar^2\left[\vec{n}_\varphi\frac{\partial}{\partial\vartheta} - \vec{n}_\vartheta\frac{1}{\sin\vartheta}\frac{\partial}{\partial\varphi}\right]\cdot\left[\vec{n}_\varphi\frac{\partial}{\partial\vartheta} - \vec{n}_\vartheta\frac{1}{\sin\vartheta}\frac{\partial}{\partial\varphi}\right] =$$

$$= -\hbar^2\Bigg[\underbrace{\vec{n}_\varphi\frac{\partial}{\partial\vartheta}\cdot\vec{n}_\varphi\frac{\partial}{\partial\vartheta}}_{\frac{\partial^2}{\partial\vartheta^2}} - \underbrace{\vec{n}_\varphi\frac{\partial}{\partial\vartheta}\cdot\vec{n}_\vartheta\frac{1}{\sin\vartheta}\frac{\partial}{\partial\varphi}}_{0} - \underbrace{\vec{n}_\vartheta\frac{1}{\sin\vartheta}\frac{\partial}{\partial\varphi}\cdot\vec{n}_\varphi\frac{\partial}{\partial\vartheta}}_{-\cot\vartheta\frac{\partial}{\partial\vartheta}} + \underbrace{\vec{n}_\vartheta\frac{1}{\sin\vartheta}\frac{\partial}{\partial\varphi}\cdot\vec{n}_\vartheta\frac{1}{\sin\vartheta}\frac{\partial}{\partial\varphi}}_{\frac{1}{\sin^2\vartheta}\frac{\partial^2}{\partial\varphi^2}}\Bigg]$$

$$= -\hbar^2\Bigg[\underbrace{\frac{\partial^2}{\partial\vartheta^2} + \cot\vartheta\frac{\partial}{\partial\vartheta}}_{\frac{1}{\sin\vartheta}\frac{\partial}{\partial\vartheta}\sin\vartheta\frac{\partial}{\partial\vartheta}} + \frac{1}{\sin^2\vartheta}\frac{\partial^2}{\partial\varphi^2}\Bigg] \Rightarrow \boxed{\hat{L}^2 = -\hbar^2\left[\frac{1}{\sin\vartheta}\frac{\partial}{\partial\vartheta}\sin\vartheta\frac{\partial}{\partial\vartheta} + \frac{1}{\sin^2\vartheta}\frac{\partial^2}{\partial\varphi^2}\right]}$$

▶ Eigenequation $\hat{L}^2\Omega_{\lambda m}(\vartheta,\varphi) = \lambda^2\Omega_{\lambda m}(\vartheta,\varphi)$

solved with a factorized function $\Omega_{\lambda m}(\vartheta,\varphi) \equiv f_{\lambda m}(\vartheta)\,e^{im\varphi}$

$$\left[\frac{1}{\sin\vartheta}\frac{\partial}{\partial\vartheta}\sin\vartheta\frac{\partial}{\partial\vartheta} - \frac{m^2}{\sin^2\vartheta} + \frac{\lambda^2}{\hbar^2}\right] f_{\lambda m}(\vartheta) = 0 \xrightarrow[\xi=\cos\vartheta]{\text{subst.}} \left[\frac{\partial}{\partial\xi}(1-\xi^2)\frac{\partial}{\partial\xi} - \frac{m^2}{1-\xi^2} + \frac{\lambda^2}{\hbar^2}\right] f_{\lambda m}(\xi) = 0$$

The solution known in the form (for derivation see elsewhere):

$$f_{\lambda m}(\xi) \equiv P_{lm}(\xi) \propto (1-\xi^2)^{\frac{m}{2}}\frac{d^{l+m}}{d\xi^{l+m}}(\xi^2-1)^l \quad \textit{associated Legendre polynomial}$$

Eigenvalues $\boxed{\lambda^2 = l(l+1)\hbar^2 \text{ with } \begin{cases} l = 0, 1, 2\ldots \\ m = -l, (-l{+}1)\ldots 0\ldots(+l{-}1), +l \end{cases}}$

Eigenfunctions

$$\Omega_{\lambda m}(\vartheta,\varphi) = \boxed{\underbrace{\mathcal{N}_{lm}}_{\text{normalization}} P_{lm}(\cos\vartheta)\,e^{im\varphi} \equiv Y_{lm}(\vartheta,\varphi)} \quad \textbf{spherical harmonics}$$

Relation between l and m quantum numbers is represented by the following diagram:

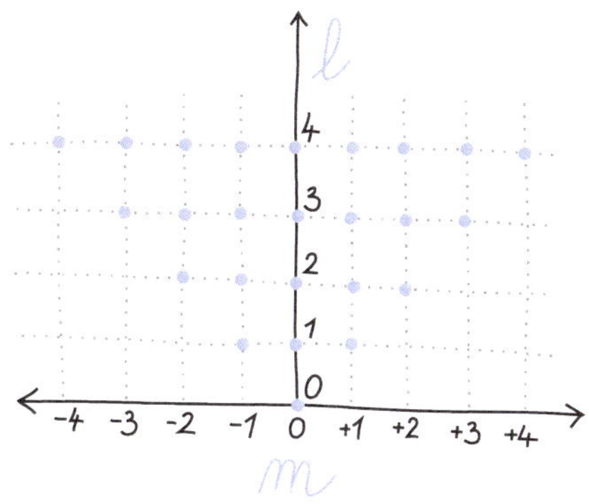

Note: The existence of simultaneous eigenstates of $\hat{L}^2$ and $\hat{L}_z$ is not accidental. It follows from the fact that both operators commute (see Sec. 1.3). The selection rules for m and l will be derived in Sec. 4.1.

Hamiltonian with isotropic potential

Equipped with the angular momentum operators, we can return to the Hamiltonian of a particle moving in a spherically symmetric potential field $\boxed{V(\vec{x}) = V(r)}$ This is a rather important situation in general since nature likes rotational invariance. We will briefly report three well known examples, assuming a certain degree of the reader's acquaintance with these elementary results.

▶ Hamiltonian written in **spherical coordinates**

$$\boxed{\hat{H} = -\frac{\hbar^2}{2M}\Delta + V(r)} = \frac{1}{2M}\left[\underbrace{\frac{-\hbar^2}{r^2}\frac{\partial}{\partial r}r^2\frac{\partial}{\partial r}}_{\hat{p}_r^2} + \underbrace{\frac{-\hbar^2}{r^2\sin\vartheta}\frac{\partial}{\partial\vartheta}\sin\vartheta\frac{\partial}{\partial\vartheta} + \frac{-\hbar^2}{r^2\sin^2\vartheta}\frac{\partial^2}{\partial\varphi^2}}_{r^{-2}\hat{L}^2}\right] + V(r)$$

can be decomposed into three parts:

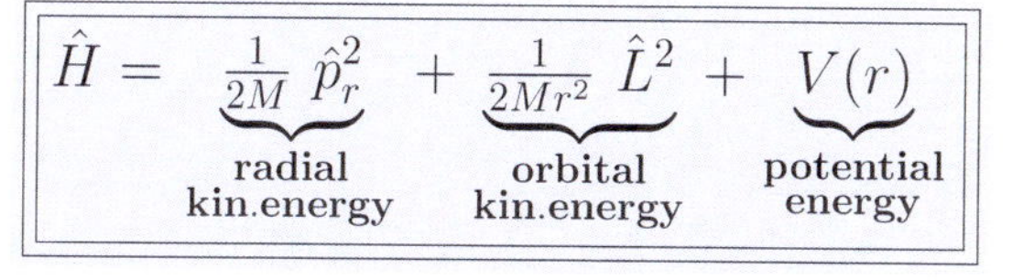

$$\hat{H} = \underbrace{\frac{1}{2M}\hat{p}_r^2}_{\text{radial kin.energy}} + \underbrace{\frac{1}{2Mr^2}\hat{L}^2}_{\text{orbital kin.energy}} + \underbrace{V(r)}_{\text{potential energy}}$$

with $\boxed{\hat{p}_r \equiv -i\hbar\left(\frac{\partial}{\partial r} + \frac{1}{r}\right)}$ **radial momentum**

▶ **Separation of variables**

The isotropic form of the Hamiltonian enables one to separate radial and angular variables through the wavefunction ansatz:

$$\boxed{\psi_{nlm}(r,\vartheta,\varphi) \equiv \underbrace{R_{nl}(r)}_{\frac{u_{nl}(r)}{r}} Y_{lm}(\vartheta,\varphi)}$$

$\hat{L}^2 Y_{lm}(\vartheta,\varphi) = \hbar^2 l(l+1) Y_{lm}(\vartheta,\varphi)$

The equation for R_{nl} reads: $\left[-\frac{\hbar^2}{2M}\frac{1}{r^2}\frac{d}{dr}r^2\frac{d}{dr} + \frac{\hbar^2 l(l+1)}{2Mr^2} + V(r)\right] R_{nl}(r) = E_{nl}R_{nl}(r)$

$$\boxed{\left[-\frac{\hbar^2}{2M}\frac{d^2}{dr^2} + \underbrace{\frac{\hbar^2 l(l+1)}{2Mr^2} + V(r)}_{V_{\rm eff}^{(l)}(r)}\right] u_{nl}(r) = E_{nl}u_{nl}(r)}$$ **radial Schrödinger eq.**

▶ **Unbound-state asymptotics** (eigenfunctions of radial momentum)

For $V(r) \xrightarrow{r\to\infty} 0$ we write down an $E > 0$ asymptotic radial solution for $l = 0$:

Spherical wave (for $r \gg 0$): $\boxed{R_{p_r}(r) \propto \frac{e^{i\frac{p_r r}{\hbar}}}{r}}$ $\equiv$ plane wave of $u(r)$

$-i\hbar\left(\frac{\partial}{\partial r}+\frac{1}{r}\right)\frac{e^{ip_r r/\hbar}}{r} = p_r\frac{e^{ip_r r/\hbar}}{r}$

▶ Bound state close to the origin $r{=}0$

Approximate equation $\frac{d^2u}{dr^2} - \frac{l(l+1)}{r^2}u \approx 0$ can be solved with $u(r) \propto r^k$

$k(k-1) = l(l+1) \quad\Rightarrow\quad k = \begin{cases} l+1 \\ -l \text{ (nonphysical)}\end{cases} \quad\Rightarrow\quad \boxed{u_{nl}(r)|_{r\sim 0} \approx r^{l+1} \xrightarrow{r\to 0} 0}$

▶ Example (a): finite spherical square well

$$\boxed{V(r) = \begin{cases} -V_0<0 & \text{for } r<R \\ 0 & \text{for } r\geq R\end{cases}}$$

Radial equation: $\left[\frac{d^2}{dr^2} - \frac{l(l+1)}{r^2} + \frac{2M(E-V)}{\hbar^2}\right]u_{nl}(r) = 0$ with $V = \begin{cases} -V_0 \\ 0\end{cases}$

Discrete spectrum $E_i \in (-V_0, 0)$, continuous spectrum $E \in (0, +\infty)$

$\kappa = \frac{\sqrt{2M(E+V_0)}}{\hbar} \qquad k = \frac{\sqrt{2ME}}{\hbar}\begin{cases} >0 & \text{for } E\geq 0 \\ = i\varkappa & \text{for } E<0\end{cases} \qquad r \to \rho \equiv \begin{cases} \kappa r & \text{for } r<R \\ kr & \text{for } r\geq R\end{cases}$

The eigenfunctions in a general case can expressed through Bessel & Neumann functions, or alternatively through Hankel functions:

$$R_{nl}(\rho) = \frac{u_{nl}(\rho)}{\rho} = \begin{cases} \textit{Bessel} & j_l(\rho) \propto_{\rho\to 0} \rho^l \\ \textit{Neumann} & n_l(\rho) \propto_{\rho\to 0} \rho^{-(l+1)} \\ \textit{Hankel} & h_l^+(\rho) = j_l(\rho) + in_l(\rho) \propto_{r\to\infty} \frac{e^{i(\rho - l\pi/2)}}{i\rho} \\ \quad\textit{functions} & h_l^-(\rho) = j_l(\rho) - in_l(\rho) \propto_{r\to\infty} \frac{e^{-i(\rho-l\pi/2)}}{i\rho}\end{cases}$$

For bound states ($E<0$) the $r{=}0$ & $r{\to}\infty$ conditions restrict the solution to:

$$\boxed{R_{nl}(r) = \begin{cases} Aj_l(\kappa r) & \text{for } r<R \\ B\Re h_l^+(i\varkappa r) & \text{for } r\geq R\end{cases}} \quad (\Re \equiv \mathrm{Re})$$

$$\text{Conditions}\left\{\begin{aligned} \frac{\frac{d}{d\rho}j_l(\kappa R)}{j_l(\kappa R)} &= \frac{\frac{d}{d\rho}\Re h_l^{(1)}(i\varkappa R)}{\Re h_l^+(i\varkappa R)} \\ \kappa^2 + \varkappa^2 &= \frac{2MV_0}{\hbar^2}\end{aligned}\right\} \Rightarrow \begin{cases}\text{numerical determination} \\ \text{of energy levels } E_{nl}\end{cases}$$

▶ Example (b): isotropic harmonic oscillator (revisited) $\boxed{V(r) = \frac{M\omega^2}{2}r^2}$

From the solution of the separable problem we know:

$$E_N = \hbar\omega(N+\tfrac{3}{2}) \text{ where } N = n_1+n_2+n_3$$

Solution in spherical coordinates (for the derivation see elsewhere):

$R_{nl}(\xi) \propto \xi^l L_n^{l+1/2}(\xi^2) \qquad$ with $\xi = \sqrt{\frac{M\omega}{\hbar}}r$ and $L_i^a(\rho) \equiv \rho^{-a}e^{\rho}\frac{d^i}{d\rho^i}(\rho^{i+a}e^{-\rho})$

$\equiv$ *generalized Laguerre polynomial*

Relation between quantum numbers from both solutions:

$\boxed{\underbrace{N+1}_{1,2,3\ldots} = 2n_r + l + 1}$ $\quad n_r = 0, 1, 2, \ldots$
radial quantum number $=$ number of nodes of $R_{nl}(r)$

► **Example (c): attractive Coulomb field** (hydrogen atom) $\boxed{V(r) = -\frac{\alpha}{r}}$

For hydrogen $\alpha = \frac{e^2}{4\pi\epsilon_0}$ and $\frac{M\alpha^2}{2\hbar^2} \doteq 13.6\,\text{eV}$

Bound states energies & wavefunctions (for the derivation see elsewhere):

$\boxed{E_n = -\frac{M\alpha^2}{2\hbar^2}\frac{1}{n^2}}$ $n = 1, 2, 3, \cdots \equiv$ principal quantum number : $\boxed{n = n_r + l + 1}$

$n_r = 0, 1, 2 \cdots \equiv$ radial q. number $=$ num. of nodes of $R_{nl}(r)$

Level n degeneracy $\begin{cases} l = 0, 1, \ldots (n-1) \\ m = -l, \cdots + l \end{cases} \Rightarrow d_n = \sum_{l=0}^{n-1}(2l+1) = n^2$

$\boxed{R_{nl}(r) \propto \rho^l e^{-\rho/2} L_{n-l-1}^{2l+1}(\rho)}$ with $\rho \equiv \frac{2}{na}r$, where $\boxed{a = \frac{\hbar^2}{\alpha M}}$ $\equiv$ **Bohr radius** and

$L_i^j(\rho) \equiv \frac{d^j}{d\rho^j} e^\rho \frac{d^i}{d\rho^i}(\rho^i e^{-\rho}) \equiv$ *associated Laguerre polynomials*

► Graphical expression of oscillator and hydrogen selection rules for quantum numbers

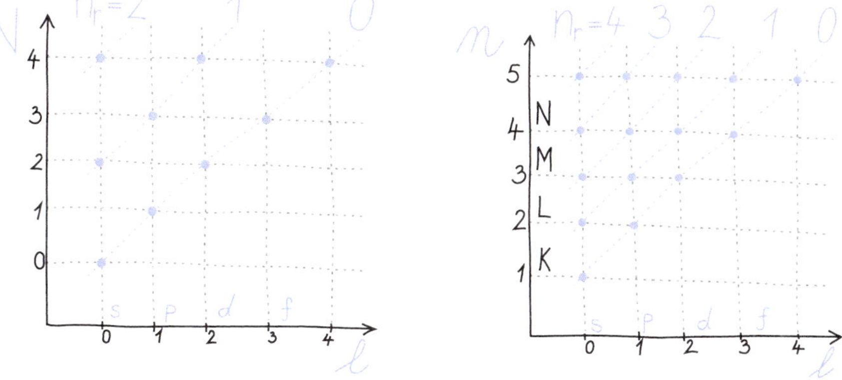

◄ **Historical remark**

1926: Erwin Schrödinger presents a series of 4 papers introducing wavefunction and explaining the quantization of energy in terms of the eigenvalue problem. The solutions of hydrogen-like atom and harmonic oscillator are given with the aid of the orbital angular momentum

1928-30: Application to molecules and solids; L. Pauling explains chemical bond, P.M. Morse describes vibrations of diatomic molecules (Morse potential), F. Bloch and L. Brillouin analyze eigenstates in periodic potentials

1929: First numerical algorithm for solving the eigenvalue problem

1932-1949: Formation of the shell model of atomic nuclei (D. Ivanenko, M. Goeppert-Mayer, J.H.D. Jensen)

■ Hamiltonian of a spin-$\frac{1}{2}$ particle in static electromagnetic field

At last, we look at the Hamiltonian of an electron-like particle moving in static electric and magnetic fields. This is an important example! We will discuss the invariance of the Schrödinger equation under the gauge transformation—the concept that becomes essential in the relativistic quantum theory.

▶ In analogy with the classical expression, the quantum Hamiltonian reads as:

$$\hat{H} = \frac{1}{2M}\left[\hat{\vec{p}} - q\vec{A}(\hat{\vec{x}})\right]^2 + qV(\hat{\vec{x}}) - \hat{\vec{\mu}}\cdot\vec{B}(\hat{\vec{x}})$$

$q \equiv$ particle charge
$V(\vec{x}) \equiv$ **scalar potential**
$\vec{A}(\vec{x}) \equiv$ **vector potential**

$$\left.\begin{array}{l}\vec{B}(\vec{x}) = \vec{\nabla}\times\vec{A}(\vec{x}) \equiv \text{ magnetic induction} \\ \vec{E}(\vec{x}) = -\vec{\nabla}V(\vec{x}) - \underbrace{\frac{\partial}{\partial t}\vec{A}(\vec{x})}_{0} \equiv \text{ electric intensity}\end{array}\right\} \text{ of stationary field}$$

▶ Operator of the particle **magnetic moment** $\hat{\vec{\mu}}$ is proportional to its spin:

$$\hat{\vec{\mu}} = g\,\mu\,\frac{1}{\hbar}\hat{\vec{S}} = g\,\frac{q\hbar}{2M}\,\frac{1}{2}\hat{\vec{\sigma}}$$

$g \equiv$ gyromagnetic ratio $\begin{pmatrix}\text{electron} & g{=}2 \\ \text{proton} & g{=}5.5856 \\ \text{neutron} & g{=}-3.8263\end{pmatrix}$

$\mu = \frac{q\hbar}{2M} = \left\{\begin{smallmatrix}\text{Bohr or} \\ \text{nuclear}\end{smallmatrix}\right\}$ magneton, $q = e$, $M = \left\{\begin{smallmatrix}m_e \\ m_p\end{smallmatrix}\right\}$

▶ Evaluation of the kinetic term [mind that $\hat{\vec{p}}$ and $\vec{A}(\hat{\vec{x}})$ do *not* commute]

$$\left[\hat{\vec{p}} - q\vec{A}(\hat{\vec{x}})\right]^2 \equiv \underbrace{\hat{\vec{p}}^2}_{-\hbar^2\Delta} \underbrace{-q\left[\hat{\vec{p}}\cdot\vec{A}(\hat{\vec{x}}) + \vec{A}(\hat{\vec{x}})\cdot\hat{\vec{p}}\right]}_{\substack{+i\hbar q\left[\vec{\nabla}\cdot\vec{A} + \vec{A}\cdot\vec{\nabla}\right] = \\ +i\hbar q\left[(\vec{\nabla}\cdot\vec{A}) + 2\vec{A}\cdot\vec{\nabla}\right]}} + q^2\vec{A}(\hat{\vec{x}})^2 \qquad \text{Hermitian}$$

▶ Spinor $\mathbf{\Psi}(\vec{x}) = \begin{pmatrix}\psi_\uparrow(\vec{x}) \\ \psi_\downarrow(\vec{x})\end{pmatrix}$ $\Rightarrow$ $\hat{H}\mathbf{\Psi} = E\mathbf{\Psi}$ yields **Pauli equation**:

$$-\frac{\hbar^2}{2M}\Delta\mathbf{\Psi} + \frac{i\hbar q}{2M}\underbrace{(\vec{\nabla}\cdot\vec{A})}_{0 \text{ in Lorentz calibration}}\mathbf{\Psi} + \frac{i\hbar q}{M}(\vec{A}\cdot\vec{\nabla}\mathbf{\Psi}) + \frac{q^2}{2M}\vec{A}^2\mathbf{\Psi} + qV\mathbf{\Psi} - g\frac{q\hbar}{2M}\frac{1}{2}(\hat{\vec{\sigma}}\cdot\vec{B})\mathbf{\Psi} = E\mathbf{\Psi}$$

▶ Special case: **homogeneous magnetic field**

$\vec{B}(\vec{x}) \equiv (0,0,B) \quad \Leftarrow \quad \vec{A}(\vec{x}) = \frac{B}{2}(-y,+x,0) \qquad$ satisfies $\vec{\nabla}\cdot\vec{A} = 0$

The term $\frac{i\hbar q}{M}(\vec{A}\cdot\vec{\nabla}\mathbf{\Psi}) = \frac{qB}{2M}\underbrace{i\hbar\left[-y\frac{\partial}{\partial x} + x\frac{\partial}{\partial y}\right]}_{-\hat{L}_z}\mathbf{\Psi}$ yields orbital ang. momentum

$\Rightarrow$ the whole Pauli eq.

$$\left[\underbrace{-\frac{\hbar^2}{2M}\Delta}_{\text{kin. en.}} \underbrace{+qV}_{\text{electrostat. energy}} \underbrace{-\frac{q}{2M}B\left(\hat{L}_z + g\hat{S}_z\right)}_{\substack{\text{mag. moment interaction} \\ \Rightarrow\textbf{Zeeman splitting}}} + \underbrace{\underbrace{\frac{q^2}{2M}\frac{B^2}{4}(x^2+y^2)}_{\frac{1}{2}M\omega_{\rm L}^2}}_{\text{kin. en. of \textbf{precessional motion}} \approx 0}\right]\mathbf{\Psi} = E\mathbf{\Psi}$$

For electron in hydrogen $\omega_{\rm L} \equiv \frac{qB}{2M}$ (Larmor frequency) $\lesssim \omega_{\rm orbital}$ for $B \lesssim 10^5$ T
$\Rightarrow$ the last term can be neglected unless the field is extremely large

▶ **Invariance under gauge transformations**

The action of classical elmg. field is invariant under gauge transformations generated by $f(\vec{x})$:

$$\vec{A} \mapsto \vec{A}' = \vec{A} - \vec{\nabla}f \qquad V \mapsto V' = V + \underbrace{\frac{\partial}{\partial t}f}_{=0 \text{ in stac. case}}$$

These transformations do not change $\vec{E}$ and $\vec{B}$, but they *change* Pauli equation. However, gauge transformation of $\vec{A}$ in Pauli equation is compensated by a local phase transformation of the wavefunction:

$$\boxed{\Psi(\vec{x}) \mapsto \Psi'(\vec{x}) \equiv \Psi(\vec{x}) e^{-i\frac{q}{\hbar}f(\vec{x})}}$$

Direct verification:
$(-i\hbar\vec{\nabla} - q\vec{A}') \cdot (-i\hbar\vec{\nabla} - q\vec{A}')\Psi' = e^{-i\frac{q}{\hbar}f}(-i\hbar\vec{\nabla} - q\vec{A})^2\Psi$
Therefore:

$$\left.\begin{array}{l} \hat{H}\Psi = E\Psi \quad \Rightarrow \quad \hat{H}'\Psi' = E\Psi' \\ |\psi(\vec{x}, m_s)|^2 = |\psi'(\vec{x}, m_s)|^2 \end{array}\right\} \Rightarrow$$

energy & probability density unchanged (but not all mystery is gone, see p. 120)

◀ **Historical remark**
1918: Hermann Weyl introduces the local gauge invariance of the metric tensor
1927: Wolfgang Pauli writes down the spinor equation for particle in mag.field
1928: H. Weyl concludes (also based on earlier works of other authors) that gauge transformation in QM is related to changing the phase of wavefunction

1.3 Compatible and incompatible observables

Operators, in contrast to ordinary numbers and functions used in classical physics, have one revolutionary property: they *may not* be commuting. The product $\hat{A}\hat{B}$ does not have to be the same operator as $\hat{B}\hat{A}$. This property turns out to be of essential importance for physics. For instance, we will see that it is responsible for the key feature of the quantum world: uncertainty.

We introduce the **commutator** of operators, $\boxed{[\hat{A}, \hat{B}] \equiv \hat{A}\hat{B} - \hat{B}\hat{A}}$, which is zero if $\hat{A}\hat{B} = \hat{B}\hat{A}$ and nonzero if $\hat{A}\hat{B} \neq \hat{B}\hat{A}$, and rise a classification of observables among each other:

$$\begin{cases} \text{compatible observables with} & \left[\hat{A}, \hat{B}\right] = 0 \\ \text{incompatible observables with} & \left[\hat{A}, \hat{B}\right] \neq 0 \end{cases}$$

■ Compatible observables

We first explore the case $\hat{A}\hat{B} = \hat{B}\hat{A}$. We show that such operators can be diagonalized simultaneously. A maximal set of commuting operators selects a unique basis in the Hilbert space and in this way create a particular representation.

► $[\hat{A}, \hat{B}] = 0 \Rightarrow$ eigenspaces of $\hat{B}$ invariant under $\hat{A}$ and vice versa

$$\hat{B}|\psi\rangle = b|\psi\rangle \Rightarrow \hat{B}\underbrace{(\hat{A}|\psi\rangle)}_{|\psi'\rangle} = b\underbrace{(\hat{A}|\psi\rangle)}_{|\psi'\rangle}$$

► Commuting operators have a **complete set of common eigenvectors**

Intuitively, this is obvious from the invariance of the eigenspaces $\mathcal{H}_a$ of $\hat{A}$ under the action of $\hat{B}$. The subspace $\mathcal{H}_a$ can therefore be considered as the Hilbert space where operator $\hat{B}$ finds eigenvectors $|b\rangle$.

A more rigorous proof:

$\{|a_i^{(k)}\rangle\}_{i,k}$ and $\{|b_j^{(l)}\rangle\}_{j,l} \equiv$ orthonormal eigenbases of $\hat{A}$ and $\hat{B}$, respectively

Unique expansion: $|a_i^{(k)}\rangle = \sum_j \underbrace{\sum_l \alpha_{ij}^{(kl)} |b_j^{(l)}\rangle}_{|\psi_{ij}^{(k)}\rangle} = \sum_j |\psi_{ij}^{(k)}\rangle$, where: $\hat{B}|\psi_{ij}^{(k)}\rangle = b_j|\psi_{ij}^{(k)}\rangle$

Eigenstate condition reads as:

$$(\hat{A}-a_i\hat{I})|a_i^{(k)}\rangle = 0 = \sum_j \overbrace{(\hat{A}-a_i\hat{I})|\psi_{ij}^{(k)}\rangle}^{|\tilde{\psi}_{ij}^{(k)}\rangle}$$

where: $\hat{B}|\tilde{\psi}_{ij}^{(k)}\rangle = b_j|\tilde{\psi}_{ij}^{(k)}\rangle$ (from invariance of $\mathcal{H}_{b_j}$ under $\hat{A}$)

$|\tilde{\psi}_{ij}^{(k)}\rangle$ with different j orthogonal $\Rightarrow$ the condition satisfied *iff* $|\tilde{\psi}_{ij}^{(k)}\rangle = 0\ \forall j$

$\Rightarrow |\psi_{ij}^{(k)}\rangle$ is a simultaneous eigenvector of $\hat{A}$ and $\hat{B}$ (eigenvalues a_i and b_j)

The same procedure repeated $\forall\ |a_i^{(k)}\rangle \Rightarrow$ the resulting set $\{|\psi_{ij}^{(k)}\rangle\}_{i,j,k}$ of simultaneous eigenvectors is complete since it allows one to expand the basis $\{|a_i^{(k)}\rangle\}_{i,k}$

$\Rightarrow \exists$ a *simultaneous orthonormal eigenbasis* $\{|a_ib_j{}^{(k)}\rangle\}_{i,j,k}$ of both $\hat{A}$ and $\hat{B}$, where ${}^{(k)}$ enumerates the states with the same combination of eigenvalues a_i, b_j

$\Rightarrow$ Observables A, B are "compatible"

▶ $\boxed{[\hat{A},\hat{B}] = 0 \quad \Leftrightarrow \quad [\hat{P}_{a_i},\hat{P}_{b_j}] = 0 \quad \forall\ i,j}$

$\Leftarrow$ follows from spectral decompositions: $\hat{A} = \sum_i a_i\hat{P}_{a_i}$ and $\hat{B} = \sum_j b_j\hat{P}_{b_j}$

$\Rightarrow$ follows from $\hat{P}_{a_i} = \sum_{j'\in\mathcal{S}_B^{a_i}} \sum_k |a_ib_{j'}{}^{(k)}\rangle\langle a_ib_{j'}{}^{(k)}|$, $\hat{P}_{b_j} = \sum_{i'\in\mathcal{S}_A^{b_j}} \sum_l |a_{i'}b_j{}^{(l)}\rangle\langle a_{i'}b_j{}^{(l)}|$

$\left\{ {\mathcal{S}_B^{a_i} \atop \mathcal{S}_A^{b_j}} \right\} \equiv$ the set of eigenvalues $\left\{ {b_{j'} \atop a_{i'}} \right\}$ contained in the eigenspace of $\left\{ {a_i \atop b_j} \right\}$

$$\hat{P}_{a_i}\hat{P}_{b_j} = \sum_{i',j'}\sum_{k,l} |a_ib_{j'}{}^{(k)}\rangle \underbrace{\langle a_ib_{j'}{}^{(k)}|a_{i'}b_j{}^{(l)}\rangle}_{\delta_{ii'}\delta_{jj'}\delta_{kl}} \langle a_{i'}b_j{}^{(l)}| = \sum_k |a_ib_j{}^{(k)}\rangle\langle a_ib_j{}^{(k)}| = \hat{P}_{b_j}\hat{P}_{a_i}$$

▶ **Complete set of commuting operators (of compatible observables)**

The above conclusions concerning 2 commuting operators can be generalized to an arbitrary number n of mutually commuting operators:

$n=3$: operators $\hat{A},\hat{B},\hat{C}$ satisfying $[\hat{A},\hat{B}]=[\hat{A},\hat{C}]=[\hat{B},\hat{C}]=0 \Rightarrow \exists$ simultaneous orthonormal eigenbasis $\{|a_ib_jc_k{}^{(l)}\rangle\}_{i,j,k,l}$ such that $\left.\begin{matrix}\hat{A}\\ \hat{B}\\ \hat{C}\end{matrix}\right\} |a_ib_jc_k{}^{(l)}\rangle = \left.\begin{matrix}a_i\\ b_j\\ c_k\end{matrix}\right\} |a_ib_jc_k{}^{(l)}\rangle$

...and analogously for $n>3$

A set of mutually commuting operators $\overbrace{\hat{A},\hat{B},\hat{C}\ldots}^{n}$ is **complete** if eigenvalues $\underbrace{a_i, b_j, c_k\ldots}_{n}$ uniquely determine a *single* eigenvector $|a_ib_jc_k\ldots\rangle$ (no ${}^{(l)}$ needed)

Consider $\hat{X}$ commuting with all operators $\hat{A},\hat{B},\hat{C}\ldots$ of a complete set. Then $\hat{X}|a_ib_jc_k\ldots\rangle = x|a_ib_jc_k\ldots\rangle$ and the eigenvalue x is determined by $a_i, b_j, c_k\ldots$.

$\Rightarrow x = f(a,b,c\ldots) \Rightarrow$ spectral decomposition $\hat{X} = \sum_{a_i,b_j,c_k\ldots} f(a_i,b_j,c_k\ldots)\hat{P}_{a_i,b_j,c_k\ldots}$

$\Rightarrow \boxed{\hat{X} = f(\hat{A},\hat{B},\hat{C}\ldots)}$ $\quad \hat{X}$ is a function of $\hat{A},\hat{B},\hat{C}\ldots$

The number n of operators in a complete set is usually identified with the number f of **quantum degrees of freedom**. Examples: Spin- & structureless particle in 3D has $f=3 \Rightarrow$ we need 3 commuting operators to uniquely determine a basis in $\mathcal{H}$. Note: the number n is fixed only within a certain algebra of pre-selected operators.

▶ **Combining complete sets in a product spaces**

Consider a composite system: $\mathcal{H} \equiv \mathcal{H}_1 \otimes \mathcal{H}_2$

$\underbrace{\{\hat{A}_1, \hat{B}_1, \hat{C}_1 \ldots\}}_{n_1} \equiv$ complete set in $\mathcal{H}_1$ $\quad \underbrace{\{\hat{A}_2, \hat{B}_2, \hat{C}_2 \ldots\}}_{n_2} \equiv$ complete set in $\mathcal{H}_2$

$$\Rightarrow \underbrace{\left\{\left\{(\hat{A}_1 \otimes \hat{I}), (\hat{B}_1 \otimes \hat{I}), (\hat{C}_1 \otimes \hat{I}) \ldots\right\}, \left\{(\hat{I} \otimes \hat{A}_2), (\hat{I} \otimes \hat{B}_2), (\hat{I} \otimes \hat{C}_2) \ldots\right\}\right\}}_{n_1+n_2} \equiv \textbf{complete set} \text{ in } \mathcal{H} \equiv \mathcal{H}_1 \otimes \mathcal{H}_2$$

$\boxed{[\hat{X}_1 \otimes \hat{I}, \hat{I} \otimes \hat{Y}_2] = 0}\ \forall \hat{X}_1, \hat{Y}_2$ (the same eigenvalues as the original sets)

$\Rightarrow$ addition of freedom-degree numbers for composite systems: $n = n_1 + n_2$

■ Incompatible observables

We turn to the case $\hat{A}\hat{B} \neq \hat{B}\hat{A}$. Such observables cannot be simultaneously diagonalized and exhibit a mutual uncertainty: increasing precision of one observable reduces precision of the other.

▶ Nonzero commutator expressed as: $\boxed{[\hat{A}, \hat{B}] = i\hat{C}}$

$(i\hat{C})^\dagger = (\hat{A}\hat{B} - \hat{B}\hat{A})^\dagger = \hat{B}^\dagger \hat{A}^\dagger - \hat{A}^\dagger \hat{B}^\dagger = -[\hat{A}, \hat{B}] = -i\hat{C} \Rightarrow \boxed{\hat{C} = \hat{C}^\dagger}$ for $\left\{ \begin{smallmatrix} \hat{A}=\hat{A}^\dagger \\ \hat{B}=\hat{B}^\dagger \end{smallmatrix} \right\}$

▶ **Uncertainty relation**

$$\boxed{\underbrace{[\langle A^2 \rangle_\psi - \langle A \rangle_\psi^2]}_{\langle\langle A^2 \rangle\rangle_\psi} \underbrace{[\langle B^2 \rangle_\psi - \langle B \rangle_\psi^2]}_{\langle\langle B^2 \rangle\rangle_\psi} \geq \tfrac{1}{4} \langle \psi | \hat{C} | \psi \rangle^2}$$

lower bound of the product of dispersions depends on $|\psi\rangle$

Proof:

$[\langle A^2 \rangle_\psi - \langle A \rangle_\psi^2] = \langle \psi | [\hat{A} - \langle A \rangle_\psi \hat{I}]^2 | \psi \rangle = \langle \varphi | \varphi \rangle$ with $|\varphi\rangle = [\hat{A} - \langle A \rangle_\psi \hat{I}] |\psi\rangle$

$[\langle B^2 \rangle_\psi - \langle B \rangle_\psi^2] = \langle \psi | [\hat{B} - \langle B \rangle_\psi \hat{I}]^2 | \psi \rangle = \langle \chi | \chi \rangle$ with $|\chi\rangle = [\hat{B} - \langle B \rangle_\psi \hat{I}] |\psi\rangle$

$$\langle\langle A^2 \rangle\rangle_\psi \langle\langle B^2 \rangle\rangle_\psi = \langle \varphi | \varphi \rangle \langle \chi | \chi \rangle \geq |\langle \varphi | \chi \rangle|^2 = \left| \langle \psi | [\hat{A} - \langle A \rangle_\psi \hat{I}][\hat{B} - \langle B \rangle_\psi \hat{I}] | \psi \rangle \right|^2 =$$

$$\left| \langle \psi | \hat{A}\hat{B} | \psi \rangle - \langle A \rangle_\psi \langle B \rangle_\psi \right|^2 = \left| \langle \psi | \tfrac{\hat{A}\hat{B} + \hat{B}\hat{A}}{2} | \psi \rangle + \langle \psi | \underbrace{\tfrac{\hat{A}\hat{B} - \hat{B}\hat{A}}{2}}_{\frac{i}{2}\hat{C}} | \psi \rangle - \langle A \rangle_\psi \langle B \rangle_\psi \right|^2$$

$$\geq \tfrac{1}{4} \langle \psi | \hat{C} | \psi \rangle^2$$

$\Rightarrow$ Non-commuting operators $\hat{A}, \hat{B}$ cannot be diagonalized simultaneously:

$\Rightarrow$ Observables A, B are "incompatible"

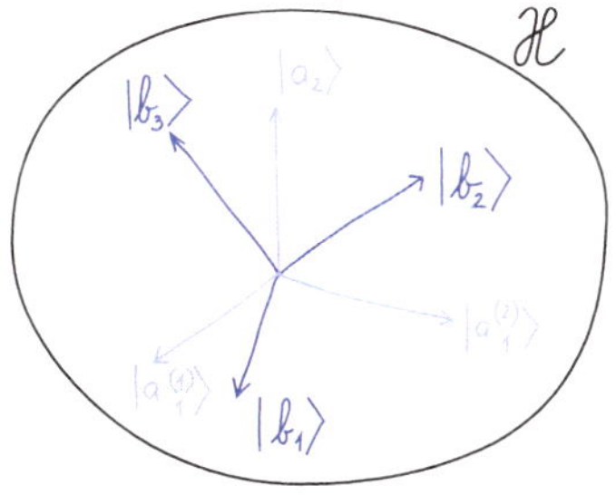

$\mathcal{H}$

$|b_3\rangle$ $|b_2\rangle$ $|b_1\rangle$

Analogy with Poisson brackets

Although incompatible ("non-commuting") observables are genuinely quantum invention, there exists a surprising parallel of this behavior in classical mechanics. It is based on the properties of Poisson brackets.

▶ Some properties of commutators

(a) $[\hat{A},\hat{B}]=-[\hat{B},\hat{A}]$ $\quad[\hat{A},\mathrm{const}\,\hat{I}]=0$

(b) Sums: $[a\hat{A}+a'\hat{A}',\hat{B}]=a[\hat{A},\hat{B}]+a'[\hat{A}',\hat{B}]$, $\quad[\hat{A},b\hat{B}+b'\hat{B}']=b[\hat{A},\hat{B}]+b'[\hat{A},\hat{B}']$

(c) Products: $[\hat{A}\hat{A}',\hat{B}]=\hat{A}[\hat{A}',\hat{B}]+[\hat{A},\hat{B}]\hat{A}'$, $\quad[\hat{A},\hat{B}\hat{B}']=\hat{B}[\hat{A},\hat{B}']+[\hat{A},\hat{B}]\hat{B}'$

(d) Jacobi identity: $[\hat{A},[\hat{B},\hat{C}]]+[\hat{B},[\hat{C},\hat{A}]]+[\hat{C},[\hat{A},\hat{B}]]=0$

▶ **Poisson bracket** for 2 classical observables in f degrees of freedom

$$\{A,B\}\equiv\sum_{i=1}^{f}\left(\frac{\partial A}{\partial p_i}\frac{\partial B}{\partial q_i}-\frac{\partial B}{\partial p_i}\frac{\partial A}{\partial q_i}\right)$$

$A\equiv A(p_1...p_f,q_1...q_f)$, $B\equiv B(p_1...p_f,q_1...q_f)$

Note: another definition, obtained by an exchange $p_i\leftrightarrow q_i$, yields the opposite sign of the Poisson bracket

▶ Properties of Poisson brackets are analogous to those of commutators:

$$\left\{\begin{array}{l}\{A,B\}=-\{B,A\},\qquad \{A,\mathrm{const}\}=0\\ \{aA+a'A',B\}=a\{A,B\}+a'\{A',B\},\qquad \{AA',B\}=A\{A',B\}+\{A,B\}A'\\ \{A,\{B,C\}\}+\{B,\{C,A\}\}+\{C,\{A,B\}\}=0\end{array}\right\}\Leftrightarrow[\hat{A},\hat{B}]$$

▶ **Geometrical meaning** of Poisson bracket

$$\{A,B\}=\underbrace{\left(-\frac{\partial A}{\partial q_1},...-\frac{\partial A}{\partial q_f},+\frac{\partial A}{\partial p_1},...+\frac{\partial A}{\partial p_f}\right)}_{\mathbb{J}_{2f}\vec{\nabla}_{2f}A\text{ vector}\perp\text{to gradient}}\cdot\underbrace{\left(\frac{\partial B}{\partial p_1},...\frac{\partial B}{\partial p_f},\frac{\partial B}{\partial q_1},...\frac{\partial B}{\partial q_f}\right)}_{\vec{\nabla}_{2f}B\text{ gradient}}$$

ordinary scalar product of two $2f$-dim vectors in the phase space

$\mathbb{J}_{2f}\equiv\begin{pmatrix}0 & -I_f\\ +I_f & 0\end{pmatrix}$ is the *symplectic matrix* in dim. $2f$ ($I_f\equiv$ unit matrix in dim. f)

$$\{A,A\}=0\;\Leftrightarrow\;\underbrace{(\mathbb{J}_{2f}\vec{\nabla}_{2f}A)}_{\text{one of the tangent vectors to }A=\text{const}}\perp\underbrace{(\vec{\nabla}_{2f}A)}_{\text{normal vector to }A=\text{const}}$$

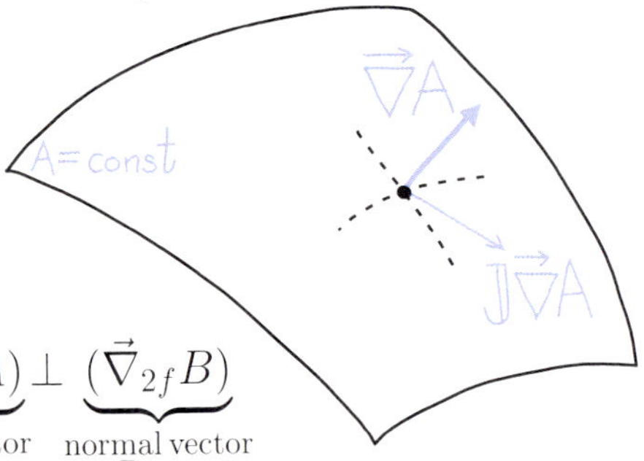

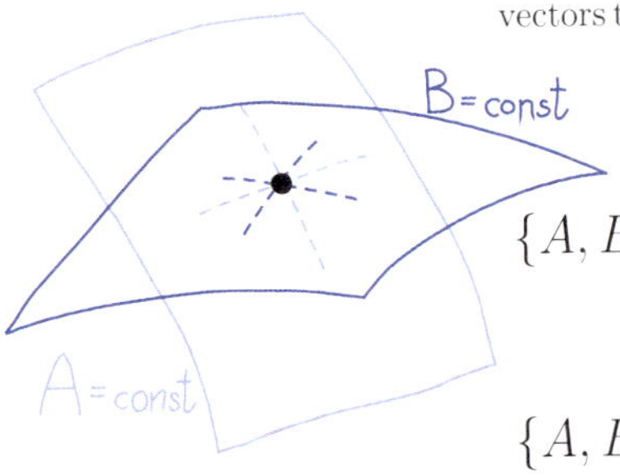

A=const, $\vec{\nabla}A$, $\mathbb{J}\vec{\nabla}A$, B=const, A=const

$$\{A,B\}=0\Leftrightarrow\underbrace{(\mathbb{J}_{2f}\vec{\nabla}_{2f}A)}_{\text{a tangent vector to }A=\text{const}}\perp\underbrace{(\vec{\nabla}_{2f}B)}_{\text{normal vector to }B=\text{const}}$$

$$\{A,B\}\neq0\Leftrightarrow(\mathbb{J}_{2f}\vec{\nabla}_{2f}A)\angle(\vec{\nabla}_{2f}B)$$

$\Rightarrow$ $\{A,B\}=0\;\Rightarrow$ hypersurfaces A=const & B=const locally coincide
$\{A,B\}\neq0\;\Rightarrow$ hypersurfaces A=const & B=const locally deviate

Consequences for classical statistical physics: Consider a statistical ensemble with probability distribution $\rho(p_1...,q_1...)$ spread around a point $(p_1^{(0)}...,q_1^{(0)}...)\equiv\mathbf{r}^{(0)}$ in the phase space. From the above geometrical considerations it follows that in this ensemble quantities A, B *cannot* have sharp values simultaneously if $\{A, B\} \neq 0$ at $\mathbf{r}^{(0)}$. This is a *classical analogue of uncertainty*!

All these analogies justify the following requirement:

▶ **Dirac quantization assumption** for observables $A, B, C \leftrightarrow \hat{A}, \hat{B}, \hat{C}$

Postulate: $\{A, B\} = C$ (classical) $\Rightarrow$ $[\hat{A}, \hat{B}] = -i\hbar\hat{C}$ (quantum)

■ Equivalent representations of quantum mechanics

A fascinating feature of physical description is that it can be cast in infinitely many equivalent ways. In other words, there exists a multitude of mathematical representations yielding the same observable output. In classical mechanics, this feature is anchored in the concept of canonical transformations. In quantum mechanics, the equivalent descriptions follow from the use of various Hilbert-space bases, which may be generated by alternative complete sets of observables.

▶ Any **complete set of commuting operators** $\{\hat{A}, \hat{B}, \dots\}$ with **discrete spectra** generates a countable orthonormal basis $\{|i\rangle\}_{i=1}^{d_{\mathcal{H}}}$ of $\mathcal{H}$:

$$\sum_{i=1}^{d_{\mathcal{H}}} |i\rangle\langle i| = \hat{I}_{\mathcal{H}} \qquad \langle i|j\rangle = \delta_{ij}$$

State vectors: $|\psi\rangle = \sum_i |i\rangle\langle i|\psi\rangle = \sum_i \underbrace{\langle i|\psi\rangle}_{\psi_i} |i\rangle \;\Leftrightarrow\; \boxed{|\psi\rangle \equiv \begin{pmatrix} \psi_1 \\ \psi_2 \\ \vdots \end{pmatrix}}$ (in)finite "columns"

Operators: $|\psi'\rangle = \hat{A}|\psi\rangle \Rightarrow \sum_i |i\rangle \underbrace{\langle i|\psi'\rangle}_{\psi'_i} = \sum_i |i\rangle \sum_j \underbrace{\langle i|\hat{A}|j\rangle}_{A_{ij}} \underbrace{\langle j|\psi\rangle}_{\psi_j} \Rightarrow \psi'_i = \sum_j A_{ij}\psi_j$

$$\boxed{\begin{pmatrix} \psi'_1 \\ \psi'_2 \\ \vdots \end{pmatrix} = \begin{pmatrix} A_{11} & A_{12} & \dots \\ A_{21} & A_{22} & \\ \vdots & & \ddots \end{pmatrix} \begin{pmatrix} \psi_1 \\ \psi_2 \\ \vdots \end{pmatrix}}$$ $\Rightarrow$ lin. operators $\equiv$ (in)finite matrices

▶ For a complete set $\{\hat{A}, \hat{B}, \dots\}$ with **continuous spectra** there $\exists$ a continuous "orthonormal basis" $\{|x\rangle\}_{x\in\mathcal{D}} \in \overline{\mathcal{H}}$

$$\int_{x\in\mathcal{D}} |x\rangle\langle x|dx = \hat{I}_{\overline{\mathcal{H}}} \qquad \langle x|x'\rangle = \delta(x-x')$$

State vectors: $|\psi\rangle = \int |x\rangle\langle x|\psi\rangle\, dx = \int \underbrace{\langle x|\psi\rangle}_{\psi(x)} |x\rangle\, dx \;\Leftrightarrow\; \boxed{|\psi\rangle \equiv \psi(x)}$ wavefuncs.

Operators: $|\psi'\rangle = \hat{A}|\psi\rangle \Rightarrow \int |x\rangle \underbrace{\langle x|\psi'\rangle}_{\psi'(x)}\, dx = \int |x\rangle \int \underbrace{\langle x|\hat{A}|x'\rangle}_{A(x,x')} \underbrace{\langle x'|\psi\rangle}_{\psi(x')}\, dx'\, dx$

$$\boxed{\psi'(x) = \int A(x,x')\,\psi(x')\,dx'}$$ $\Rightarrow$ lin. operators $\equiv$ integral kernels

▶ Complete set $\{\hat{A}, \hat{B}, \dots\}$ with mixed **discrete & continuous spectra** $\Rightarrow$ combined discrete-continuous "orthonormal basis" $\{|i, x\rangle\}_{\substack{i\in\mathcal{D}_i \\ x\in\mathcal{D}_x}} \in \overline{\mathcal{H}}$

$$\sum_{i\in\mathcal{D}_i}\int_{\mathcal{D}_x}|i,x\rangle\langle i,x|\,dx=\hat{I}_{\mathcal{H}}\qquad\langle i,x|i',x'\rangle=\delta_{ii'}\delta(x-x')$$

State vectors: $|\psi\rangle=\sum_{i\in\mathcal{D}_i}\int_{\mathcal{D}_x}|i,x\rangle\underbrace{\langle i,x|\psi\rangle}_{\psi_i(x)}dx\Leftrightarrow\boxed{|\psi\rangle\equiv\begin{pmatrix}\psi_1(x)\\\psi_2(x)\\\vdots\end{pmatrix}}$ (in)finite wavefunc. "columns"

Operators: $\boxed{\begin{pmatrix}\psi'_1(x)\\\psi'_2(x)\\\vdots\end{pmatrix}=\int\begin{pmatrix}A_{11}(x,x')&A_{12}(x,x')&\dots\\A_{21}(x,x')&A_{22}(x,x')&\\\vdots&&\ddots\end{pmatrix}\begin{pmatrix}\psi_1(x')\\\psi_2(x')\\\vdots\end{pmatrix}dx'}$ matrix integral kernels

◄ **Historical remark**
1925-6: M. Born, W. Heisenberg, P. Jordan write commutation relations between various observables (matrix mechanics) and introduce the concept of compatibility
1927: P. Jordan, P. Dirac attempt to introduce canonical transformations to QM
1927: John von Neumann formulates the concept of complete sets of observables and associates "canonical transformations" with different choices of this set
1927: Werner Heisenberg writes down the $\Delta x\Delta p$ uncertainty relation
1928: E.H. Kennard and H. Weyl derive the uncertainty relation from the commutator, generalization ∀ incompatible observables by H.P. Robertson in 1929
1930: P. Dirac relates commutators to Poisson brackets (⇒ canonical quantization)

2.3 Examples of commuting & noncommuting operators

We now apply the results of the previous section to the single-particle operators introduced in Sec. 2.2. In particular, the algebra of coordinate & momentum operators and that of angular momentum operators will be investigated. Representations of the single-particle Hilbert space will be built using these operators.

■ Coordinate & momentum

Coordinate & momentum operators jointly form the commonly known commutation relation—twin of the canonical Poisson bracket of classical mechanics. It leads to the familiar form of the uncertainty principle but also to the problems of coordinate & momentum in the ordinary Hilbert space (see Sec. 2.1).

► Canonical commutation relations

$$\hat{x}_i\equiv x_i\cdot\qquad\hat{p}_i\equiv-i\hbar\frac{\partial}{\partial x_i}\quad\Rightarrow\quad\boxed{[\hat{x}_i,\hat{x}_j]=[\hat{p}_i,\hat{p}_j]=0\quad[\hat{x}_i,\hat{p}_j]=i\hbar\delta_{ij}\hat{I}}$$

Poisson brackets $\{x_i,x_j\}=\{p_i,p_j\}=0,\{x_i,p_j\}=-\delta_{ij}$
These relations define general **canonically conjugate quantities**

Note: The same commutation relations can also be satisfied with:
$\hat{x}_i\equiv x_i\cdot\qquad\hat{p}_i\equiv-i\hbar\frac{\partial}{\partial x_i}+f(\vec{x})$

► **Heisenberg uncertainty relation**

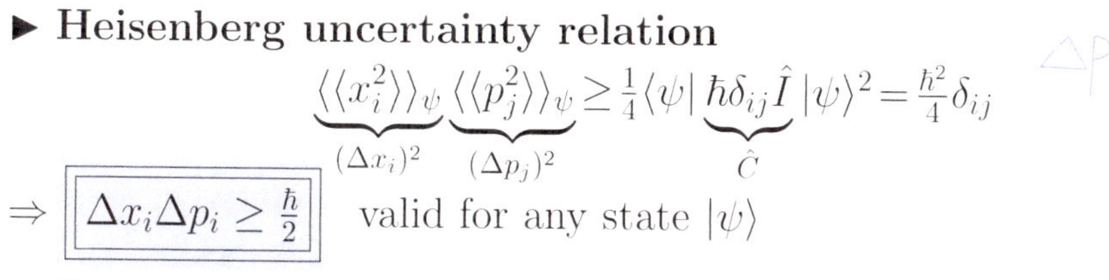

$$\underbrace{\langle\langle x_i^2\rangle\rangle_\psi}_{(\Delta x_i)^2}\underbrace{\langle\langle p_j^2\rangle\rangle_\psi}_{(\Delta p_j)^2}\geq\tfrac{1}{4}\langle\psi|\underbrace{\hbar\delta_{ij}\hat{I}}_{\hat{C}}|\psi\rangle^2=\tfrac{\hbar^2}{4}\delta_{ij}$$

$\Rightarrow$ $\boxed{\Delta x_i\Delta p_i\geq\frac{\hbar}{2}}$ valid for any state $|\psi\rangle$

► **Some general consequences** of canonical commutation relations

(a) Operators $\hat{x}_i$ and $\hat{p}_i$ *cannot* be represented in $\mathcal{H}$ of a *finite dimension d*. To show this, we introduce an important general quantity:

Trace of operator: $\boxed{\mathrm{Tr}\hat{A}=\sum_i\langle\phi_i|\hat{A}|\phi_i\rangle=\sum_i d_i a_i}\equiv\begin{cases}\text{sum of all eigenvalues (with}\\ \text{multiplicity factors } d_i\geq 1\\ \text{counting for degeneracies)}\end{cases}$

The trace is invariant under the $|\phi_i\rangle\to|\phi_i'\rangle$ transformation $\Rightarrow$ is independent of the chosen orthonormal basis. It also has the following important property:

$$\mathrm{Tr}(\hat{A}\hat{B})=\sum_i\langle\phi_i|\hat{A}\hat{B}|\phi_i\rangle=\sum_{i,j}\langle\phi_i|\hat{A}|\phi_j\rangle\langle\phi_j|\hat{B}|\phi_i\rangle=\sum_{j,i}\langle\phi_j|\hat{B}|\phi_i\rangle\langle\phi_i|\hat{A}|\phi_j\rangle=\mathrm{Tr}(\hat{B}\hat{A})$$

For coordinate and momentum operators represented in a finite dimension d, this relation yields a contradiction: $\mathrm{Tr}[\hat{x}_i,\hat{p}_i]=0\neq\mathrm{Tr}(i\hbar\hat{I}_d)=i\hbar d$

However, there $\exists$ various $d{=}\infty$ **discrete representations** of $\hat{x}$ and $\hat{p}$. For instance, the one obtained in the basis of 1D harmonic oscillator:

$$\hat{x}=\sqrt{\frac{\hbar}{2M\omega}}\begin{pmatrix}0&\sqrt{1}&0&0&0&\dots\\ \sqrt{1}&0&\sqrt{2}&0&0&\\ 0&\sqrt{2}&0&\sqrt{3}&0&\\ 0&0&\sqrt{3}&0&\sqrt{4}&\\ \vdots&&&\ddots&\ddots&\ddots\end{pmatrix}\qquad\hat{p}=\sqrt{\frac{M\hbar\omega}{2}}\begin{pmatrix}0&-i\sqrt{1}&0&0&0&\dots\\ i\sqrt{1}&0&-i\sqrt{2}&0&0&\\ 0&i\sqrt{2}&0&-i\sqrt{3}&0&\\ 0&0&i\sqrt{3}&0&-i\sqrt{4}&\\ \vdots&&&\ddots&\ddots&\ddots\end{pmatrix}$$

(b) *Eigenstates* of $\hat{x}_i$ and $\hat{p}_i$ are out of $\mathcal{H}$ (more precisely: $\nexists$ within $\mathcal{H}$)

Assume coordinate eigenstate $|x_i\rangle\in\mathcal{H}$ satisfying $\langle x_i|x_i\rangle=1\ \Rightarrow\ \frac{1}{i\hbar}\langle x_i|[\hat{x}_i,\hat{p}_i]|x_i\rangle=1$

But $\langle x_i|[\hat{x}_i,\hat{p}_i]|x_i\rangle=x_i\langle x_i|\hat{p}_i|x_i\rangle-x_i\langle x_i|\hat{p}_i|x_i\rangle=0\quad\Rightarrow$ contradiction

► **Canonical & mechanical momentum of particle in elmg. field**

$H=\frac{1}{2M}[\vec{p}-q\vec{A}(\vec{x})]^2+qV(\vec{x})$ with $\vec{p}\equiv$ canonical momentum

Mechanical momentum $\vec{\pi}$ defined through velocity: $\dot{\vec{x}}=\frac{\partial H}{\partial\vec{p}}=\frac{1}{M}\underbrace{[\vec{p}-q\vec{A}(\vec{x})]}_{\vec{\pi}}$

Operators $\boxed{\hat{\vec{p}}=-i\hbar\vec{\nabla}}$ & $\boxed{\hat{\vec{\pi}}=-i\hbar\vec{\nabla}-q\vec{A}(\vec{x})}$ of canonical & mechanical mom.

Commutator $[\hat{\pi}_i,\hat{\pi}_j]=\underbrace{[\hat{p}_i,\hat{p}_j]}_{0}-q[\hat{p}_i,\hat{A}_j]-q[\hat{A}_i,\hat{p}_j]+q^2\underbrace{[\hat{A}_i,\hat{A}_j]}_{0}=i\hbar q\underbrace{\left(\frac{\partial A_j}{\partial x_i}-\frac{\partial A_i}{\partial x_j}\right)}_{\varepsilon_{ijk}B_k}$

$\boxed{[\hat{\pi}_i,\hat{\pi}_j]=i\hbar q\,\varepsilon_{ijk}B_k(\vec{x})}$ $\Rightarrow$ for $\vec{B}{\neq}0$ velocity components incompatible

■ Coordinate & momentum representations

Although coordinate and momentum operators are not the nicest ones (the corresponding eigenstates dwelling somewhere outside the ordinary Hilbert space), the most familiar representations of quantum mechanics are based on these operators. For the sake of simplicity, we restrict ourselves to the 1D case.

▶ Coordinate representation in 1D

State vector $|\psi\rangle = \int \langle x|\psi\rangle |x\rangle \, dx \quad \Rightarrow \quad$ wavefunction $\boxed{\psi(x) \equiv \langle x|\psi\rangle}$

$$\left.\begin{array}{l} \text{Scalar product: } \langle\psi|\psi'\rangle = \int \langle\psi|x\rangle\langle x|\psi'\rangle dx = \int \psi(x)^* \, \psi'(x) \, dx \\ \text{Position operator: } \hat{x}\psi(x) = x\psi(x) \\ \text{Momentum operator: } \hat{p}\psi(x) = -i\hbar \frac{d}{dx}\psi(x) \end{array}\right\} \text{expressions used so far}$$

Note: Strictly, all these relations (as well as those below) should be restricted only to $|\psi\rangle \in \underline{\mathcal{H}}$ (a dense subset of $\mathcal{H}$)

▶ Momentum representation in 1D

State vector $|\psi\rangle = \int \langle p|\psi\rangle |p\rangle \, dp \quad \Rightarrow \quad$ wavefunction $\boxed{\tilde{\psi}(p) \equiv \langle p|\psi\rangle}$

One gets expressions analogous (complementary) to the x-representation:

Scalar product: $\langle\psi|\psi'\rangle = \int \langle\psi|p\rangle\langle p|\psi'\rangle dp = \int \tilde{\psi}(p)^* \, \tilde{\psi}'(p) \, dp$

Momentum operator: $\hat{p}\tilde{\psi}(p) = \langle p|\hat{p}|\psi\rangle = p\langle p|\psi\rangle \quad \Rightarrow \quad \boxed{\hat{p}\,\tilde{\psi}(p) = p\,\tilde{\psi}(p)}$

Position operator: $\hat{x}\tilde{\psi}(p) = \langle p|\hat{x}|\psi\rangle = \int \underbrace{\langle p|\hat{x}|p'\rangle}_{X(p,p')} \underbrace{\langle p'|\psi\rangle}_{\tilde{\psi}(p')} dp' =$

$$= \iiint \underbrace{\langle p|x\rangle}_{\frac{1}{\sqrt{2\pi\hbar}}e^{-i\frac{px}{\hbar}}} \underbrace{\langle x|\hat{x}|x'\rangle}_{x\delta(x-x')} \underbrace{\langle x'|p'\rangle}_{\frac{1}{\sqrt{2\pi\hbar}}e^{+i\frac{p'x'}{\hbar}}} \tilde{\psi}(p') \, dx \, dx' \, dp' = \frac{1}{2\pi\hbar} \iint \underbrace{x e^{i\frac{(p'-p)x}{\hbar}}}_{i\hbar\frac{d}{dp}e^{i\frac{(p'-p)x}{\hbar}}} \tilde{\psi}(p') \, dx \, dp'$$

$$= \frac{i}{2\pi}\frac{d}{dp} \int \underbrace{\int e^{i\frac{(p'-p)x}{\hbar}} dx}_{2\pi\hbar\delta(p'-p)} \tilde{\psi}(p') \, dp' = i\hbar\frac{d}{dp}\tilde{\psi}(p) \quad \Rightarrow \quad \boxed{\hat{x}\tilde{\psi}(p) = +i\hbar\frac{d}{dp}\tilde{\psi}(p)}$$

▶ Relation between x- & p-representations: Fourier transformations

Relation between eigenstates:

	coordinate rep.	momentum rep.
$\lvert x'\rangle$	$\delta(x-x')$	$\frac{1}{\sqrt{2\pi\hbar}}\, e^{-ix'p/\hbar}$
$\lvert p'\rangle$	$\frac{1}{\sqrt{2\pi\hbar}}\, e^{+ip'x/\hbar}$	$\delta(p-p')$

Relation between general states:

$$\langle p|\psi\rangle = \int_{-\infty}^{+\infty} \underbrace{\langle p|x\rangle}_{\frac{1}{\sqrt{2\pi\hbar}}e^{-i\frac{px}{\hbar}}} \underbrace{\langle x|\psi\rangle}_{\psi(x)} dx = \boxed{\frac{1}{\sqrt{2\pi\hbar}} \int_{-\infty}^{+\infty} e^{-i\frac{px}{\hbar}} \, \psi(x) \, dx = \tilde{\psi}(p)}$$

$$\langle x|\psi\rangle = \int_{-\infty}^{+\infty} \underbrace{\langle x|p\rangle}_{\frac{1}{\sqrt{2\pi\hbar}}e^{+i\frac{px}{\hbar}}} \underbrace{\langle p|\psi\rangle}_{\tilde{\psi}(p)} dp = \boxed{\frac{1}{\sqrt{2\pi\hbar}} \int_{-\infty}^{+\infty} e^{+i\frac{px}{\hbar}} \, \tilde{\psi}(p) \, dp = \psi(x)}$$

In transition to **3D**, one applies the following substitutions:

$$\frac{1}{\sqrt{2\pi\hbar}} \to \frac{1}{(2\pi\hbar)^{\frac{3}{2}}} \qquad \begin{array}{l} dx \to d\vec{x} \\ dp \to d\vec{p} \end{array} \qquad p\,x \to \vec{p}\cdot\vec{x}$$

▶ Gaussian wavepackets

A family of wavefunctions $\in \mathcal{H}$ suitable for the description of particles partially localized in both coordinate & momentum spaces. They are defined as states whose probability density $\rho(p) \equiv |\tilde{\psi}(p)|^2$ in momentum space has the Gaussian form with average p_0 and dispersion σ_p^2:

$$\boxed{\tilde{\psi}(p) = \frac{1}{(2\pi\sigma_p^2)^{\frac{1}{4}}} e^{-\frac{(p-p_0)^2}{4\sigma_p^2}}} \qquad \text{normalization: } \int\limits_{-\infty}^{+\infty} |\tilde{\psi}(p)|^2 \, dp = 1$$

Coordinate representation:

$$\psi(x) = \frac{1}{\sqrt{2\pi\hbar}} \int\limits_{-\infty}^{+\infty} e^{+i\frac{px}{\hbar}} \, \tilde{\psi}(p) \, dp = \underbrace{\frac{1}{(8\pi^3\hbar^2\sigma_p^2)^{\frac{1}{4}}}}_{C} \int\limits_{-\infty}^{+\infty} \underbrace{e^{+i\frac{px}{\hbar} - \frac{p^2 - pp_0 + p_0^2}{4\sigma_p^2}}}_{e^{-\frac{1}{4\sigma_p^2}p^2 + (\frac{p_0}{2\sigma_p^2} + \frac{ix}{\hbar})p - \frac{p_0^2}{4\sigma_p^2}} \equiv e^{ap^2+bp+c}} dp = C\sqrt{\frac{\pi}{|a|}} \, e^{c - \frac{b^2}{4a}} =$$

$$\boxed{\frac{1}{(2\pi\sigma_x^2)^{\frac{1}{4}}} e^{-\frac{x^2}{4\sigma_x^2}} \, e^{+i\frac{p_0 x}{\hbar}} = \psi(x)}$$

with σ_x satisfying $\boxed{\sigma_x \sigma_p = \frac{\hbar}{2}}$

$\Rightarrow$ Heisenberg relation minimized

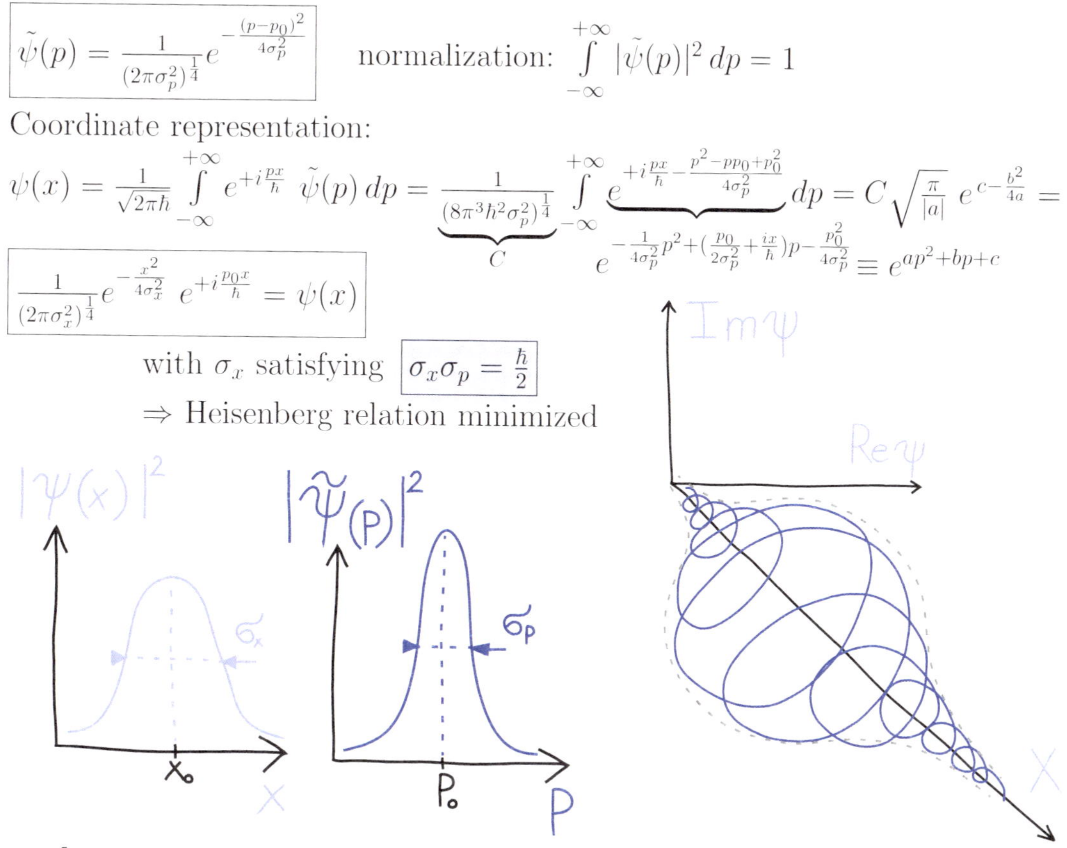

■ Angular momentum operators

Let us analyze commutation relations of angular-momentum operators. In fact, it is these relations what allows us to recognize that a given set of operators (like Pauli matrices) represents an angular momentum. In other words: what commutes like angular momentum *is* angular momentum.

▶ Components of **spin** $\frac{1}{2}$

$$\left\{ \begin{array}{l} [\hat{S}_x, \hat{S}_y] = \frac{\hbar^2}{4}[\hat{\sigma}_x, \hat{\sigma}_y] = 2i\frac{\hbar^2}{4}\hat{\sigma}_z = i\hbar\hat{S}_z \\ [\hat{S}_y, \hat{S}_z] = \frac{\hbar^2}{4}[\hat{\sigma}_y, \hat{\sigma}_z] = 2i\frac{\hbar^2}{4}\hat{\sigma}_x = i\hbar\hat{S}_x \\ [\hat{S}_z, \hat{S}_x] = \frac{\hbar^2}{4}[\hat{\sigma}_z, \hat{\sigma}_x] = 2i\frac{\hbar^2}{4}\hat{\sigma}_y = i\hbar\hat{S}_y \end{array} \right\} \Rightarrow \boxed{[\hat{S}_i, \hat{S}_j] = i\hbar\varepsilon_{ijk}\hat{S}_k} \text{ or } [\hat{\sigma}_i, \hat{\sigma}_j] = 2i\varepsilon_{ijk}\hat{\sigma}_k$$

Uncertainty relation $\langle\langle S_x^2\rangle\rangle_\psi \langle\langle S_y^2\rangle\rangle_\psi \geq \frac{\hbar^2}{4}\langle\psi|\hat{S}_z|\psi\rangle^2$

$$\Rightarrow \quad \boxed{(\Delta S_x)_\psi (\Delta S_y)_\psi \geq \frac{\hbar}{2}\,|\langle S_z\rangle_\psi|}$$

$|\psi\rangle = \alpha|\uparrow\rangle + \beta|\downarrow\rangle$ with $|\alpha|^2 + |\beta|^2 = 1$

$\Rightarrow (\Delta S_x)_\psi (\Delta S_y)_\psi \geq \frac{\hbar^2}{2}\left|\frac{1}{2} - |\beta|^2\right|$

$(\Delta S_x)_\psi (\Delta S_y)_\psi \geq 0$ for $|\beta|^2 = \frac{1}{2}$
corresponds to spin $\in xy$ plane (see p. 28)

▶ Components of **orbital angular momentum**

$$\left.\begin{cases}[\hat{L}_x,\hat{L}_y]=[\hat{y}\hat{p}_z-\hat{z}\hat{p}_y,\hat{z}\hat{p}_x-\hat{x}\hat{p}_z]=[\hat{y}\hat{p}_z,\hat{z}\hat{p}_x]+[\hat{z}\hat{p}_y,\hat{x}\hat{p}_z]=i\hbar(\hat{x}\hat{p}_y-\hat{y}\hat{p}_x)=i\hbar\hat{L}_z\\ [\hat{L}_y,\hat{L}_z]=[\hat{z}\hat{p}_x-\hat{x}\hat{p}_z,\hat{x}\hat{p}_y-\hat{y}\hat{p}_x]=[\hat{z}\hat{p}_x,\hat{x}\hat{p}_y]+[\hat{x}\hat{p}_z,\hat{y}\hat{p}_x]=i\hbar(\hat{y}\hat{p}_z-\hat{z}\hat{p}_y)=i\hbar\hat{L}_x\\ [\hat{L}_z,\hat{L}_x]=[\hat{x}\hat{p}_y-\hat{y}\hat{p}_x,\hat{y}\hat{p}_z-\hat{z}\hat{p}_y]=[\hat{x}\hat{p}_y,\hat{y}\hat{p}_z]+[\hat{y}\hat{p}_x,\hat{z}\hat{p}_y]=i\hbar(\hat{z}\hat{p}_x-\hat{x}\hat{p}_z)=i\hbar\hat{L}_y\end{cases}\right\} \Rightarrow \boxed{[\hat{L}_i,\hat{L}_j]=i\hbar\varepsilon_{ijk}\hat{L}_k}$$

Poisson brackets $\{L_i, L_j\} = -\varepsilon_{ijk} L_k$

▶ Components of **total angular momentum** of a spin-$\frac{1}{2}$ particle

$$\boxed{\hat{J}_i = \hat{L}_i + \hat{S}_i} \quad \boxed{[\hat{L}_i, \hat{S}_j] = 0} \quad \Rightarrow \quad [\hat{J}_i, \hat{J}_j] = [\hat{L}_i, \hat{L}_j] + [\hat{S}_i, \hat{S}_j] = i\hbar\varepsilon_{ijk} \overbrace{(\hat{L}_k + \hat{S}_k)}^{\hat{J}_k}$$

▶ **General angular momentum**

Components $\{\hat{J}_i\}_{i=1}^3$ satisfying commutation relations $\boxed{[\hat{J}_i, \hat{J}_j] = i\hbar\varepsilon_{ijk}\hat{J}_k}$

$\hat{J}_i$ mutually incompatible ⇒ uncertainty relations: $\boxed{(\Delta J_x)_\psi (\Delta J_y)_\psi \geq \frac{\hbar}{2} |\langle J_z \rangle_\psi|}$

The **squared angular momentum** $\boxed{\hat{J}^2 = \sum_{j=1}^{3} \hat{J}_j^2 \equiv \hat{J}_j \hat{J}_j}$ is compatible with all components $\hat{J}_i$:

$$[\hat{J}_i, \hat{J}_j\hat{J}_j] = \hat{J}_j \underbrace{[\hat{J}_i, \hat{J}_j]}_{i\hbar\varepsilon_{ijk}\hat{J}_k} + \underbrace{[\hat{J}_i, \hat{J}_j]}_{i\hbar\varepsilon_{ijk}\hat{J}_k} \hat{J}_j = i\hbar\varepsilon_{ijk}(\hat{J}_j\hat{J}_k + \hat{J}_k\hat{J}_j) = \boxed{0 = [\hat{J}_i, \hat{J}^2]}$$

⇒ $\hat{J}^2$ can be diagonalized simultaneously with any component $\hat{J}_i$

▶ **Simultaneous eigenfunctions** of $\hat{J}_z$ and $\hat{J}^2$

Orbital momentum:
$$\boxed{\begin{array}{l}\hat{L}^2 R(r) Y_{lm}(\vartheta,\varphi) = l(l+1)\hbar^2\, R(r) Y_{lm}(\vartheta,\varphi)\\ \hat{L}_z\, R(r) Y_{lm}(\vartheta,\varphi) = m\hbar\, R(r) Y_{lm}(\vartheta,\varphi)\end{array}} \quad \forall R(r)$$

$$l = 0, 1, 2, ..., \quad m = -l, (-l+1), \ldots, (+l-1), +l.$$

Spin $\frac{1}{2}$: $$\hat{S}^2 = \frac{\hbar^2}{4}[\underbrace{\hat{\sigma}_x^2}_{\hat{I}} + \underbrace{\hat{\sigma}_y^2}_{\hat{I}} + \underbrace{\hat{\sigma}_z^2}_{\hat{I}}] = \underbrace{\tfrac{3}{4}}_{\frac{1}{2}(\frac{1}{2}+1) = s(s+1)} \hbar^2 \hat{I}$$

$$\hat{S}^2 \left(\begin{smallmatrix}\alpha\\ \beta\end{smallmatrix}\right) = \tfrac{1}{2}(\tfrac{1}{2}+1)\hbar^2 \left(\begin{smallmatrix}\alpha\\ \beta\end{smallmatrix}\right) \quad \forall \left(\begin{smallmatrix}\alpha\\ \beta\end{smallmatrix}\right) \in \mathbb{C}^2, \quad \hat{S}_z \left(\begin{smallmatrix}1\\ 0\end{smallmatrix}\right) = +\tfrac{1}{2}\hbar \left(\begin{smallmatrix}1\\ 0\end{smallmatrix}\right), \quad \hat{S}_z \left(\begin{smallmatrix}0\\ 1\end{smallmatrix}\right) = -\tfrac{1}{2}\hbar \left(\begin{smallmatrix}0\\ 1\end{smallmatrix}\right)$$

General spin: $\boxed{s = \frac{1}{2}, 1, \frac{3}{2}, 2, \frac{5}{2}, \ldots}$ represented in $\mathcal{H}_{\text{spin}} \equiv \mathbb{C}^{2s+1}$ (see Sec. 4):

$\hat{S}^2 = s(s+1)\hbar^2 \hat{I}$ $\quad$ $\hat{S}_z$ can be represented by a diagonal matrix:

(for another s=1 representation see p. 60)

$$\boxed{\hat{S}^2 \begin{pmatrix}\alpha_{+s}\\ \vdots\\ \alpha_{-s}\end{pmatrix} = s(s+1)\hbar^2 \begin{pmatrix}\alpha_{+s}\\ \vdots\\ \alpha_{-s}\end{pmatrix}} \; \forall \begin{pmatrix}\alpha_{+s}\\ \vdots\\ \alpha_{-s}\end{pmatrix} \in \mathbb{C}^{2s+1}, \quad \boxed{\hat{S}_z \begin{pmatrix}0\\ \vdots\\ \alpha_{m_s}\\ \vdots\\ 0\end{pmatrix} = m_s\hbar \begin{pmatrix}0\\ \vdots\\ \alpha_{m_s}\\ \vdots\\ 0\end{pmatrix}}$$

⇒ Any vector $\in \mathbb{C}^{2s+1}$ is an eigenvector of $\hat{S}^2$ with quantum number s

Vectors $\left(\begin{smallmatrix}1\\ 0\\ \vdots\\ 0\end{smallmatrix}\right), \left(\begin{smallmatrix}0\\ 1\\ 0\\ \vdots\end{smallmatrix}\right), ..., \left(\begin{smallmatrix}0\\ \vdots\\ 0\\ 1\end{smallmatrix}\right)$ are simultaneously eigenvectors of $\hat{S}_z$ with quantum numbers $m = +s, (+s-1), ..., -s$, respectively

◀ **Historical remark**
1926: M. Born, W. Heisenberg, P. Jordan give commutation relations for position & momentum and for the components of angular momentum
1927-8: H. Weyl analyzes algebraic properties of position & momentum operators
1930: Paul Dirac elaborates position & momentum representations and presents an algebraic derivation of angular momentum eigenvalues
1931: M. Stone & J.von Neumann prove unitary equivalence of representations conserving the canonical commutation relation (Stone - von Neumann theorem)

■ Complete sets of commuting operators for a structureless particle

Below we give several examples of the complete set of observables characterizing a single spinless particle in 3D. Such a system has $f=3$ classical degrees of freedom, and also its quantum state is completely determined by eigenvalues of 3 commuting operators. These operators can be chosen in different ways.

▶ **Coordinates** $\hat{\vec{x}} \equiv (\hat{x}_1, \hat{x}_2, \hat{x}_3)$

Eigenbasis $\boxed{\Phi_{\vec{y}}(\vec{x}) = \delta(\vec{x}-\vec{y})}$ satisfying $\langle \Phi_{\vec{y}} | \Phi_{\vec{y}'} \rangle = \delta(\vec{y}-\vec{y}')$
General wavefunction: $\psi(\vec{x}) = \int \psi(\vec{y})\, \Phi_{\vec{y}}(\vec{x})\, d\vec{y}$

Note: dimension of $\Phi_{\vec{y}}(\vec{x})$ is $[\text{length}]^{-3} \Rightarrow$ it represents an amplitude density in a **joint space** of $\vec{x}$ & $\vec{y}$ (normal wavefunction is amplitude density only in $\vec{x}$)

▶ **Momenta** $\hat{\vec{p}} \equiv (\hat{p}_1, \hat{p}_2, \hat{p}_3)$

Eigenbasis $\boxed{\Phi_{\vec{p}}(\vec{x}) = \mathcal{N}_{\vec{p}}\, e^{i\frac{\vec{p}\cdot\vec{x}}{\hbar}}}$ with coeffs. $\mathcal{N}_{\vec{p}}$ given by "normalization":

$$\langle \Phi_{\vec{p}} | \Phi_{\vec{p}'} \rangle = \mathcal{N}^*_{\vec{p}} \mathcal{N}_{\vec{p}'} \underbrace{\int e^{-i\frac{(\vec{p}-\vec{p}')\cdot\vec{x}}{\hbar}} d\vec{x}}_{(2\pi\hbar)^3 \delta(\vec{p}-\vec{p}')} \Rightarrow \boxed{\mathcal{N}_{\vec{p}} = (2\pi\hbar)^{-\frac{3}{2}}}$$

General wavefunction: $\psi(\vec{x}) = \int \tilde{\psi}(\vec{p})\, \Phi_{\vec{p}}(\vec{x})\, d\vec{p}$

Note: dimension of $\Phi_{\vec{p}}(\vec{x})$ is $[\text{length}]^{-\frac{3}{2}}[\text{momentum}]^{-\frac{3}{2}} \Rightarrow$ it represents an amplitude density in a joint space of both $\vec{x}$ & $\vec{p}$

▶ **Radial momentum** $\hat{p}_r$ **& orbital momentum** $\hat{L}^2$, $\hat{L}_z$

Eigenbasis $\boxed{\Phi_{p_r lm}(\vec{x}) = \underbrace{\mathcal{N}_{p_r} \tfrac{1}{r} e^{i\frac{p_r r}{\hbar}}}_{R_{p_r}(r)} Y_{lm}(\vartheta, \varphi)}$ with $\mathcal{N}_{p_r} = (\pi\hbar)^{-\frac{1}{2}}$

$$\langle \Phi_{p_r lm} | \Phi_{p'_r l'm'} \rangle = \mathcal{N}^*_{p_r} \mathcal{N}_{p'_r} \int_0^\infty \frac{1}{r^2} e^{-i\frac{(p_r - p'_r)r}{\hbar}} \overbrace{\left[\int_0^{2\pi}\!\!\int_0^{\pi} Y^*_{lm}(\vartheta,\varphi) Y_{l'm'}(\vartheta,\varphi) \sin\vartheta\, d\vartheta\, d\varphi \right]}^{\delta_{ll'}\delta_{mm'}} r^2 dr = \delta(p_r - p'_r)\delta_{ll'}\delta_{mm'}$$

▶ **Isotropic Hamiltonian** $\hat{H}_{\text{rot}}$ **& orbital momentum** $\hat{L}^2$, $\hat{L}_z$

$\hat{H}_{\text{rot}} = -\frac{\hbar^2}{2M}\Delta + V(r)$ with $V \equiv$ an infinite potential well (Coulomb, rectangle...)

Eigenbasis $\boxed{\Phi_{nlm}(\vec{x}) = \underbrace{R_{nl}(r)}_{\frac{1}{r}u_{nl}(r)} Y_{lm}(\vartheta,\varphi)}$ with $u_{nl}(r)$ from rad. Schrödinger eq.

$$\left[-\frac{\hbar^2}{2M}\frac{d^2}{dr^2}+\frac{\hbar^2 l(l+1)}{2Mr^2}+V\right]u_{nl}=E_{nl}u_{nl}$$

$\langle\Phi_{nlm}|\Phi_{n'l'm'}\rangle = \delta_{nn'}\delta_{ll'}\delta_{mm'}$

▶ Infinitely many other choices possible

In all cases, the number of operators = **number of degrees of freedom** $f=3$

▶ Particle with **spin** $(s=\frac{1}{2}) \Rightarrow$ the same sets + **spin projection** $\hat{S}_z$

Eigenbases $\boxed{\Phi_{\vec{y}s_z}(\vec{x})}$, $\boxed{\Phi_{\vec{p}s_z}(\vec{x})}$, $\boxed{\Phi_{p_r lms_z}(\vec{x})}$, $\boxed{\Phi_{nlms_z}(\vec{x})}$

Another possibility is to use **total angular momentum** $\boxed{\hat{\vec{J}} = \hat{\vec{L}} + \hat{\vec{S}}}$

$\Rightarrow$ eigenvalues of $\{\hat{J}^2, \hat{J}_z\} \equiv \{\hbar^2 j(j+1), \hbar m_j\}$

Commutation relations of the $\hat{J}_z, \hat{J}^2$ operators:

$[\hat{J}^2,\hat{L}^2] = [\hat{J}^2,\hat{S}^2] = 0 = [\hat{J}_z,\hat{L}^2] = [\hat{J}_z,\hat{S}^2]$ but $[\hat{J}^2,\hat{L}_i] \neq 0 \neq [\hat{J}^2,\hat{S}_i]$

New complete set: $\hat{H}_{\rm rot}$ & $\hat{L}^2, \hat{S}^2$ & total ang. momentum $\hat{J}^2, \hat{J}_z$

$\Rightarrow$ eigenbasis $\boxed{\Phi_{nljm_j}(\vec{x})}$ (for exact form see Sec. 4.1 and p. 141)

This remains valid for a particle with any value of spin s: the Hilbert space is expanded $(2s+1)$ times compared to $\mathcal{H}$ of a spinless particle

1.4 Representation of physical transformations

Representation of observables is not the only role of operators in quantum mechanics. A specific type of operators, namely the unitary ones, is used to express various kinds of transformations that lead to equivalent descriptions of the same physics.

■ Unitary operators

At first, we explore basic mathematical properties of unitary operators. In a separable Hilbert space, these operators can be introduced as transformations between different orthonormal bases.

▶ Transformations of orthonormal bases

Basis I: $\{|i\rangle\}_i \equiv \{|1\rangle, |2\rangle, \dots\}$ $\quad \langle i|j\rangle = \delta_{ij}$

Basis II: $\{|i'\rangle\}_i \equiv \{|1'\rangle, |2'\rangle, \dots\}$ $\quad \langle i'|j'\rangle = \delta_{ij}$

$|i'\rangle = \hat{U}|i\rangle$ where $\hat{U} \equiv \sum_i |i'\rangle\langle i|$ is an **unitary operator**: $\hat{U}^\dagger = \sum_i |i\rangle\langle i'| = \hat{U}^{-1}$

▶ **3 equivalent definitions** of an unitary operator:

(1) Transforms an orthonormal basis to any other orthonormal basis: $\boxed{\{|i\rangle\}_i \underset{\hat{U}^{-1}}{\overset{\hat{U}}{\rightleftarrows}} \{|i'\rangle\}_i}$

(2) Inversion = Hermitian conjugation: $\boxed{\hat{U}^{-1} = \hat{U}^\dagger}$

(3) Conserves scalar products: $\boxed{\langle\psi_1'|\psi_2'\rangle = \langle\psi_1|\psi_2\rangle}$, where $|\psi_\bullet'\rangle = \hat{U}|\psi_\bullet\rangle$

► **Eigenvalues & eigenvectors** of unitary operators

$\hat{U}|u\rangle = u|u\rangle \quad \Leftrightarrow \quad \langle u|\hat{U}^\dagger = \langle u|u^*$

$\Rightarrow \langle u|\underbrace{\hat{U}^\dagger\hat{U}}_{\hat{I}}|u\rangle = uu^*\langle u|u\rangle \Rightarrow uu^* = 1 \Rightarrow \boxed{u = e^{i\phi}}$

$\Rightarrow \langle u|\overbrace{\hat{U}^\dagger\hat{U}}^{\hat{I}}|u'\rangle = u'u^*\langle u|u'\rangle \Rightarrow \underbrace{u'u^* = 1}_{e^{i(\phi'-\phi)}=1}$ or $\langle u|u'\rangle = 0$

$\Rightarrow$ for $\phi' \neq \phi(\mathrm{mod}2\pi)$: $\boxed{\langle u|u'\rangle = 0}$

► **Spectral decomposition** $\boxed{\hat{U} = \sum_i \underbrace{e^{i\phi_i}}_{u_i} \underbrace{\hat{P}_{\phi_i}}_{\sum_k |u_i^{(k)}\rangle\langle u_i^{(k)}|}}$ with $\boxed{\hat{P}_{\phi_i}\hat{P}_{\phi_j} = \delta_{ij}\hat{P}_{\phi_i}}$

► Any unitary operator = **exponential of a Hermitian operator**

$\boxed{\hat{U} = e^{i\hat{A}}}$ with $\hat{A} = \hat{A}^\dagger$ and $\boxed{e^{\hat{X}} \equiv \sum_{k=0}^{\infty} \frac{\hat{X}^k}{k!}}$ ≡ operator exponential defined through the Taylor series

(a) exponential ⇒ unitary: $\hat{U}^\dagger = \sum_{k=0}^{\infty} \frac{(-i\hat{A})^k}{k!} = e^{-i\hat{A}} = \hat{U}^{-1}$

(b) exponential ⇐ unitary: $\forall \hat{U} \equiv \sum_i e^{i\phi_i}\hat{P}_{\phi_i}$ define $\boxed{\hat{A} \equiv \sum_i \phi_i \hat{P}_{\phi_i}} = \hat{A}^\dagger \Rightarrow \hat{U} = e^{i\hat{A}}$

► **Example**: $\hat{U} = \left(\begin{smallmatrix} 0 & 1 \\ 1 & 0 \end{smallmatrix}\right)$

Eigenvalues $u_1 = 1 = e^{i0}$ and $u_2 = -1 = e^{i\pi}$

Eigenvectors $|+1\rangle \equiv \frac{1}{\sqrt{2}}\left(\begin{smallmatrix} 1 \\ 1 \end{smallmatrix}\right)$ and $|-1\rangle \equiv \frac{1}{\sqrt{2}}\left(\begin{smallmatrix} +1 \\ -1 \end{smallmatrix}\right)$ (orthonormal)

$\hat{A} = 0|+1\rangle\langle+1| + \pi|-1\rangle\langle-1| = \frac{\pi}{2}\left(\begin{smallmatrix} +1 & -1 \\ -1 & +1 \end{smallmatrix}\right)$ with $\left(\begin{smallmatrix} +1 & -1 \\ -1 & +1 \end{smallmatrix}\right)^k = 2^{k-1}\left(\begin{smallmatrix} +1 & -1 \\ -1 & +1 \end{smallmatrix}\right)$ for $k \geq 1$

$$e^{i\hat{A}} = \hat{I} + \underbrace{\sum_{k=1}^{\infty} \frac{(i\pi)^k}{k!}}_{e^{i\pi}-1=-2} \underbrace{\left[\frac{1}{2}\left(\begin{smallmatrix} +1 & -1 \\ -1 & +1 \end{smallmatrix}\right)\right]^k}_{\frac{1}{2}\left(\begin{smallmatrix} +1 & -1 \\ -1 & +1 \end{smallmatrix}\right)} = \left(\begin{smallmatrix} 1 & 0 \\ 0 & 1 \end{smallmatrix}\right) - \left(\begin{smallmatrix} +1 & -1 \\ -1 & +1 \end{smallmatrix}\right) = \left(\begin{smallmatrix} 0 & 1 \\ 1 & 0 \end{smallmatrix}\right) = \hat{U}$$

► Combining **exponentials of non-commuting operators**

$e^{\hat{X}}e^{\hat{Y}} = e^{\hat{Y}}e^{\hat{X}} = e^{\hat{X}+\hat{Y}}$ for $[\hat{X},\hat{Y}] = 0$, $\boxed{e^{\hat{X}}e^{\hat{Y}} \neq e^{\hat{Y}}e^{\hat{X}} \neq e^{\hat{X}+\hat{Y}}}$ for $[\hat{X},\hat{Y}] \neq 0$

Baker-Campbell-Hausdorff (BCH) formula (one of its forms):

$$\boxed{e^{\hat{X}}\hat{A}e^{-\hat{X}} = \underbrace{\hat{A}}_{[\hat{X},\hat{A}]_0} + \frac{1}{1!}\underbrace{[\hat{X},\hat{A}]}_{[\hat{X},\hat{A}]_1} + \frac{1}{2!}\underbrace{[\hat{X},[\hat{X},\hat{A}]]}_{[\hat{X},\hat{A}]_2} + \frac{1}{3!}\underbrace{[\hat{X},[\hat{X},[\hat{X},\hat{A}]]]}_{[\hat{X},\hat{A}]_3} \ldots + \frac{1}{k!}\left[\hat{X},\hat{A}\right]_k + \ldots}$$

$$e^{\hat{X}}e^{\hat{Y}}e^{-\hat{X}} = e^{\hat{Y}} + \tfrac{1}{1!}[\hat{X}, e^{\hat{Y}}] + \tfrac{1}{2!}[\hat{X}, [\hat{X}, e^{\hat{Y}}]] + \tfrac{1}{3!}[\hat{X}, [\hat{X}, [\hat{X}, e^{\hat{Y}}]]] + \cdots = \sum_{k=0}^{\infty} \tfrac{1}{k!} \left[\hat{X}, e^{\hat{Y}}\right]_k$$

$$= e^{\hat{Y}} + \sum_{k,l=1}^{\infty} \tfrac{1}{k!\,l!} \left[\hat{X}, \hat{Y}^l\right]_k$$

Special case: $[\hat{X}, [\hat{X}, \hat{Y}]] = [\hat{Y}, [\hat{X}, \hat{Y}]] = \cdots = 0 \quad \Rightarrow \quad e^{\hat{X}+\hat{Y}} = e^{\hat{X}} e^{\hat{Y}} e^{-\frac{1}{2}[\hat{X}, \hat{Y}]}$

■ Unitary transformations as "quantum canonical transformations"

Unitary operators materialize transitions between alternative QM representations, defined by distinct bases in the system's Hilbert space (see p. 43). They also express transformations between state vectors of the same system as seen from various reference frames, differing, e.g., by translations, rotations, or Galilean boosts. Physical descriptions in all these representations or reference frames must be fully equivalent. In this sense, the unitary transformations are analogues of classical canonical transformations.

▶ Diagonalization of an operator:

Transformation $\{|i\rangle\}_i$ (general basis) $\xrightarrow{\hat{U}} \{|a_j^{(k)}\rangle\}_{j,k}$ (eigenbasis)

$$\underbrace{\begin{pmatrix} U_{11} & U_{12} & \cdots \\ U_{21} & U_{22} & \\ \vdots & & \ddots \end{pmatrix}}_{\hat{U}} \underbrace{\begin{pmatrix} A_{11} & A_{12} & \cdots \\ A_{21} & A_{22} & \\ \vdots & & \ddots \end{pmatrix}}_{\hat{A}} \underbrace{\begin{pmatrix} U_{11}^* & U_{21}^* & \cdots \\ U_{12}^* & U_{22}^* & \\ \vdots & & \ddots \end{pmatrix}}_{\hat{U}^\dagger} = \underbrace{\begin{pmatrix} a_1 & 0 & \cdots \\ 0 & a_2 & \\ \vdots & & \ddots \end{pmatrix}}_{\hat{A}_{\rm diag}} \Rightarrow \boxed{\hat{A}_{\rm diag} = \hat{U}\hat{A}\hat{U}^\dagger}$$

eigenvectors of $\hat{A}$: $\begin{pmatrix} U_{i1}^* \\ U_{i2}^* \\ \vdots \end{pmatrix}$

▶ Link between equivalent representations

Postulate: Various representations of quantum state vectors & operators are equivalent *iff* they are connected by a unitary transformation

General transformation of bases: $\{|i\rangle\}_i \xrightarrow{\hat{U}} \{|i'\rangle\}_i$

Transformation of vectors: $|\psi\rangle = \sum_i \alpha_i |i\rangle \;\mapsto\; |\psi'\rangle = \sum_i \alpha_i |i'\rangle$

$\Rightarrow \boxed{|\psi'\rangle = \hat{U}|\psi\rangle}$

Transformation of **operators**: $\hat{A} = \sum_i a_i \hat{P}_{a_i} \;\mapsto\; \hat{A}' = \sum_i a_i \hat{U}\hat{P}_{a_i}\hat{U}^\dagger$

$\Rightarrow \boxed{\hat{A}' = \hat{U}\hat{A}\hat{U}^\dagger} = \hat{U}\hat{A}\hat{U}^{-1}$

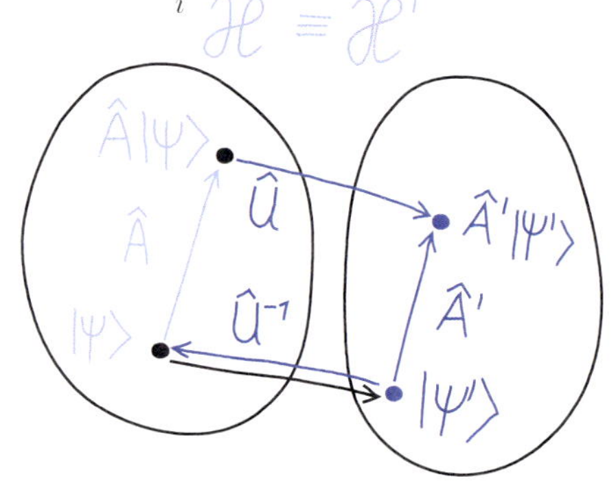

Interpretation of this relation using the identity $\hat{A}'|\psi'\rangle = \hat{U}\big(\hat{A}\underbrace{|\psi\rangle}_{\hat{U}^{-1}|\psi'\rangle}\big)$

$\hat{A}'|\psi'\rangle$ defined via the "detour path"; see the sketch on the right, where the target space of $\hat{U}$ is formally denoted as $\mathcal{H}'$ ($\equiv \mathcal{H}$)

◄ **Historical remark**
1897-1906: Independent derivations of the Baker-Campbell-Hausdorff formula
1900-10: David Hilbert elaborates the theory of (unitary) operators in $\mathcal{H}$
1927-32: Unitary operators and representation theory appear in the mathematical formulation of QM (P. Dirac, J. von Neumann)

■ Symmetry in quantum mechanics

It is often repeated that symmetry represents the most powerful concept in physics. The famous "Weyl's definition":[§] *"A thing is symmetrical if there is something you can do to it so that after you have finished doing it it looks the same as before."* can be always applied. For us, the *thing* means either a given physical system (its most essential attributes) or a general form of its QM description. *To do something to it* then means to look at the system from another reference frame or through a different representation.

► Symmetry in a wider sense (*sensu lato*)

Frameworks S & S' ≡ alternative "observer frames" or "ways of description" (representations)

$$\text{Transf. S}\to\text{S'} \Leftrightarrow \text{Operator } \hat{U}_{\text{S}\to\text{S}'}\equiv\hat{U}: \left\{\begin{array}{lccc} \text{states} & |\psi\rangle & \mapsto & |\psi'\rangle=\hat{U}|\psi\rangle \\ \text{observables} & \hat{A} & \mapsto & \hat{A}'=\hat{U}\hat{A}\hat{U}^{\dagger} \end{array}\right\}$$

The structure and all predictions of quantum mechanics remain the same:

$$\langle\psi_1'|\psi_2'\rangle=\langle\hat{U}\psi_1|\hat{U}\psi_2\rangle=\langle\psi_1|\underbrace{\hat{U}^{\dagger}\hat{U}}_{\hat{I}}|\psi_2\rangle=\langle\psi_1|\psi_2\rangle \qquad \text{...scalar products}$$

$$\langle i'|\hat{A}'|j'\rangle=\langle i|\overbrace{\hat{U}^{\dagger}\hat{U}}^{\hat{I}}\hat{A}\overbrace{\hat{U}^{\dagger}\hat{U}}^{\hat{I}}|j\rangle=\langle i|\hat{A}|j\rangle \qquad \text{...matrix elements}$$

$$\hat{A}|a\rangle=a|a\rangle\Rightarrow\underbrace{\hat{U}\hat{A}\hat{U}^{\dagger}}_{\hat{A}'}\underbrace{\hat{U}|a\rangle}_{|a'\rangle}=a\underbrace{\hat{U}|a\rangle}_{|a'\rangle} \qquad \text{...eigenvalues}$$

$$\overbrace{[\hat{A}',\hat{B}']}^{i\hat{C}}=\hat{A}'\hat{B}'-\hat{B}'\hat{A}'=\hat{U}\hat{A}\hat{U}^{\dagger}\hat{U}\hat{B}\hat{U}^{\dagger}-\hat{U}\hat{B}\hat{U}^{\dagger}\hat{U}\hat{A}\hat{U}^{\dagger}=\overbrace{\hat{U}[\hat{A},\hat{B}]\hat{U}^{\dagger}}^{i\hat{C}'} \qquad \text{...commutators}$$

Hence: " $\boxed{\text{QM}_{\text{S}'}=\text{QM}_{\text{S}}}$ " that is: the structure of QM description is the same for both S & S'

► Symmetry in a narrower sense (*sensu stricto*)

A *system is invariant* under the transformation S → S'
iff its Hamiltonian does not change: $\boxed{\hat{H}'=\hat{H}}$

$$\hat{H}'=\hat{U}\hat{H}\hat{U}^{\dagger}=\hat{H} \quad\Rightarrow\quad \hat{U}\hat{H}=\hat{H}\hat{U} \quad\Rightarrow\quad \boxed{[\hat{H},\hat{U}]=0}$$

Reasoning: Hamiltonian $\hat{H}$ represents the most important physical operator (describing e.g. the system's dynamics), the symmetry is therefore associated with the invariance of $\hat{H}$ under the unitary transformation S → S'.

[§]In fact, this is Feynman's informal transcription of Weyl's original formulation.

Usual consequences of this meaning of symmetry:

(a) **degeneracy** of energy levels: $\hat{H}|\psi\rangle = E|\psi\rangle \quad \Rightarrow \boxed{\hat{H}(\hat{U}|\psi\rangle) = E(\hat{U}|\psi\rangle)}$

$\Rightarrow$ if $\hat{U}|\psi\rangle \neq |\psi\rangle$, the level E is degenerate

(b) **conservation laws** ($\exists$ integrals of motions, see p. 67)

"Flight over the group theory nest"

Group theory represents a superb example of "the unreasonable effectiveness of mathematics in the natural sciences", which was pointed out by Wigner. Initiated as a purely theoretical discipline, it grew into one of the most commonly applied branches of mathematics today. Here we just summarize the basic concepts of the theory that are of immediate importance for QM.

▶ Unitary transformations do not typically come alone but in groups!

Group $\mathcal{G}$ = set of elements $\{g\}$ closed under a binary operation $\circ$ (group multiplication) $\boxed{g_1, g_2 \in \mathcal{G} \Rightarrow \underbrace{(g_1 \circ g_2)}_{g_1 g_2} \in \mathcal{G}}$ satisfying the following properties:

(1) $(g_1 g_2)g_3 = g_1(g_2 g_3)$ associativity

(2) $\exists\ e \in \mathcal{G}:\ ge = eg = g\ \forall g \in \mathcal{G}$ unit element

(3) $\forall\ g \in \mathcal{G}\ \exists\ g^{-1}:\ gg^{-1} = g^{-1}g = e$ inverse elements

Note: commutativity not required!
If $g_1 g_2 = g_2 g_1\ \forall\ g_1, g_2 \in \mathcal{G}$, the group is called **Abelian**

▶ **Unitary representation** of group $\mathcal{G}$:

Mapping to unitary operators: $\boxed{g \mapsto \hat{U}_g}$, $\boxed{g_1 \circ g_2 \mapsto \hat{U}_2 \hat{U}_1}$

Group properties naturally satisfied: $(\hat{U}_2\hat{U}_1)^\dagger = \hat{U}_1^{-1}\hat{U}_2^{-1} = (\hat{U}_2\hat{U}_1)^{-1}$ closure

$(\hat{U}_3\hat{U}_2)\hat{U}_1 = \hat{U}_3(\hat{U}_2\hat{U}_1)$ associativity

$e \mapsto \hat{I} = \hat{I}^\dagger = \hat{I}^{-1}$ unit element

$\hat{U} = e^{i\hat{A}} \Rightarrow \hat{U}^{-1} = e^{-i\hat{A}}$ inverse elements

Hilbert space $\mathcal{H}$ where $\hat{U}$ act $\equiv$ **carrier space** of $\mathcal{G}$
QM works with $\mathcal{H} \Rightarrow$ it provides a direct physical "arena" for group theory

Invariant subspace: a subspace $\mathcal{H}_{\mathcal{G}} \subset \mathcal{H}$ is invariant under $\mathcal{G}$ if $\hat{U}|\psi\rangle \in \mathcal{H}_{\mathcal{G}}$ $\forall \hat{U} \in \mathcal{G}$ and $\forall\,|\psi\rangle \in \mathcal{H}_{\mathcal{G}}$

$$\text{Matrix representation: } \hat{U} \equiv \begin{pmatrix} \ddots & 0 & \cdots & & \\ 0 & \boxed{\in \mathcal{H}_{\mathcal{G}}^{(1)}} & 0 & & \\ \vdots & 0 & \ddots & 0 & \\ & & 0 & \boxed{\in \mathcal{H}_{\mathcal{G}}^{(2)}} & \\ & & & & \ddots \end{pmatrix} \quad \text{block diagonal structure of all } \hat{U}$$

Irreducible representation (irrep) of group $\mathcal{G}$: $\nexists$ invariant subspace $\mathcal{H}_{\mathcal{G}} \subset \mathcal{H}$

▶ Finite (discrete) groups

Groups with a finite (or at least discrete) number of elements (describe, e.g., spatial symmetries of crystals or reflection transformations): $\boxed{\mathcal{G} \equiv \{g_i\}_{i\in\mathbb{N}}}$

Example: **cyclic group** $\boxed{\mathcal{Z}_2 \equiv \{\hat{P}, \hat{I}\}}$ with a generalized **parity transformation** $\hat{P} = \hat{P}^{-1} = \hat{P}^{\dagger}$ ($\Rightarrow \hat{P}^2 = \hat{I}$) $\equiv$ spatial inversion, 2-particle exchange, particle-antiparticle or particle-hole transformation...

▶ Continuous (Lie) groups

Groups with elements parametrized by a d-dimensional real vector $\vec{s} \Rightarrow$ the group elements (e.g., spatial translations) form a continuum: $\boxed{\mathcal{G} \equiv \{g(\vec{s})\}_{\vec{s}\in\mathbb{R}^d}}$

$$\left.\begin{array}{lll} g(\vec{s}_1)g(\vec{s}_2)=g(\vec{s}_3) & \Rightarrow & \vec{s}_3 = \vec{f}(\vec{s}_1,\vec{s}_2) \\ g(\vec{s})^{-1} = g(\vec{s}\,') & \Rightarrow & \vec{s}\,' = \vec{h}(\vec{s}) \end{array}\right\} \text{functions } \vec{f} \text{ \& } \vec{h} \text{ differentiable}$$

$\Rightarrow \mathcal{G} \equiv$ **Lie group**

Unitary representation = mapping of a given Lie group to a family of unitary operators acting in a suitable Hilbert space: $\boxed{g(\vec{s}) \mapsto \hat{U}(\vec{s})}$

▶ dim=1 Lie group $\boxed{\mathcal{G} \equiv \{g(s)\}_{s\in\mathbb{R}}}$

$$\text{Requirements:} \begin{cases} \hat{U}(0)=\hat{I} & \text{choice of origin} \\ \hat{U}(s+ds)=\hat{U}(s)\hat{U}(ds) & \text{local additivity} \Leftarrow f(s,ds)=s+\overbrace{\tfrac{\partial f}{\partial s_2}}^{\overset{!}{=}1}ds \end{cases}$$

$$\left.\begin{array}{l} \hat{U}(s) = \hat{I} + \left(\frac{d\hat{U}}{ds}\right)_0 s + \frac{1}{2}\left(\frac{d^2\hat{U}}{ds^2}\right)_0 s^2 + \dots \\ \hat{U}(s)\hat{U}(s)^{\dagger} = \hat{I} + \underbrace{\left[\left(\frac{d\hat{U}}{ds}\right)_0 + \left(\frac{d\hat{U}}{ds}\right)_0^{\dagger}\right]}_{0} s + \underbrace{\Big[\dots\Big]}_{0} s^2 + \dots \end{array}\right\} \Rightarrow \begin{array}{l} \left(\frac{d\hat{U}}{ds}\right)_0 = i\hat{G} \\ \text{with } \boxed{\hat{G} = \hat{G}^{\dagger}} \end{array}$$

$$\Rightarrow \text{condition} \quad \left(\frac{d\hat{U}}{ds}\right)_s = \lim_{ds\to 0}\frac{\hat{U}(s+ds)-\hat{U}(s)}{ds} = \hat{U}(s)\left(\frac{d\hat{U}}{ds}\right)_0 = i\hat{U}(s)\hat{G}$$

$\boxed{\hat{U}(s) = e^{i\hat{G}s}}$ is the most general solution, where $\hat{G} \equiv$ **generator** of $\mathcal{G}$

$\Rightarrow$ the group is **Abelian**: $\hat{U}(s_1)\hat{U}(s_2) = \hat{U}(s_1+s_2) = \hat{U}(s_2)\hat{U}(s_1)$

▶ dim>1 Lie group $\boxed{\mathcal{G} \equiv \{g(\vec{s})\}_{\vec{s}\in\mathbb{R}^d}}$

$$\boxed{\hat{U}(\vec{s}) = e^{i\hat{\vec{G}}\cdot\vec{s}}} = e^{i\sum_{k=1}^{d}\hat{G}_k s_k} \text{ with } \hat{\vec{G}} \equiv \left\{\hat{G}_k = \frac{1}{i}\left(\frac{\partial U(\vec{s})}{\partial s_k}\right)_{\vec{s}=0}\right\}_{k=1}^{d} \equiv \text{set of generators}$$

$$\neq \prod_{k=1}^{d} e^{i\hat{G}_k s_k} \quad \text{in the } \textbf{non-Abelian} \text{ case: } [\hat{G}_k,\hat{G}_l] \neq 0$$

$$\hat{U}(\vec{s}_1)\hat{U}(\vec{s}_2) \neq \hat{U}(\vec{s}_1+\vec{s}_2) \neq \hat{U}(\vec{s}_2)\hat{U}(\vec{s}_1)$$

$$\overbrace{\hat{U}(\vec{s})}^{e^{i\hat{A}}} \approx \hat{I} + i\overbrace{\sum_i \hat{G}_i s_i}^{\hat{A}} - \frac{1}{2}\overbrace{\sum_j\sum_k \hat{G}_j\hat{G}_k s_j s_k}^{\hat{A}^2} + \dots \overset{\text{sum.conv.}}{\equiv} \hat{I} + i\hat{G}_i s_i - \frac{1}{2}\hat{G}_j\hat{G}_k s_j s_k + \dots$$

▶ **Closure relation for infinitesimal transformations** (dim>1)

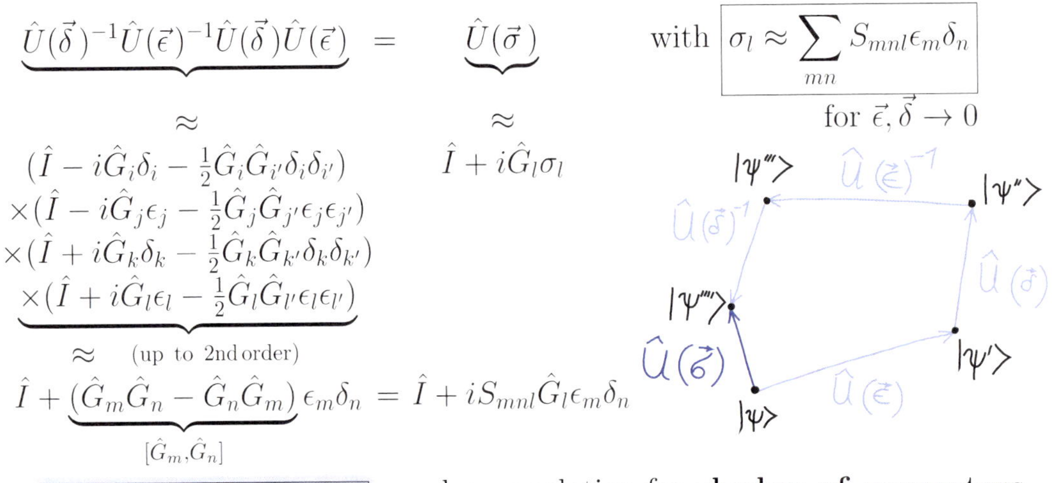

$$\underbrace{\hat{U}(\vec{\delta})^{-1}\hat{U}(\vec{\epsilon})^{-1}\hat{U}(\vec{\delta})\hat{U}(\vec{\epsilon})} = \underbrace{\hat{U}(\vec{\sigma})} \qquad \text{with } \boxed{\sigma_l \approx \sum_{mn} S_{mnl}\epsilon_m\delta_n} \text{ for } \vec{\epsilon},\vec{\delta}\to 0$$

$$\approx \qquad\qquad \approx$$

$$\underbrace{\begin{array}{l}(\hat{I} - i\hat{G}_i\delta_i - \frac{1}{2}\hat{G}_i\hat{G}_{i'}\delta_i\delta_{i'})\\ \times(\hat{I} - i\hat{G}_j\epsilon_j - \frac{1}{2}\hat{G}_j\hat{G}_{j'}\epsilon_j\epsilon_{j'})\\ \times(\hat{I} + i\hat{G}_k\delta_k - \frac{1}{2}\hat{G}_k\hat{G}_{k'}\delta_k\delta_{k'})\\ \times(\hat{I} + i\hat{G}_l\epsilon_l - \frac{1}{2}\hat{G}_l\hat{G}_{l'}\epsilon_l\epsilon_{l'})\end{array}} \qquad \hat{I} + i\hat{G}_l\sigma_l$$

$\approx$ (up to 2nd order)

$$\hat{I} + \underbrace{(\hat{G}_m\hat{G}_n - \hat{G}_n\hat{G}_m)}_{[\hat{G}_m,\hat{G}_n]}\epsilon_m\delta_n = \hat{I} + iS_{mnl}\hat{G}_l\epsilon_m\delta_n$$

$$\Rightarrow \boxed{[\hat{G}_m,\hat{G}_n] = i\sum_l S_{mnl}\hat{G}_l}$$

closure relation for **algebra of generators**
$S_{mnl} \equiv$ structure constants

▶ **Invariant (Casimir) operator**

An operator $\hat{C}_{\mathcal{G}} \equiv \hat{C}(\vec{\hat{G}})$ associated with group $\mathcal{G}$ such that $\boxed{[\hat{C}_{\mathcal{G}},\hat{G}_i] = 0}$ $\forall i$

Eigenspaces of $\hat{C}_{\mathcal{G}}$ within the space $\mathcal{H}$ are invariant under the action of all generators $\{\hat{G}_i\}$ $\Rightarrow$ these subspaces often carry irreducible representations of $\mathcal{G}$

◀ **Historical remark**

1830 (approx.): dawn of the group theory (the name given by É. Galois)
1873: Sophus Lie introduces continuous groups (later work of W. Killing, E. Cartan)
1928-32: M.H. Stone and J.von Neumann obtain QM-related results on Lie groups
1928: Hermann Weyl: *Gruppentheorie und Quantenmechanik*—book placing the group theory to the foundations of QM
1927-37: Eugene Wigner elaborates group techniques in the classification of atomic and later nuclear spectra; the 1931 book *Group Theory and Its Application to the Quantum Mechanics of Atomic Spectra*
1929: Hans Bethe applies point groups in polyatomic molecules
1931: Hendrik Casimir introduces the invariant operator
1940's-50's: Giulio Racah refines group methods in the theory of complex spectra

2.4 Fundamental spatio-temporal symmetries

We are going to describe basic spatial and spatio-temporal transformations of non-relativistic physical systems. We will see that elementary physical operators in QM can be naturally introduced as generators of the corresponding Lie groups. Extrapolating this path, one may seek the very origin of the quantum uncertainty (incompatibility) in the non-Abelian character of some of these groups.

Space translation

Translations in the coordinate space form an Abelian group generated by momentum operators. This is a nice playground to exercise work with generators.

▶ **Coordinate translation operator**

Required action of translations on coordinate operators: $\boxed{\hat{T}_{\vec{a}}\hat{\vec{x}}\hat{T}_{\vec{a}}^{-1} = \hat{\vec{x}} - \vec{a}\hat{I}}$

$\Rightarrow$ commutation relations $[\hat{x}_i, \hat{T}_{\vec{a}}] = a_i\hat{T}_{\vec{a}}$

$\Rightarrow \quad \hat{x}_i|\vec{x}\rangle = x_i|\vec{x}\rangle \quad \Rightarrow \quad \hat{x}_i(\hat{T}_{\vec{a}}|\vec{x}\rangle) = (x_i+a_i)(\hat{T}_{\vec{a}}|\vec{x}\rangle) \quad \Rightarrow \quad \boxed{\hat{T}_{\vec{a}}|\vec{x}\rangle = |\vec{x}+\vec{a}\rangle}$

(required unitarity of $\hat{T}_{\vec{a}} \Rightarrow$ proportionality coeff. is unity)

$\Rightarrow \quad \langle\vec{x}|\hat{T}_{\vec{a}}\psi\rangle = \langle\hat{T}_{\vec{a}}^{-1}\vec{x}|\psi\rangle = \langle\vec{x}-\vec{a}|\psi\rangle = \boxed{\psi(\vec{x}-\vec{a}) = \hat{T}_{\vec{a}}\psi(\vec{x})}$

▶ Remark: **general translation operator** for an arbitrary operator $\hat{O}$

Assume operator $\hat{T}_{\Delta o}$ satisfying $\boxed{[\hat{O}, \hat{T}_{\Delta o}] = \Delta o\,\hat{T}_{\Delta o}}$ with $\Delta o \equiv$ number

$\Rightarrow \quad \hat{O}|o\rangle = o|o\rangle \quad \Rightarrow \quad \hat{O}(\hat{T}_{\Delta o}|o\rangle) = (o+\Delta o)(\hat{T}_{\Delta o}|o\rangle)$

$\Rightarrow \quad \hat{T}_{\Delta o}|o\rangle = |o+\Delta o\rangle$ (operator $\hat{T}_{\Delta o}$ shifts eigenstates of $\hat{O}$ by value Δo)

▶ **Generators of x, y, z translations**

Translation $\vec{a} = a\vec{n}_j$ along $j = 1, 2, 3$ axes: $[\hat{x}_i, \hat{T}_{a\vec{n}_j}] = \delta_{ij}a\hat{T}_{a\vec{n}_j}$

Infinitesimal translations: $\hat{T}_{(\delta a)\vec{n}_j} \approx \hat{I} + i\hat{G}_j(\delta a) \Rightarrow \boxed{[\hat{x}_i, \hat{G}_j] = -i\delta_{ij}\hat{I}}$

$\Rightarrow$ we can set $\boxed{\hat{G}_j = -\frac{1}{\hbar}\hat{p}_j}$ (generators $\propto$ momentum components)

$[\hat{G}_i, \hat{G}_j] = 0 \Rightarrow$ **Abelian group**, $[\hat{G}_i, \hat{p}_j] = 0 \Rightarrow \boxed{\hat{T}_{\vec{a}}\hat{\vec{p}}\hat{T}_{\vec{a}}^{-1} = \hat{\vec{p}}}$

▶ **Finite translations in any direction**

j^{th}axis: $\hat{T}_{a\vec{n}_j} = \lim_{n\to\infty}\left(\hat{I} - \frac{i}{\hbar}\hat{p}_j\frac{a}{n}\right)^n = e^{-i\frac{a\hat{p}_j}{\hbar}} \quad \Rightarrow$ general direction: $\boxed{\hat{T}_{\vec{a}} = e^{-i\frac{\vec{a}\cdot\hat{\vec{p}}}{\hbar}}}$

Direct verification of wavefunction transformation:

$$\hat{T}_{\vec{a}}\psi(\vec{x}) = \sum_{k=0}^{\infty}\frac{1}{k!}(-\vec{a}\cdot\vec{\nabla})^k\psi(\vec{x}) = \psi(\vec{x}-\vec{a})$$

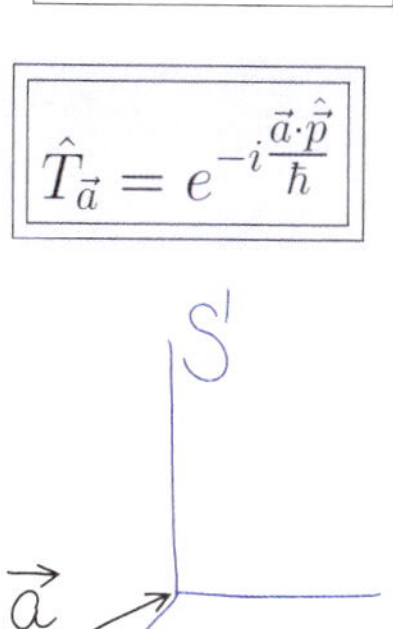

▶ **Translation for $N > 1$ particle systems**

$$\hat{\vec{P}} = \sum_{k=1}^{N}\hat{\vec{p}}_k \;\Rightarrow\; \hat{T}_{\vec{a}} = e^{-i\frac{\vec{a}\cdot\hat{\vec{P}}}{\hbar}} = \bigotimes_{k=1}^{N}\hat{T}_{\vec{a},k}$$

$$\Rightarrow \; (\hat{T}_{\vec{a}}\Psi)(\vec{x}_1, \ldots, \vec{x}_N) = \Psi(\vec{x}_1-\vec{a}, \ldots, \vec{x}_N-\vec{a})$$

Particles with spin: translation does not affect the spin variables

▶ **Translational invariance** *sensu stricto* $[\hat{T}_{\vec{a}}, \hat{H}] = 0$

For 1 particle this means that $\hat{H} = H(\hat{\vec{p}}, \hat{\vec{S}}) \Rightarrow$ Hamiltonian must not depend on spatial coordinates, just on momentum and spin components

For $N>1$ particles: $\boxed{\hat{H} = H\left(\{\hat{\vec{p}}_k\}, \{\hat{\vec{x}}_k - \hat{\vec{x}}_l\}, \{\hat{\vec{S}}_k\}\right)}$ $\Rightarrow$ Hamiltonian must depend only on *relative coordinates*, e.g.: $\hat{H} = \sum_k \frac{1}{2M_k}\hat{\vec{p}}_k^{\,2} + \sum_{k\geq l} V\left(\{\hat{\vec{x}}_k - \hat{\vec{x}}_l\}\right)$

▶ **Discrete translations** (invariance of a crystal lattice)

Discrete set of translation vectors $\vec{a}_{\vec{n}} = (n_x L_x, n_y L_y, n_z L_z) = \vec{n} \cdot \vec{L}$
with $n_i = 0, \pm 1, \pm 2, \ldots$

$[\hat{T}_{\vec{a}_{\vec{n}}}, \hat{H}] = 0 \quad \Rightarrow \quad \hat{T}_{\vec{a}_{\vec{n}}}$ and $\hat{H}$ have a common set of eigenfunctions

General $\hat{T}_{\vec{a}_{\vec{n}}}$ eigenfunction has the form: $\boxed{\psi(\vec{x}) = u(\vec{x})\, e^{i\frac{\vec{\pi}\cdot\vec{x}}{\hbar}}}$ **Bloch theorem**

where $u(\vec{x})$ is any $\vec{L}$-periodic function: $u(\vec{x}+\vec{a}_{\vec{n}}) = u(\vec{x})$, and $\vec{\pi}$ is a vector called **quasimomentum**

Proof: $\underbrace{e^{-i\frac{\hat{\vec{p}}\cdot\vec{a}_{\vec{n}}}{\hbar}}}_{\hat{T}_{\vec{a}_{\vec{n}}}} u(\vec{x}) e^{i\frac{\vec{\pi}\cdot\vec{x}}{\hbar}} = u(\vec{x}-\vec{a}_{\vec{n}}) e^{i\frac{\vec{\pi}\cdot(\vec{x}-\vec{a}_{\vec{n}})}{\hbar}} = \underbrace{e^{-i\frac{\vec{\pi}\cdot\vec{a}_{\vec{n}}}{\hbar}}}_{\text{eigenvalue}} u(\vec{x}) e^{i\frac{\vec{\pi}\cdot\vec{x}}{\hbar}}$

■ Space rotation

Generators of rotations are angular momentum operators. The group is therefore non-Abelian. If working with scalar wavefunctions, one will manage just with the orbital angular momentum. However, to describe rotations of more complicated types of wavefunction (like spinors or vectors), we need to introduce an additional angular momentum—spin.

▶ **Coordinate transformation**

Rotation about axis $\vec{n}$ by angle ϕ in ordinary space expressed by: $\boxed{\vec{x}\,' = \overbrace{\mathbf{R}_{\vec{n}\phi}^{-1}}^{3\times 3 \text{ rotation matrix}} \vec{x}}$

Radius conserved $\Rightarrow$ orthogonality: $\boxed{\mathbf{R}_{\vec{n}\phi}^{\mathrm{T}} \mathbf{R}_{\vec{n}\phi} = \mathbf{I}} \Rightarrow \sum_i r_{ij} r_{ik} = \delta_{jk}$

$$\mathbf{R}_{\vec{n}_z\phi}^{-1} = \begin{pmatrix} \cos\phi & \sin\phi & 0 \\ -\sin\phi & \cos\phi & 0 \\ 0 & 0 & 1 \end{pmatrix} \Rightarrow \mathbf{R}_{\vec{n}_z\delta\phi}^{-1} \approx \mathbf{I} + \underbrace{\begin{pmatrix} 0 & 1 & 0 \\ -1 & 0 & 0 \\ 0 & 0 & 0 \end{pmatrix}}_{i\mathbf{G}_3} \delta\phi$$

$$\mathbf{R}_{\vec{n}_y\phi}^{-1} = \begin{pmatrix} \cos\phi & 0 & -\sin\phi \\ 0 & 1 & 0 \\ \sin\phi & 0 & \cos\phi \end{pmatrix} \Rightarrow \mathbf{R}_{\vec{n}_y\delta\phi}^{-1} \approx \mathbf{I} + \underbrace{\begin{pmatrix} 0 & 0 & -1 \\ 0 & 0 & 0 \\ 1 & 0 & 0 \end{pmatrix}}_{i\mathbf{G}_2} \delta\phi$$

$$\mathbf{R}_{\vec{n}_x\phi}^{-1} = \begin{pmatrix} 1 & 0 & 0 \\ 0 & \cos\phi & \sin\phi \\ 0 & -\sin\phi & \cos\phi \end{pmatrix} \Rightarrow \mathbf{R}_{\vec{n}_x\delta\phi}^{-1} \approx \mathbf{I} + \underbrace{\begin{pmatrix} 0 & 0 & 0 \\ 0 & 0 & 1 \\ 0 & -1 & 0 \end{pmatrix}}_{i\mathbf{G}_1} \delta\phi$$

Any finite rotation is expressed via generators of infinitesimal rotations: $\boxed{\mathbf{R}_{\vec{n}\phi}^{-1} = e^{i(\vec{\mathbf{G}}\cdot\vec{n})\phi}}$

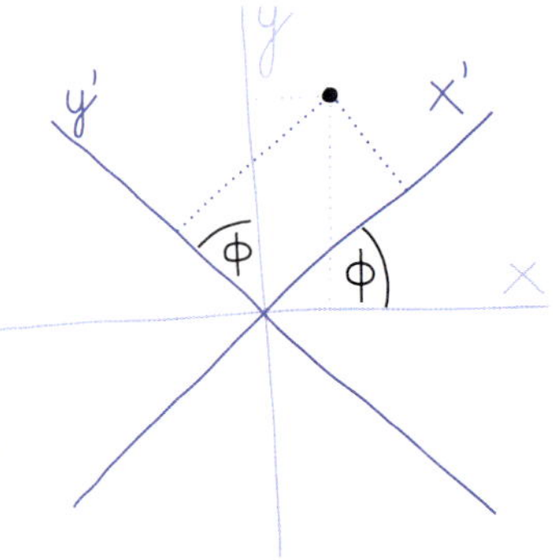

Commutators of the 3×3 generator matrices:

$$\boxed{[\mathbf{G}_i, \mathbf{G}_j] = i\varepsilon_{ijk}\mathbf{G}_k}$$

These are the commutators of angular momentum $/\hbar$
The same can be assumed for the QM generators $\hat{G}_i$

▶ **Quantum rotation operator:** $\boxed{\hat{R}_{\vec{n}\phi} = e^{-i(\hat{\vec{G}}\cdot\vec{n})\phi}}$ with $\boxed{\hat{\vec{G}} = \frac{1}{\hbar}\underbrace{(\hat{\vec{L}}+\hat{\vec{S}})}_{\hat{\vec{J}}}}$

Generators for 1 particle $\propto$ orbital + spin ang. momentum

Generators for any system $\propto$ general operators of total angular momentum

Postulate: Angular momentum operators of an arbitrary quantum system $= \hbar\times$ generators of rotation ($\Rightarrow$ spin $\leftrightarrow$ transformation properties)

▶ **Transformation of coordinates & momenta**

(a) rotation around z:

$$\hat{x}'_i \equiv \hat{R}_{\vec{n}_z\phi}\hat{x}_i\hat{R}^{-1}_{\vec{n}_z\phi} = \underbrace{e^{-i(\hat{L}_3+\hat{S}_3)\phi/\hbar}}_{e^{-i\hat{L}_3\phi/\hbar}e^{-i\hat{S}_3\phi/\hbar}}\hat{x}_i\underbrace{e^{+i(\hat{L}_3+\hat{S}_3)\phi/\hbar}}_{e^{+i\hat{S}_3\phi/\hbar}e^{+i\hat{L}_3\phi/\hbar}} = e^{-i\hat{L}_3\phi/\hbar}\hat{x}_i e^{+i\hat{L}_3\phi/\hbar}$$

Infinitesimal rotation:

$$\hat{x}'_i \approx \left(\hat{I}-\tfrac{i}{\hbar}\hat{L}_3\delta\phi\right)\hat{x}_i\left(\hat{I}+\tfrac{i}{\hbar}\hat{L}_3\delta\phi\right) \approx \hat{x}_i - \tfrac{i}{\hbar}\underbrace{[\hat{L}_3,\hat{x}_i]}_{-i\hbar(\delta_{i2}\hat{x}_1-\delta_{i1}\hat{x}_2)}\delta\phi = \begin{cases}\hat{x}_1+\hat{x}_2\delta\phi\\ \hat{x}_2-\hat{x}_1\delta\phi\\ \hat{x}_3\end{cases}$$

The same for momentum:

$$\hat{p}'_i \approx \left(\hat{I}-\tfrac{i}{\hbar}\hat{L}_3\delta\phi\right)\hat{p}_i\left(\hat{I}+\tfrac{i}{\hbar}\hat{L}_3\delta\phi\right) \approx \hat{p}_i - \tfrac{i}{\hbar}\underbrace{[\hat{L}_3,\hat{p}_i]}_{+i\hbar(\delta_{i1}\hat{p}_2-\delta_{i2}\hat{p}_1)}\delta\phi = \begin{cases}\hat{p}_1+\hat{p}_2\delta\phi\\ \hat{p}_2-\hat{p}_1\delta\phi\\ \hat{p}_3\end{cases}$$

(b) general rotation:

$$\boxed{\underbrace{\hat{\vec{x}}'}_{\begin{pmatrix}\hat{x}'_1\\ \hat{x}'_2\\ \hat{x}'_3\end{pmatrix}} \equiv \hat{R}_{\vec{n}\phi}\,\hat{\vec{x}}\,\hat{R}^{-1}_{\vec{n}\phi} = \underbrace{\mathbf{R}^{-1}_{\vec{n}\phi}}_{\begin{pmatrix}r_{11}&r_{12}&r_{13}\\ r_{21}&r_{22}&r_{23}\\ r_{31}&r_{32}&r_{33}\end{pmatrix}^{-1}}\hat{\vec{x}}_{\begin{pmatrix}\hat{x}_1\\ \hat{x}_2\\ \hat{x}_3\end{pmatrix}}} \quad \boxed{\hat{\vec{p}}' \equiv \hat{R}_{\vec{n}\phi}\,\hat{\vec{p}}\,\hat{R}^{-1}_{\vec{n}\phi} = \mathbf{R}^{-1}_{\vec{n}\phi}\,\hat{\vec{p}}}$$

$$\Rightarrow \quad \hat{\vec{x}}\hat{R}_{\vec{n}\phi} = \mathbf{R}_{\vec{n}\phi}\hat{R}_{\vec{n}\phi}\hat{\vec{x}}$$

▶ **Transformation of angular momentum**

(a) z-rotation:
$$\hat{J}'_i \approx \left(\hat{I}-\tfrac{i}{\hbar}\hat{J}_3\delta\phi\right)\hat{J}_i\left(\hat{I}+\tfrac{i}{\hbar}\hat{J}_3\delta\phi\right) \approx \hat{J}_i - \tfrac{i}{\hbar}\underbrace{[\hat{J}_3,\hat{J}_i]}_{i\hbar\varepsilon_{3ij}\hat{J}_j}\delta\phi = \begin{cases}\hat{J}_1+\hat{J}_2\delta\phi\\ \hat{J}_2-\hat{J}_1\delta\phi\\ \hat{J}_3\end{cases}$$

(b) general rotation: $\boxed{\hat{\vec{J}}' \equiv \hat{R}_{\vec{n}\phi}\hat{\vec{J}}\hat{R}^{-1}_{\vec{n}\phi} = \mathbf{R}^{-1}_{\vec{n}\phi}\hat{\vec{J}}}$ $\quad \hat{\vec{S}}' = \mathbf{R}^{-1}_{\vec{n}\phi}\hat{\vec{S}} \quad \hat{\vec{L}}' = \mathbf{R}^{-1}_{\vec{n}\phi}\hat{\vec{L}}$

▶ **Action on wavefunctions** (coordinate & momentum representation)

$$\underbrace{\hat{\vec{x}}\,(\hat{R}_{\vec{n}\phi}}_{\mathbf{R}_{\vec{n}\phi}\hat{R}_{\vec{n}\phi}\hat{\vec{x}}}|\vec{x}\rangle) = (\mathbf{R}_{\vec{n}\phi}\vec{x})(\hat{R}_{\vec{n}\phi}|\vec{x}\rangle) \quad \Rightarrow \quad \hat{R}_{\vec{n}\phi}|\vec{x}\rangle = |\mathbf{R}_{\vec{n}\phi}\vec{x}\rangle$$

$$\left.\begin{aligned}\langle\vec{x}|\hat{R}_{\vec{n}\phi}\psi\rangle &= \langle\mathbf{R}^{-1}_{\vec{n}\phi}\vec{x}|\psi\rangle\\ \langle\vec{p}|\hat{R}_{\vec{n}\phi}\psi\rangle &= \langle\mathbf{R}^{-1}_{\vec{n}\phi}\vec{p}|\psi\rangle\end{aligned}\right\} \Rightarrow \boxed{\begin{aligned}\hat{R}_{\vec{n}\phi}\psi(\vec{x}) &= \psi(\mathbf{R}^{-1}_{\vec{n}\phi}\vec{x})\\ \hat{R}_{\vec{n}\phi}\tilde{\psi}(\vec{p}) &= \tilde{\psi}(\mathbf{R}^{-1}_{\vec{n}\phi}\vec{p})\end{aligned}}$$

▶ **Transformation of scalar wavefunction** $\psi(\vec{x})$

Only the argument of $\psi(\vec{x})$ affected by the transformation:

$$\boxed{\hat{R}_{\vec{n}\phi}\psi(\vec{x}) = \psi(\underbrace{\mathbf{R}^{-1}_{\vec{n}\phi}\vec{x}}_{\vec{x}'})} \quad \Rightarrow \quad \hat{\vec{J}} \equiv \hat{\vec{L}} \quad \Rightarrow \quad \textbf{spin 0}$$

$$\text{Example:}\quad \underbrace{\hat{R}_{\vec{n}_3\delta\phi}}_{e^{-i\hat{L}_3\delta\phi/\hbar}}\psi(\vec{x})\approx\left[\hat{I}-\left(x_1\frac{\partial}{\partial x_2}-x_2\frac{\partial}{\partial x_1}\right)\delta\phi\right]\psi(\vec{x})=\psi\left[\overbrace{\begin{pmatrix}1&+\delta\phi&0\\-\delta\phi&1&0\\0&0&1\end{pmatrix}}^{\mathbf{R}^{-1}_{\vec{n}_3\delta\phi}}\begin{pmatrix}x_1\\x_2\\x_3\end{pmatrix}\right]$$

▶ Transformation of vector wavefunction $\Psi(\vec{x})\equiv\begin{pmatrix}\psi_1(\vec{x})\\\psi_2(\vec{x})\\\psi_3(\vec{x})\end{pmatrix}$

Besides the argument, also the direction of the vector $\Psi(\vec{x})$ affected by rotation. Defining transformation property:

$$\hat{R}_{\vec{n}\phi}\Psi(\vec{x})=\underbrace{\begin{pmatrix}r_{11}&r_{12}&r_{13}\\r_{21}&r_{22}&r_{23}\\r_{31}&r_{32}&r_{33}\end{pmatrix}_{\vec{n}\phi}}_{\hat{\mathbf{R}}_{\vec{n}\phi}\equiv\mathbf{R}_{\vec{n}\phi}}\begin{pmatrix}\psi_1\\\psi_2\\\psi_3\end{pmatrix}(\mathbf{R}^{-1}_{\vec{n}\phi}\vec{x})=\underbrace{e^{-i\frac{\hat{\vec{S}}\cdot\vec{n}}{\hbar}\phi}\Psi}_{\Psi'}(\underbrace{\mathbf{R}^{-1}_{\vec{n}\phi}\vec{x}}_{\vec{x}'})$$

Generators of $\hat{\mathbf{R}}^{-1}_{\vec{n}\phi}$ (cf. p. 58): $\hat{S}_1=\hbar\begin{pmatrix}0&0&0\\0&0&-i\\0&+i&0\end{pmatrix}$ $\hat{S}_2=\hbar\begin{pmatrix}0&0&+i\\0&0&0\\-i&0&0\end{pmatrix}$ $\hat{S}_3=\hbar\begin{pmatrix}0&-i&0\\+i&0&0\\0&0&0\end{pmatrix}$

$$\text{Example (}z\text{-rotation):}\quad \underbrace{\hat{R}_{\vec{n}_3\delta\phi}}_{e^{-i[\hat{S}_3+\hat{L}_3]\delta\phi/\hbar}}\begin{pmatrix}\psi_1(\vec{x})\\\psi_2(\vec{x})\\\psi_3(\vec{x})\end{pmatrix}\approx\underbrace{\left[\hat{I}-\begin{pmatrix}0&+1&0\\-1&0&0\\0&0&0\end{pmatrix}\delta\phi\right]}_{\begin{pmatrix}1&-\delta\phi&0\\+\delta\phi&1&0\\0&0&1\end{pmatrix}}\underbrace{e^{-i\frac{\hat{L}_3\delta\phi}{\hbar}}\begin{pmatrix}\psi_1(\vec{x})\\\psi_2(\vec{x})\\\psi_3(\vec{x})\end{pmatrix}}_{\begin{pmatrix}\psi_1(\mathbf{R}^{-1}_{\vec{n}_3\delta\phi}\vec{x})\\\psi_2(\mathbf{R}^{-1}_{\vec{n}_3\delta\phi}\vec{x})\\\psi_3(\mathbf{R}^{-1}_{\vec{n}_3\delta\phi}\vec{x})\end{pmatrix}}$$

$$\hat{S}_1^2+\hat{S}_2^2+\hat{S}_3^2=\overbrace{s(s+1)}^{2}\hbar^2\hat{I}\ \Rightarrow\ \boxed{s=1}$$

$$\hbar\,\mathrm{Det}(\hat{S}_i-\lambda\hat{I})=0\ \Rightarrow\lambda=\pm1,0\ \Rightarrow\ \boxed{s_i=\hbar\begin{cases}-1\\0\\+1\end{cases}}\ \Rightarrow\ \textbf{spin 1}$$

⇒ 3-component wavefuctions $\Psi(\vec{x})$ with vector transformation properties describe particles with spin 1. We now find the link of Cartesian components $\psi_i(\vec{x})$ with i=1,2,3 to the probability amplitudes $\psi_{m_s}(\vec{x})$ for individual spin projections m_s=0, ±1 to the z-axis direction:

Eigenvectors of $\hat{S}_3$: $\xi_{+1}=\frac{1}{\sqrt{2}}\begin{pmatrix}-1\\-i\\0\end{pmatrix}$ $\xi_0=\begin{pmatrix}0\\0\\1\end{pmatrix}$ $\xi_{-1}=\frac{1}{\sqrt{2}}\begin{pmatrix}+1\\-i\\0\end{pmatrix}$

$$\overbrace{\begin{pmatrix}\psi_1(\vec{x})\\\psi_2(\vec{x})\\\psi_3(\vec{x})\end{pmatrix}}^{\psi_{+1}(\vec{x})\xi_{+1}+\psi_0(\vec{x})\xi_0+\psi_{-1}(\vec{x})\xi_{-1}}=\begin{pmatrix}\frac{1}{\sqrt{2}}[\psi_{-1}(\vec{x})-\psi_{+1}(\vec{x})]\\-\frac{i}{\sqrt{2}}[\psi_{-1}(\vec{x})+\psi_{+1}(\vec{x})]\\\psi_0(\vec{x})\end{pmatrix}\Rightarrow\boxed{\begin{pmatrix}\psi_{+1}(\vec{x})\\\psi_0(\vec{x})\\\psi_{-1}(\vec{x})\end{pmatrix}=\begin{pmatrix}-\frac{1}{\sqrt{2}}[\psi_1(\vec{x})-i\psi_2(\vec{x})]\\\psi_3(\vec{x})\\+\frac{1}{\sqrt{2}}[\psi_1(\vec{x})+i\psi_2(\vec{x})]\end{pmatrix}}$$

▶ Transformation of spinor wavefunction $\Psi(\vec{x})\equiv\begin{pmatrix}\psi_\uparrow(\vec{x})\\\psi_\downarrow(\vec{x})\end{pmatrix}$ spin $\frac{1}{2}$

For spinors we will proceed the opposite way. In this case we know spin matrices. Assuming the standard form of the transformation:

$$\boxed{\hat{R}_{\vec{n}\phi}\Psi(\vec{x})=\underbrace{[\hat{\mathbf{S}}_{\vec{n}\phi}\Psi]}_{\Psi'}(\mathbf{R}^{-1}_{\vec{n}\phi}\vec{x})=e^{-i\frac{\hat{\vec{S}}\cdot\vec{n}}{\hbar}\phi}\Psi(\mathbf{R}^{-1}_{\vec{n}\phi}\vec{x})}\quad\text{we find the unknown matrix }\hat{\mathbf{S}}_{\vec{n}\phi}$$

$$\hat{\mathbf{S}}_{\vec{n}\phi}=e^{-i\frac{\hat{\vec{S}}\cdot\vec{n}}{\hbar}\phi}=\sum_{k=0}^{\infty}\frac{1}{k!}\left(-\frac{i\phi}{2}\right)^k(\hat{\vec{\sigma}}\cdot\vec{n})^k=\dots\qquad\text{with }(\hat{\vec{\sigma}}\cdot\vec{n})^k=\begin{cases}\hat{I}&\text{for }k\text{=even}\\\hat{\vec{\sigma}}\cdot\vec{n}&\text{for }k\text{=odd}\end{cases}$$

$$(\hat{\vec{\sigma}}\cdot\vec{n})^2=\sum_{i,j=1}^{3}n_in_j\hat{\sigma}_i\hat{\sigma}_j=\tfrac{1}{2}\sum_{i,j=1}^{3}n_in_j\underbrace{(\hat{\sigma}_i\hat{\sigma}_j+\hat{\sigma}_j\hat{\sigma}_i)}_{2\delta_{ij}\hat{I}}+\tfrac{1}{2}\underbrace{\sum_{i,j=1}^{3}n_in_j(\hat{\sigma}_i\hat{\sigma}_j-\hat{\sigma}_j\hat{\sigma}_i)}_{0}=\underbrace{\sum_{i=1}^{3}n_i^2}_{1}\hat{I}$$

$$\cdots=\underbrace{\sum_{k=0,2,4,\dots}\tfrac{1}{k!}\left(-\tfrac{i\phi}{2}\right)^k}_{\cos\frac{\phi}{2}}\hat{I}+\underbrace{\sum_{k=1,3,5,\dots}\tfrac{1}{k!}\left(-\tfrac{i\phi}{2}\right)^k}_{-i\sin\frac{\phi}{2}}(\hat{\vec{\sigma}}\cdot\vec{n})=\boxed{\left(\cos\tfrac{\phi}{2}\right)\hat{I}-i\left(\sin\tfrac{\phi}{2}\right)(\hat{\vec{\sigma}}\cdot\vec{n})=\hat{\mathbf{S}}_{\vec{n}\phi}}$$

spinor transformation

$$\Rightarrow\boxed{\hat{\mathbf{S}}_{\vec{n}(2\pi)}=-\hat{I},\quad\hat{\mathbf{S}}_{\vec{n}(4\pi)}=+\hat{I}}$$

Special case: $\hat{\mathbf{S}}_{\vec{n}_z\phi}=\begin{pmatrix}e^{-i\phi/2} & 0\\ 0 & e^{+i\phi/2}\end{pmatrix}$

▶ **Rotational invariance** *sensu stricto*

Hamiltonian must depend only on rotational invariants: vector squares (sizes), scalar products (e.g. $|\hat{\vec{x}}_k-\hat{\vec{x}}_l|$ or $\{\hat{\vec{S}}_k\cdot\hat{\vec{S}}_l\}$ for particles k and l) etc.

◀ **Historical remark**

1913: Élie Cartan discovered complex "tensors" with spinor transform. properties
1927: Wolfgang Pauli introduces spinors to QM

■ Space inversion

Spatial inversion (taking mirror images of all 3 spatial axes, therefore replacing "right" by "left" and vice versa) is just a discrete transformation. Nevertheless, there exists an observable associated with it—the spatial parity. In contrast to the above cases, space inversion is not a valid symmetry of this world.

▶ **Coordinate, momentum & angular momentum transformation**

$$\boxed{\hat{P}\hat{\vec{x}}\hat{P}^{-1}=-\hat{\vec{x}}}\qquad\boxed{\hat{P}\hat{\vec{p}}\hat{P}^{-1}=-\hat{\vec{p}}}\qquad\Rightarrow\qquad\hat{P}\hat{\vec{L}}\hat{P}^{-1}=\hat{P}(\hat{\vec{x}}\times\hat{\vec{p}})\hat{P}^{-1}=+\hat{\vec{L}}$$

$$\hat{P}\hat{\vec{S}}\hat{P}^{-1}=+\hat{\vec{S}}$$

Cartesian coordinates:
$x_i\to-x_i\quad(i{=}1,2,3)$

Spherical coordinates:
$r\to r$
$\vartheta\to(\pi-\vartheta)$
$\varphi\to(\varphi+\pi)$

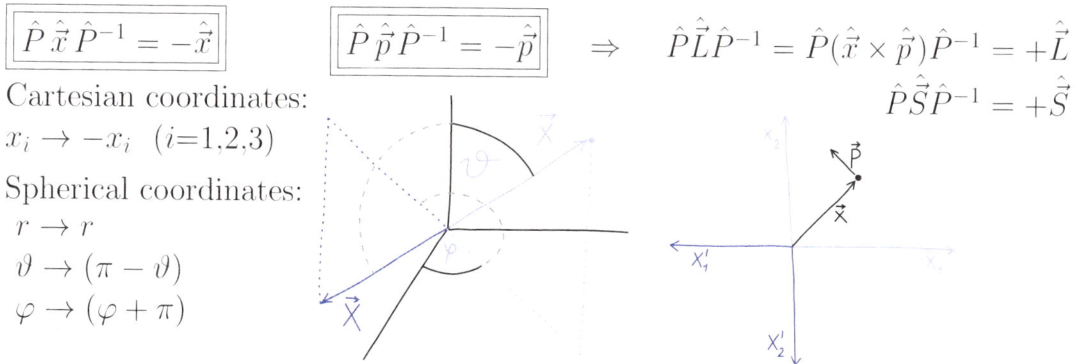

▶ **Classification of observables** with respect to space inversion:

$\hat{P}\hat{\vec{V}}\hat{P}^{-1}=-\hat{\vec{V}}$ **vector** $\qquad\hat{P}\hat{\vec{A}}\hat{P}^{-1}=+\hat{\vec{A}}$ **pseudovector** (axial-vector)

$\hat{P}\hat{S}\hat{P}^{-1}=+\hat{S}$ **scalar** $\qquad\hat{P}\hat{O}\hat{P}^{-1}=-\hat{O}$ **pseudoscalar**

▶ **Invariance** *sensu stricto* of a system under space inversion: $\hat{H}=\hat{P}\hat{H}\hat{P}^{-1}=\dots$

$$=\sum_k\tfrac{1}{2M_k}\underbrace{\hat{P}\hat{\vec{p}}_k^{\,2}\hat{P}^{-1}}_{(-\hat{\vec{p}}_k)\cdot(-\hat{\vec{p}}_k)}+\underbrace{\hat{P}V(\{\hat{\vec{x}}_k\},\{\hat{\vec{S}}_k\})\hat{P}^{-1}}_{V(\{-\hat{\vec{x}}_k\},\{\hat{\vec{S}}_k\})}\Rightarrow\boxed{V(\{\hat{\vec{x}}_k\},\{\hat{\vec{S}}_k\})=V(\{-\hat{\vec{x}}_k\},\{\hat{\vec{S}}_k\})}$$

potential must be even

▶ Physical observable associated with space inversion: **parity**

$\hat{P}^2 = \hat{I}$ $\boxed{\hat{P} = \hat{P}^\dagger = \hat{P}^{-1}}$ $\Rightarrow$ eigenvalues $\boxed{\pi = \pm 1}$

$\hat{P}\Psi_{\rm even}(\vec{x}) = +\Psi_{\rm even}(\vec{x})$ $\hat{P}\Psi_{\rm odd}(\vec{x}) = -\Psi_{\rm odd}(\vec{x})$

$$\hat{P}[R(r)Y_{lm}(\vartheta,\varphi)] = R(r)\underbrace{Y_{lm}(\pi-\vartheta,\varphi+\pi)}_{P_{lm}(-\cos\vartheta)e^{im\varphi}e^{im\pi}} = \underbrace{(-)^{l-m}(-)^m}_{(-)^l}[R(r)Y_{lm}(\vartheta,\varphi)]$$

■ Time translation

We come to transformations involving time. The most important specimen of this type represents a shift of the time-axis origin—time translation. The unitary operator expressing the transition between observers with different time settings is closely related to the evolution operator, which describes the dynamics. Motions of quantum systems generated by this operator will be in our main focus starting from Sec. 1.5.

▶ "Young" & "old" observers (different time origins)

$|\psi_{\rm young}(t')\rangle \equiv$ state of system as seen by "young"

$|\psi_{\rm old}(\underbrace{t'+\tau}_{t})\rangle \equiv$ state of system as seen by "old"

Uniqueness requirement: $|\psi_{\rm young}(t')\rangle \stackrel{!}{=} |\psi_{\rm old}(t)\rangle$

"Generational" transformation: $|\psi_{\rm young}(t')\rangle = \hat{U}(t,t')|\psi_{\rm old}(t')\rangle$

$\Rightarrow$ evolution transformation: $\boxed{|\psi_{\rm old}(t)\rangle = \hat{U}(t,t')|\psi_{\rm old}(t')\rangle}$

▶ Properties of $\hat{U}(t,t')$ in case the system is invariant under time translation:

$$\left.\begin{array}{ll} \text{(a)} & \hat{U}(t,t') \equiv \hat{U}(\overbrace{t-t'}^{\tau}) \\ \text{(b)} & \hat{U}(\tau)^{-1} = \hat{U}(\tau)^\dagger \\ \text{(c)} & \hat{U}(0) = \hat{I} \\ \text{(d)} & \hat{U}(\tau_1+\tau_2) = \hat{U}(\tau_2)\hat{U}(\tau_1) \end{array}\right\} \Rightarrow \left\{\begin{array}{l} \hat{U}(\tau) = e^{i\hat{\chi}\tau} \\ \text{Consistent choice}: \quad \hat{\chi} = -\frac{1}{\hbar}\hat{H} \\ \boxed{\hat{U}(\tau) = e^{-i\frac{\hat{H}\tau}{\hbar}}} \quad \textbf{evolution operator} \end{array}\right.$$

Association of generator $\hat{\chi}$ with the **Hamiltonian** $\hat{H}$ is equivalent to the non-stationary Schrödinger equation (see Sec. 1.5). Invariance *sensu stricto* of the system under time translation $\Leftrightarrow$ $\boxed{\hat{H}(t) \equiv \hat{H}}$ (Hamilt. independent of time)

▶ Regardless of the time-translation invariance we may assume:

Postulate: Hamiltonian $\hat{H}(t)$ of the system at time t
$= \hbar\times$ generator of *infinitesimal* time translation $\equiv$ evolution from t to $t+dt$

■ Time reversal

Time reversal means an inversion of the time arrow: going from future to past. Like the inversion of space, it is just a discrete transformation, but a more difficult one. In quantum physics it cannot be represented by a unitary operator and there is no physical observable (like parity) associated with it.

► We seek for operator $\hat{\mathcal{T}}$ satisfying: $\boxed{\hat{U}(t)\hat{\mathcal{T}}|\psi(0)\rangle = \hat{\mathcal{T}}\hat{U}(-t)|\psi(0)\rangle}$ $\forall|\psi(0)\rangle$

This means: forward evolution of time-reversed state = time reversal of backward-evolved state:

For infinitesimal time δt this implies:

$$(1 - i\tfrac{\hat{H}\delta t}{\hbar})\hat{\mathcal{T}} = \hat{\mathcal{T}}(1 + i\tfrac{\hat{H}\delta t}{\hbar})$$

$\Rightarrow$ We require: $\boxed{(-i\hat{H})\hat{\mathcal{T}} = \hat{\mathcal{T}}(i\hat{H})}$

$|\psi(0)\rangle$ 0 -t $\hat{U}(-t)|\psi(0)\rangle$ $\hat{T}\hat{U}(-t)|\psi(0)\rangle$ $\hat{T}|\psi(0)\rangle$ 0 -t $\hat{U}(t)\hat{T}|\psi(0)\rangle$

► For $\hat{\mathcal{T}}$ unitary this would mean:

$$\hat{H}\hat{\mathcal{T}} + \hat{\mathcal{T}}\hat{H} \equiv \underbrace{\{\hat{H}, \hat{\mathcal{T}}\}}_{\text{anticommutator}} = 0$$

$\Rightarrow$ $\hat{H}|E\rangle = E|E\rangle$ $\Rightarrow$ $\hat{H}(\hat{\mathcal{T}}|E\rangle) = -E(\hat{\mathcal{T}}|E\rangle)$ ...nonphysical ($\nexists$ lower bound of energy)

$\Rightarrow \hat{\mathcal{T}}$ is *not* a unitary operator

► **Operator $\hat{\mathcal{T}}$ is antiunitary**: $\hat{\mathcal{T}}(\alpha\hat{A}) = \alpha^*\hat{\mathcal{T}}\hat{A}$ $\Rightarrow$ $\boxed{[\hat{H}, \hat{\mathcal{T}}] = 0}$

$$\boxed{\hat{\mathcal{T}} \equiv \hat{U}\hat{K}} \text{ where } \begin{cases} \hat{U} \equiv & \text{a unitary operator (depends on convention)} \\ \hat{K} \equiv & \text{comp. conjugation operator : } \hat{K}\sum_i \alpha_i|i\rangle = \sum_i \alpha_i^*|i\rangle \\ & \text{with respect to basis } \{|i\rangle\}_i \text{ (basis dependent op.)} \end{cases}$$

► **Some properties**:

$$\langle\psi_1|\psi_2\rangle = \langle\hat{U}\psi_1|\hat{U}\psi_2\rangle = \textstyle\sum_i \alpha'^*_{1i}\alpha'_{2i} = (\sum_i \alpha'_{1i}\alpha'^*_{2i})^* = \langle\hat{\mathcal{T}}\psi_1|\hat{\mathcal{T}}\psi_2\rangle^* = \langle\hat{\mathcal{T}}\psi_2|\hat{\mathcal{T}}\psi_1\rangle$$

$\langle\psi_1|\hat{O}|\psi_2\rangle = \langle\hat{\mathcal{T}}\psi_2|\hat{\mathcal{T}}\hat{O}^\dagger\hat{\mathcal{T}}^{-1}|\hat{\mathcal{T}}\psi_1\rangle$ (e.g., transition matrix elements)

► **Classification of observables** with respect to time reversal:

(1) $\hat{\mathcal{T}}\hat{A}\hat{\mathcal{T}}^{-1} = +\hat{A}$ **even observables** ($\hat{H}, \hat{\vec{x}}, \dots$) ... $\hat{\mathcal{T}}$ keeps eigenvalues

(2) $\hat{\mathcal{T}}\hat{A}\hat{\mathcal{T}}^{-1} = -\hat{A}$ **odd observables** ($\hat{\vec{p}}, \hat{\vec{L}}, \hat{\vec{S}}, \dots$) ... $\hat{\mathcal{T}}$ inverts eigenvalues

► **Invariance** of a common Hamiltonian under time inversion: $\hat{H} = \hat{\mathcal{T}}\hat{H}\hat{\mathcal{T}}^{-1} = \dots$

$$= \sum_k \frac{1}{2M_k} \underbrace{\hat{\mathcal{T}}\hat{\vec{p}}_k^{\,2}\hat{\mathcal{T}}^{-1}}_{(-\hat{\vec{p}}_k)\cdot(-\hat{\vec{p}}_k)} + \underbrace{\hat{\mathcal{T}}V(\{\hat{\vec{x}}_k\}, \{\hat{\vec{S}}_k\})\hat{\mathcal{T}}^{-1}}_{V^*(\{\hat{\vec{x}}_k\}, \{-\hat{\vec{S}}_k\})} \Rightarrow \boxed{V(\{\hat{\vec{x}}_k\}, \{\hat{\vec{S}}_k\}) = V^*(\{\hat{\vec{x}}_k\}, \{-\hat{\vec{S}}_k\})}$$

potential must be real

◄ **Historical remark**

1924: O. Laporte introduces spatial parity of electron wavefunctions in atoms

1931: E. Wigner shows that time reversal is represented by an antiunitary operator

■ Galilean transformations

Nonrelativistic quantum mechanics must be invariant under transformations between inertial frames with relative speed $\vec{v}$.

▶ Galilean transformation of $\vec{x}$ & t: $\binom{\vec{x}}{t} \mapsto \binom{\vec{x}'}{t'} \equiv \boldsymbol{G}_{\vec{v}} \binom{\vec{x}}{t} = \binom{\vec{x}-\vec{v}t}{t}$
$\exists$ the corresponding family of unitary operators $\hat{G}_{\vec{v}}$ acting in $\mathcal{H}$

▶ **Quantum operator** of the Galilean transformation for a general wavefunction of single particle in the coordinate representation (the derivation is not presented here): $\hat{G}_{\vec{v}} \underbrace{\boldsymbol{\Psi}\binom{\vec{x}}{t}}_{\equiv \boldsymbol{\Psi}(\vec{x},t)} = e^{-i(M\vec{v}\cdot\vec{x}-\frac{1}{2}Mv^2t)/\hbar} \underbrace{\boldsymbol{\Psi}\left[\boldsymbol{G}_{\vec{v}}\binom{\vec{x}}{t}\right]}_{\hat{T}_{-\vec{v}t}\boldsymbol{\Psi}(\vec{x},t)}$

◀ **Historical remark**
1925: Erwin Schrödinger attempts to create a Lorenz-invariant wave equation, but because of problems he remains with the non-relativistic formulation
1926: Oskar Klein and Walter Gordon (simultaneously V. Fock *et al.*) develop a relativistic wave equation for spinless particles
1927: Paul Dirac initiates quantum field theory (the right unification of relativity with QM), in 1928 he creates a relativistic wave equation for spin-$\frac{1}{2}$ particle

■ Symmetry & degeneracy

Degeneracy of energy levels is an important signature of symmetry of Hamiltonian $\hat{H}$ under transformations $\hat{U}_g \equiv e^{i\hat{\vec{G}}\cdot\vec{s}} \in \mathcal{G}$ (since $\hat{U}_g|E\rangle$ remains an eigenstate with the same energy). However, some symmetries cause no degeneracy, and some degeneracies are not due to usual geometric symmetries.

▶ $\mathcal{G} \equiv$ **Abelian** (translations, space inversion) $\Rightarrow$ eigenstates of $\hat{H}$ are simultaneous eigenstates of all $\hat{G}_i \Rightarrow e^{i\hat{\vec{G}}\cdot\vec{s}}|E\rangle = e^{i\varphi}|E\rangle$ (vector differing just by a phase factor) $\Rightarrow$ in general **no degeneracy**

▶ $\mathcal{G} \equiv$ **non-Abelian** (rotation) $\Rightarrow \exists\ \hat{G}_i$ which acts nontrivially on the eigenstates of $\hat{H} \Rightarrow e^{i\hat{\vec{G}}\cdot\vec{s}}|E\rangle = |E'\rangle$ (eigenvector with the same energy E but in general not collinear with the initial $|E\rangle$) $\Rightarrow$ **degeneracy** occurs

Example: for rotationally invariant $\hat{H}$, the states with the same angular-momentum quantum number j and different projections m degenerated

▶ Some $\hat{H}$ have symmetries induced by groups $\mathcal{G} \supset$ standard spatio-temporal groups (e.g., groups employing both coordinates and momenta) $\Rightarrow$ **dynamical symmetry** $\Rightarrow$ occurence of "accidental degeneracies" (beyond rotational ones)
Example: **harmonic oscillator**

$$\hat{H} = \frac{1}{2M}\hat{\vec{p}}^{\,2} + \frac{M\omega^2}{2}\hat{\vec{x}}^{\,2} = \hbar\omega\left[\overbrace{\left(\sqrt{\frac{M\omega}{2\hbar}}\hat{\vec{x}} + \frac{i}{\sqrt{2M\hbar\omega}}\hat{\vec{p}}\right)}^{\hat{\vec{b}}^\dagger} \cdot \overbrace{\left(\sqrt{\frac{M\omega}{2\hbar}}\hat{\vec{x}} - \frac{i}{\sqrt{2M\hbar\omega}}\hat{\vec{p}}\right)}^{\hat{\vec{b}}} + \frac{3}{2}\hat{I}\right]$$

$\hat{H}$ invariant under transformations conserving $\hat{\vec{b}}^\dagger\hat{\vec{b}}$ (classically the norm $|\vec{b}|^2$ of a complex vector depending on $\vec{x}$ & $\vec{p}$) $\Rightarrow$ symmetry group $\mathcal{G} \equiv \mathrm{U}(3)$ (unitary group in 3D) $\supset \mathrm{O}(3)$ (orthonormal group in 3D describing rotations)

◄ **Historical remark**
1926: Wolfgang Pauli associates the accidental degeneracy in the hydrogen atom with the additional symmetry (using Lenz result from 1924)
1935-6: V. Fock and V. Bargmann analyze the "dynamical symmetry" in hydrogen

1.5 Unitary evolution of quantum systems

After all, we come to the dynamics of quantum systems. There are two types of quantum evolution: the spontaneous one—motions signifying perpetual flow of time, and an induced one—changes provoked by quantum measurements. Here we will focus on the spontaneous type of evolution.

■ Nonstationary Schrödinger equation for stationary Hamiltonian

For quantum mechanics, the dynamical Schrödinger equation means the same as what the Newton equation means for a classical mechanics. We have already introduced the evolution operator from the time translation (Sec. 2.4), so we need not make a special postulate on the spontaneous dynamics.

► **Spontaneous evolution** of a quantum system

$$\boxed{|\psi(t)\rangle = e^{-i\frac{\hat{H}t}{\hbar}}|\psi(0)\rangle} \quad \Leftrightarrow \quad \boxed{i\hbar\tfrac{d}{dt}|\psi(t)\rangle = \hat{H}|\psi(t)\rangle}$$

evolution operator **nonstationary Schrödinger equation**

► Spinless particle in a potential: $\boxed{i\hbar\frac{\partial}{\partial t}\psi(\vec{x},t) = \left[-\frac{\hbar^2}{2M}\Delta + V(\vec{x})\right]\psi(\vec{x},t)}$

► **Stationary states** $\hat{H}|E_i\rangle{=}E_i|E_i\rangle \Rightarrow |\psi(0)\rangle \equiv |E_i\rangle \xrightarrow{t} |\psi(t)\rangle = e^{-i\frac{E_i t}{\hbar}}|\psi(0)\rangle$

Eigenstates of $\hat{H}$ evolve just by changing the phase factor ⇒ "stationary"

⇒ evolution of a general state expressed by expansion into eigenstates of $\hat{H}$

$$|\psi(0)\rangle \equiv \boxed{\sum_i \underbrace{\alpha_i}_{\alpha_i(0)}|E_i\rangle \xrightarrow{t} \sum_i \underbrace{\alpha_i e^{-i\frac{E_i t}{\hbar}}}_{\alpha_i(t)}|E_i\rangle} \equiv |\psi(t)\rangle$$

■ Continuity equation & probability flow

If the dynamical Schrödinger equation is applied to the scalar wavefunction of a particle in external fields, the resulting dependence $\psi(\vec{x},t)$ describes how the probability density $\rho(\vec{x},t) = |\psi(\vec{x},t)|^2$ flows in space. This process can be described in terms of ordinary fluid dynamics.

► **Continuity equation**

Particle in scalar potential $V(\vec{x},t)$ & vector potential $\vec{A}(\vec{x},t)$:

$$\frac{\partial}{\partial t}\underbrace{|\psi|^2}_{\rho} = \underbrace{\frac{\partial\psi}{\partial t}\psi^* + \psi\frac{\partial\psi^*}{\partial t}}_{\text{from Schrödinger eq.}} = \frac{1}{i\hbar}\psi^*\left[\frac{1}{2M}(-i\hbar\vec{\nabla} - q\vec{A})^2 + V\right]\psi + \text{C.C.} =$$

$$= \frac{1}{M\hbar}\,\text{Im}\underbrace{\left[\psi^*(-\hbar^2\vec{\nabla}\cdot\vec{\nabla} + i\hbar q\vec{\nabla}\cdot\vec{A} + i\hbar q\vec{A}\cdot\vec{\nabla} + q^2\vec{A}^2)\psi\right]}_{\substack{-\hbar^2\vec{\nabla}\cdot(\psi^*\vec{\nabla}\psi) + i\hbar q\vec{\nabla}\cdot(\psi^*\vec{A}\psi) \\ +\hbar^2(\vec{\nabla}\psi^*)\cdot(\vec{\nabla}\psi) - i\hbar q\left[(\vec{\nabla}\psi^*)\cdot\vec{A}\psi - \text{C.C.}\right] + q^2\vec{A}^2|\psi|^2}} = -\vec{\nabla}\cdot\underbrace{\left[\frac{\hbar}{M}\text{Im}(\psi^*\vec{\nabla}\psi) - \frac{q}{M}\psi^*\vec{A}\psi\right]}_{\vec{j}}$$

We obtain the familiar continuity equation: $\boxed{\frac{\partial}{\partial t}\rho(\vec{x},t) + \vec{\nabla}\cdot\vec{j}(\vec{x},t) = 0}$

The change of probability in an infinitesimal volume is in balance with the incoming/outgoing flux of probability. The probability "field" $\rho(\vec{x},t)$ behaves like a fluid: its substance is locally conserved.

Conservation of total probability: Take a sphere of radius R with volume V_R and surface S_R: $\frac{d}{dt}\int\limits_{V_R}|\psi(\vec{x},t)|^2\,d\vec{x} = -\int\limits_{V_R}\vec{\nabla}\cdot\vec{j}(\vec{x},t)d\vec{x} = -\int\limits_{S_R}\vec{j}(\vec{x},t)\cdot d\vec{S}_R \xrightarrow[R\to\infty]{} 0$ (since for normalizable wavefunctions $\vec{j}\to 0$ faster than $1/S_R$) $\Rightarrow$ norm $\langle\psi|\psi\rangle = \int|\psi(\vec{x},t)|^2\,d\vec{x} = 1$ conserved in time, as is also clear from unitarity of $\hat{U}(t)$

▶ **Probability flow**

$$\vec{j}(\vec{x},t) = \frac{\hbar}{M}\underbrace{\text{Im}\left[\psi^*\vec{\nabla}\psi\right]}_{-\frac{i}{2}(\psi^*\vec{\nabla}\psi - \psi\vec{\nabla}\psi^*)} - \frac{q}{M}\psi^*\vec{A}\psi = \boxed{\frac{1}{2M}\left[\psi^*(\hat{\vec{\pi}}\psi) + \psi(\hat{\vec{\pi}}\psi)^*\right] = \vec{j}}$$

$\hat{\vec{\pi}} = (-i\hbar\vec{\nabla} - q\vec{A}) \equiv$ mechanical momentum

Parametrization: $\boxed{\psi(\vec{x},t) = R(\vec{x},t)e^{iS(\vec{x},t)/\hbar}}$ $\Rightarrow$ $\boxed{\vec{j} = \underbrace{R^2}_{\rho}\underbrace{\frac{1}{M}\left[\vec{\nabla}S - q\vec{A}\right]}_{\vec{v}}}$

This helps to understand the complex character of wavefunctions:
(a) squared absolute value $|\psi(\vec{x},t)|^2 \equiv$ probability density
(b) gradient of phase $\vec{\nabla}S(\vec{x},t) \propto$ velocity field (in absence of $\vec{A}$)

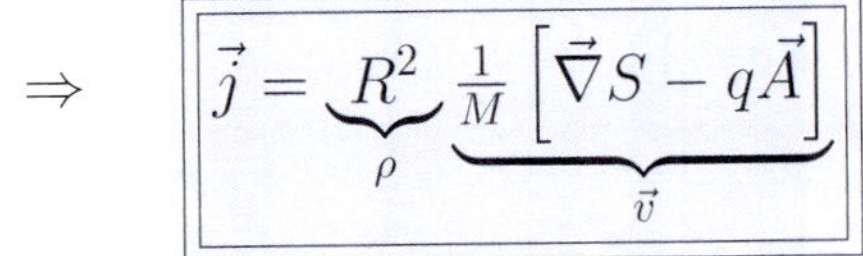

In this way, the wavefunction can be visualized by a mesh of arrows $\propto \vec{v} = \frac{1}{M}\vec{\nabla}S$, the local density of the mesh being proportional to the density $\rho = |\psi|^2$:

▶ **Flow for simple wavefunctions**

(a) Planar wave :	$\psi = \mathcal{N}e^{i\frac{\vec{p}\cdot\vec{x}}{\hbar}}$	$\vec{j} = \lvert\mathcal{N}\rvert^2\frac{\vec{p}}{M}$
(b) Spherical wave :	$\psi = \mathcal{N}\frac{1}{r}e^{i\frac{p_r r}{\hbar}}$	$\vec{j} = \lvert\mathcal{N}\rvert^2\frac{p_r}{Mr^2}\vec{n}_r$
(c) Eigenstate of orbital momentum :	$\psi = \underbrace{R(r)}_{\lvert R\rvert e^{iS_r(r)/\hbar}}\underbrace{Y_{lm}(\vartheta,\varphi)}_{P_{lm}(\cos\vartheta)e^{im\varphi}}$	$\vec{j} \propto \frac{dS_r}{dr}\vec{n}_r + \frac{m\hbar}{r\sin\vartheta}\vec{n}_\varphi$

We obtain
(a) translational
(b) divergent
(c) rotational
flows

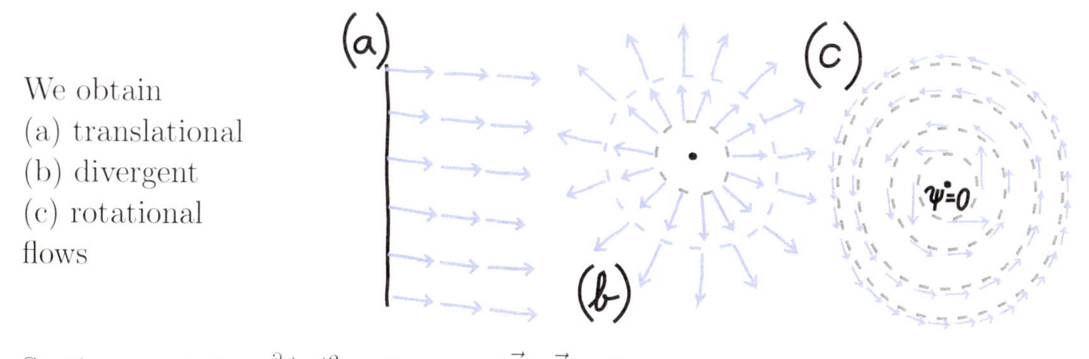

Stationary state: $\frac{\partial}{\partial t}|\psi|^2 = 0 \quad \Rightarrow \quad \vec{\nabla}\cdot\vec{j} = 0$

This follows from the existence of degenerate solutions $\left\{ \begin{smallmatrix} \psi_E \\ \psi_E^* \end{smallmatrix} \right\}$ with flows $\left\{ \begin{smallmatrix} \vec{j} \\ -\vec{j} \end{smallmatrix} \right\}$

▶ **Vorticity** of the probability flow: $\boxed{\vec{\mathfrak{v}} \equiv \vec{\nabla}\times\vec{v}}$ (standard definition)

(a) $\vec{A}{=}0 \quad\Rightarrow\quad \vec{\mathfrak{v}} = \vec{\nabla}\times\left(\frac{1}{M}\vec{\nabla}S\right) = 0$ except points $\psi{=}0$, where phase S is not determined ⇒ in absence of mg. field, the probability flow may produce vortices only in $\psi{=}0$ points

(b) $\vec{A}{\neq}0 \quad\Rightarrow\quad \vec{\mathfrak{v}} = -\frac{q}{M}\overbrace{(\vec{\nabla}\times\vec{A})}^{\vec{B}}$ flow vortical in all points where $\vec{B}\neq 0$

◀ **Historical remark**

1926: Max Born introduces probabilistic interpretation of Sch.eq. & probab. flow

■ Conservation laws & symmetries

We are ready now to appreciate the deepest dynamical consequence of symmetry. According to the famous theorem by Emmy Noether, the symmetry of a given system under an n-parameter Lie group generates n conserved quantities. But what the conservation means in the indeterministic environment of QM, where physical quantities yield just statistical values?

▶ Conservation laws in QM

Evolution of the probability distribution for measurement outcomes a of quantity A for a system in initial state $|\psi(0)\rangle$: $\quad p_\psi(a,t) \equiv \langle\psi(t)|\hat{P}_a|\psi(t)\rangle$

Quantity A conserved ⇔ $\boxed{\frac{\partial}{\partial t}p_\psi(a,t) = 0}$ $\forall|\psi(0)\rangle$ & $\forall a$

Statistical moments: $\underbrace{\langle\psi(t)|\hat{A}^k|\psi(t)\rangle}_{\langle\psi(0)|e^{i\frac{\hat{H}t}{\hbar}}\hat{A}^k e^{-i\frac{\hat{H}t}{\hbar}}|\psi(0)\rangle} = \langle\psi(0)|\hat{A}^k|\psi(0)\rangle \quad\Rightarrow\quad e^{i\frac{\hat{H}t}{\hbar}}\hat{A}e^{-i\frac{\hat{H}t}{\hbar}} = \hat{A}$

$\Rightarrow \quad \boxed{[\hat{A},\hat{H}] = 0}$

⇒ probability distribution $p_\psi(a,t)$ does not depend on time for quantities that commute with Hamiltonian

▶ Equation for the **average value**

$$i\hbar\frac{d}{dt}\langle\psi(t)|\hat{A}|\psi(t)\rangle = \overbrace{-\langle\psi(0)|e^{i\frac{\hat{H}t}{\hbar}}\hat{H}\hat{A}e^{-i\frac{\hat{H}t}{\hbar}}|\psi(0)\rangle + \langle\psi(0)|e^{i\frac{\hat{H}t}{\hbar}}\hat{A}\hat{H}e^{-i\frac{\hat{H}t}{\hbar}}|\psi(0)\rangle}^{\langle\psi(t)|[\hat{A},\hat{H}]|\psi(t)\rangle}$$

Time-derivative "operator": $\boxed{\hat{\dot{A}} \equiv \frac{1}{i\hbar}[\hat{A}, \hat{H}]}$ $\boxed{\frac{d}{dt}\langle A\rangle_\psi \equiv \langle\psi(t)|\hat{\dot{A}}|\psi(t)\rangle}$

Analogy with **Poisson bracket**: $\dot{A} = \sum_i \big(\frac{\partial A}{\partial p_i} \underbrace{\dot{p}_i}_{-\frac{\partial H}{\partial q_i}} + \frac{\partial A}{\partial q_i} \underbrace{\dot{q}_i}_{+\frac{\partial H}{\partial p_i}}\big) = -\{A, H\}$

$\Rightarrow \quad [\hat{A}, \hat{H}] \leftrightarrow -i\hbar\{A, H\}$

Example: **particle speed "operator"** for $\hat{H} = \frac{1}{2M}(\hat{\vec{p}} - q\vec{A})^2 + V$

$$\hat{\dot{\vec{x}}} = \frac{1}{i\hbar}[\hat{\vec{x}}, \hat{H}] = \frac{1}{2iM\hbar}[\hat{\vec{x}}, (\hat{\vec{p}} - q\vec{A})^2] = \frac{\hat{\vec{p}}-q\vec{A}}{M} = \frac{\hat{\vec{\pi}}}{M}$$

▶ Conservation laws generated by **symmetries**

Quantity $\hat{A} \propto \hat{G} \equiv$ Hermitian generator of group $\mathcal{G} \equiv \{e^{i\hat{G}s}\}_{s\in\mathbb{R}}$

$\Rightarrow \quad A$ conserved $\quad\Leftrightarrow\quad [e^{i\hat{G}s}, \hat{H}] = 0 \quad\Leftrightarrow\quad \mathcal{G}$ is **symmetry group** of $\hat{H}$

Generalizing to higher dimensional Lie groups, we obtain the QM version of the **Noether theorem**: invariance of $\hat{H}$ under a Lie group with n generators implies conservation of quantities associated with all generators

Standard spatio-temporal symmetries of $\hat{H}$ and related **conservation laws**:

translational invariance	$\Leftrightarrow$	**linear momentum** $\hat{\vec{p}}$
rotational invariance	$\Leftrightarrow$	**angular momentum** $\hat{\vec{J}}$
time translation invariance	$\Leftrightarrow$	**energy** $\hat{H}$
space reflection invariance	$\Leftrightarrow$	**parity** $\hat{P}$

Note: Space reflection is not a continuous transformation; parity conservation follows from an "accidental" Hermiticity of the reflection operator $\hat{P}$

◀ **Historical remark**

1915: Emmy Noether proves theorem relating conservation laws with symmetries
1924: N. Bohr, H. Kramers & J. Slater propose that in QM the conservation laws (energy, momentum) hold only "statistically" (not in every event); this is disproved in experiments of W. Bothe & H Geiger and A.H. Compton & A.W. Simon
1927: Eugene Wigner writes about symmetry & conservation laws in QM, he relates parity conservation in elmag. decays with reflection symmetry of interaction
1956: C.N. Yang & T.D. Lee propose that parity is not conserved in weak interactions; this is verified experimentally in 1957 by C.S. Wu *et al.*
1951-8: Various proofs of the CPT symmetry & conservation

■ Energy × time uncertainty relation

In physics, time is not a standard observable—it is just "a parameter" whose only role is "to fly" (and we all have to fly with it!). There is no QM operator associated with time. Nevertheless, it is often stated that time and energy form a pair of conjugated quantities similar to coordinate and momentum. This can be valid only in a limited sense, which we explore in the following.

▶ **Survival amplitude & probability**

The amplitude/probability to find the system in its initial state $|\psi(0)\rangle$ after time t: $\boxed{A_0(t) \equiv \langle\psi(0)|\psi(t)\rangle}\ \boxed{P_0(t) = |A_0(t)|^2}$

$$A_0(t)=\langle\psi(0)|e^{-i\frac{\hat{H}t}{\hbar}}\psi(0)\rangle=\langle e^{+i\frac{\hat{H}t}{\hbar}}\psi(0)|\psi(0)\rangle = A_0^*(-t) \quad \Rightarrow \quad \boxed{P_0(t) = P_0(-t)}$$

To evaluate $A_0(t)$, we use the completeness: $\begin{cases} \int\limits_{\mathcal{S}(\hat{H})} \sum\limits_{k\in\mathcal{D}_E} |Ek\rangle\langle Ek|\, dE = \hat{I} \\ \langle E'k'|Ek\rangle = \delta(E-E')\delta_{kk'} \end{cases}$

(for continuous E & discrete k: other possibilities analogous)

$$A_0(t)=\langle\psi(0)|\underset{\hat{I}}{\uparrow} e^{-i\frac{\hat{H}t}{\hbar}} \underset{\hat{I}}{\uparrow}|\psi(0)\rangle=\iint \sum_{k,k'} \underbrace{\langle\psi(0)|E'k'\rangle}_{\omega^*(E',k')} \underbrace{\langle E'k'|e^{-i\frac{\hat{H}t}{\hbar}}|Ek\rangle}_{e^{-i\frac{Et}{\hbar}}\delta(E-E')\delta_{kk'}} \underbrace{\langle Ek|\psi(0)\rangle}_{\omega(E,k)} dE\, dE'$$

$$= \int \underbrace{\left[\sum_k |\omega(E,k)|^2\right]}_{\Omega(E)\ \textbf{energy distribution}} e^{-i\frac{Et}{\hbar}} dE = \boxed{\int\limits_{\mathcal{S}(\hat{H})} \Omega(E) e^{-i\frac{Et}{\hbar}} dE = A_0(t)}$$

Fourier transformation

General property of the Fourier transformation:

$\left.\begin{smallmatrix}\text{wide}\\\text{narrow}\end{smallmatrix}\right\}$ energy distribution $\boxed{\Delta E \left\{\begin{smallmatrix}\text{large}\\\text{small}\end{smallmatrix}\right. \quad \Leftrightarrow \quad \left.\begin{smallmatrix}\text{small}\\\text{large}\end{smallmatrix}\right\} \Delta t}$ time evolution $\left\{\begin{smallmatrix}\text{fast}\\\text{slow}\end{smallmatrix}\right.$

$$\langle\langle E^2\rangle\rangle \sim \Delta E^2 \quad \Leftrightarrow \quad \Delta t^2 \sim \frac{1}{\int_{-\infty}^{+\infty} P_0(t)dt} \int_{-\infty}^{+\infty} t^2 P_0(t) dt$$

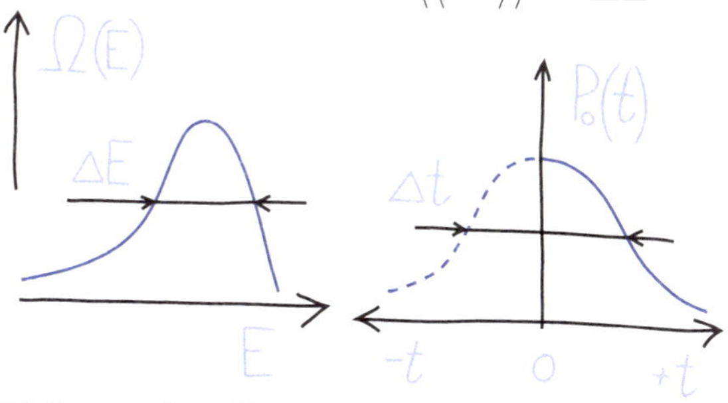

"uncertainty" relation:

$$\Delta E \cdot \Delta t \gtrsim \hbar$$

but $\Delta t \neq$ uncertainty in usual sense

This can be illustrated on the following **examples**:

▶ **Gaussian energy distribution**: $\boxed{\Omega(E) = \frac{1}{\sqrt{2\pi\sigma^2}} e^{-\frac{(E-E_0)^2}{2\sigma^2}}}$

$$A_0(t)=\frac{1}{\sqrt{2\pi\sigma^2}} \int_{-\infty}^{+\infty} \underbrace{e^{-\frac{(E-E_0)^2}{2\sigma^2} - i\frac{Et}{\hbar}}}_{e^{-\frac{1}{2\sigma^2}E^2+\left(\frac{E_0}{\sigma^2}-\frac{it}{\hbar}\right)E-\frac{E_0^2}{2\sigma^2}}} dE=e^{-\frac{\sigma^2}{2\hbar^2}t^2} e^{-i\frac{E_0 t}{\hbar}}$$

$$\boxed{P_0(t) = e^{-\left(\frac{\sigma}{\hbar}\right)^2 t^2} = e^{-\left(\frac{t}{\Delta t}\right)^2}}$$

$$\Delta t \sim \frac{\hbar}{\sigma} \qquad \sigma \equiv \Delta E \qquad \Rightarrow \Delta E \cdot \Delta t = \hbar$$

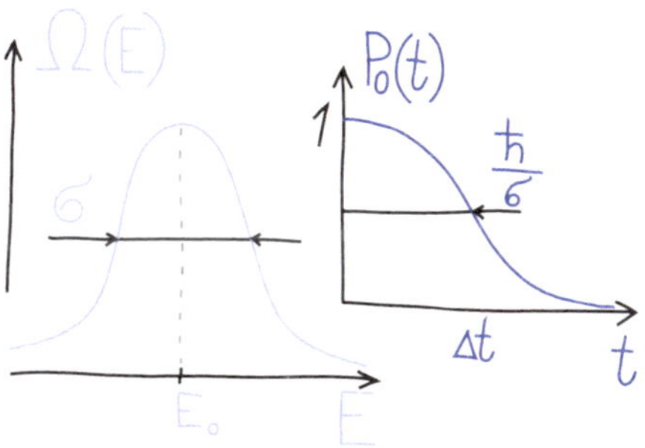

▶ **Breit-Wigner** (Cauchy) **energy distribution**: $\boxed{\Omega(E)=\frac{1}{\pi}\frac{\frac{\Gamma}{2}}{(E-E_0)^2+\left(\frac{\Gamma}{2}\right)^2}}$

Γ = finite halfwidth

$\langle\langle E^2\rangle\rangle=\infty$ infinite energy disperion because of the slow decrease of $\Omega(E)$

$\boxed{P_0(t)=e^{-\frac{t}{\tau}}}$ with $\boxed{\Gamma\tau=\hbar}$ **exponential decay** (average lifetime $\tau=\frac{\hbar}{\Gamma}$)

Inverse proof (from exponential decay to Breit-Wigner distribution):

Assume $A_0(t)=\left\{\begin{array}{c} e^{-\Gamma t/(2\hbar)}e^{-iE_0t/\hbar} \\ e^{+\Gamma t/(2\hbar)}e^{-iE_0t/\hbar}\end{array}\right\}$ for $\left\{\begin{array}{c} t\geq 0 \\ t<0\end{array}\right.$ Coherent assumption on the phase factors. The $t{>}0$ exponential decay is extended also to $t{<}0$

$$\Omega(E)=\frac{1}{2\pi\hbar}\int_{-\infty}^{+\infty}A_0(t)e^{+i\frac{Et}{\hbar}}dt=\frac{1}{2\pi\hbar}\Bigg(\underbrace{\int_{-\infty}^{0}e^{\left[\frac{\Gamma}{2\hbar}+i\frac{E-E_0}{\hbar}\right]t}dt}_{\frac{\hbar}{(\Gamma/2)+i(E-E_0)}}+\underbrace{\int_{0}^{+\infty}e^{\left[-\frac{\Gamma}{2\hbar}+i\frac{E-E_0}{\hbar}\right]t}dt}_{\frac{-\hbar}{-(\Gamma/2)+i(E-E_0)}}\Bigg)=$$

$$=\frac{1}{\pi}\frac{\Gamma/2}{(E-E_0)^2+(\Gamma/2)^2}$$

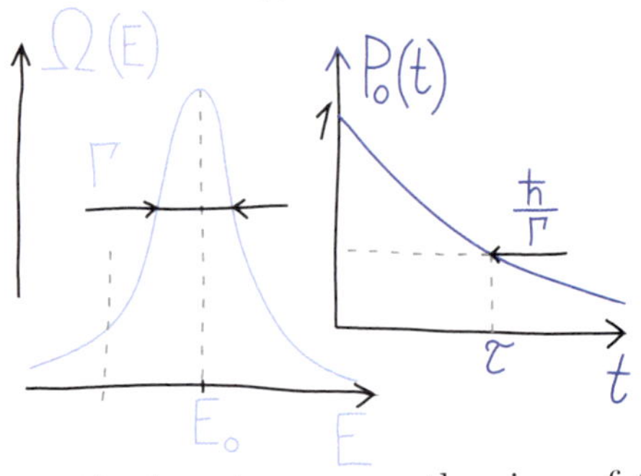

$\Rightarrow \overbrace{\Delta E}^{\sim\Gamma}\,\overbrace{\Delta t}^{\sim\tau}=\hbar$ relation obtained again

$\Rightarrow$ Low-energy cutoff of $\Omega(E)$ leads to small **deviations** from the exp. law, in particular, to a smoothening of the $t{=}0$ cusp of the extended function $P_0(t)$

▶ **Non-exponential decay**

QM always yields $\boxed{\frac{d}{dt}P_0(t)\big|_{t=0}=0}$ in contrast to exp. law: $\frac{d}{dt}P_0(t)\big|_{t=0}=-\frac{1}{\tau}$

General derivation for small times:

$|A_0(\delta t)|^2=\langle\psi(0)|e^{-i\frac{\hat{H}\delta t}{\hbar}}|\psi(0)\rangle\langle\psi(0)|e^{+i\frac{\hat{H}\delta t}{\hbar}}|\psi(0)\rangle\approx$ expand up to 2nd order in δt

$$\approx 1+\langle\psi(0)|\hat{H}|\psi(0)\rangle^2\frac{(\delta t)^2}{\hbar^2}-\langle\psi(0)|\hat{H}^2|\psi(0)\rangle\frac{(\delta t)^2}{\hbar^2}=\boxed{1-\underbrace{\frac{\langle\langle E^2\rangle\rangle}{\hbar^2}}_{\tau^{-2}}(\delta t)^2\approx P_0(\delta t)}$$

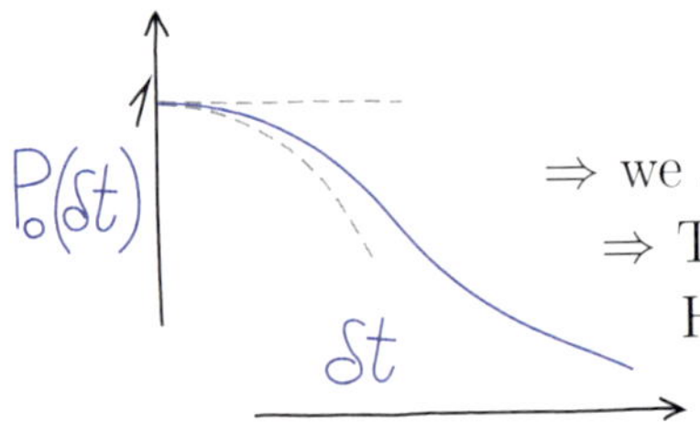

$\Rightarrow$ we again get: $\overbrace{\Delta E}^{\sqrt{\langle\langle E^2\rangle\rangle}}\;\overbrace{\Delta t}^{\tau}=\hbar$

$\Rightarrow$ The QM decay for small times is always quadratic. However, this is usually very hard to measure!

◀ **Historical remark**

1997: the first exp. detection of short-t corrections to the exponential decay law

▶ **Energy $\times$ time uncertainty in real measurements**

Let T be a quantity suitable for time determination $\Rightarrow$ **"clock" operator** $\hat{T}$

For the clock to be functioning there must be $\boxed{[\hat{T},\hat{H}]\neq 0}$ (otherwise the distribution of T for any initial state $|\psi(0)\rangle$ would be conserved in time)

$\Rightarrow$ standard $T\times E$ uncertainty relation in state $|\psi(t)\rangle$

$$\sqrt{\langle\langle E^2\rangle\rangle_{\psi(t)}\langle\langle T^2\rangle\rangle_{\psi(t)}}\geq\frac{\hbar}{2}\Big|\langle\psi(t)|\overbrace{\tfrac{1}{i\hbar}[\hat{T},H]}^{\dot{\hat{T}}}|\psi(t)\rangle\Big|$$

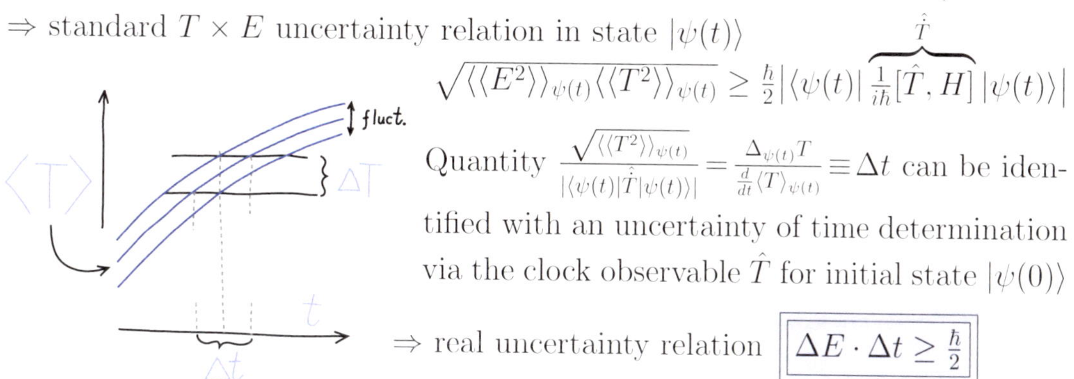

Quantity $\frac{\sqrt{\langle\langle T^2\rangle\rangle_{\psi(t)}}}{|\langle\psi(t)|\dot{\hat{T}}|\psi(t)\rangle|}=\frac{\Delta_{\psi(t)}T}{\frac{d}{dt}\langle T\rangle_{\psi(t)}}\equiv\Delta t$ can be identified with an uncertainty of time determination via the clock observable $\hat{T}$ for initial state $|\psi(0)\rangle$

$\Rightarrow$ real uncertainty relation $\boxed{\Delta E\cdot\Delta t\geq\frac{\hbar}{2}}$

Time operator in QM? For a certain *subset of initial states* of the given system, it is possible to find a suitable clock operator $\hat{T}$. However, there exists *no universal time operator* $\hat{T}$ satisfying the canonical commutation relation $[\hat{T},\hat{H}]=-i\hbar\hat{I}$, applicable for all initial states $\in\mathcal{H}$. For instance, this would imply the absence of a lower bound of energy, which is unphysical.

◀ **Historical remark**
1926, 1933: W. Pauli shows the difficulty in building a quantum operator of time
1928: N. Bohr proposes the $E\times t$ uncertainty principle, 1930's debate with Einstein
1945: L. Mandelstam & I. Tamm derive $E\times t$ uncertainty for "clock observables"
1960's–present: Discussions on the ways to formulate QM with a time operator

■ Hamiltonians depending on time

Let us have a closer look on quantum dynamics generated by a Hamiltonian which itself changes in time: $\boxed{\hat{H}=\hat{H}(t)}$. This means that, for the system under study, the time-translation invariance is violated, as is actually the case if a variable external field is applied. However, as explained in the following paragraph, time-dependent Hamiltonians naturally appear also in time-translation invariant situations—in the so-called Dirac picture of quantum dynamics.

▶ The basic form of evolution operator valid only locally: $\hat{U}(\delta t,t)=e^{-i\frac{\hat{H}(t)\delta t}{\hbar}}$

$\Rightarrow$ **Generalized Schrödinger equation** $\boxed{i\hbar\frac{d}{dt}|\psi(t)\rangle=\hat{H}(t)|\psi(t)\rangle}$

We distinguish 2 cases:

(a)	$[\hat{H}(t),\hat{H}(t')]=0\quad\forall\, t,t'$	easy but rare
(b)	$[\hat{H}(t),\hat{H}(t')]\neq 0\quad t\neq t'$	normal & difficult

▶ Equation for **generalized evolution operator** $\hat{U}(t,t_0)$

The generalized Schrödinger equation can be translated to an operator form:

$$i\hbar\frac{\partial}{\partial t}\underbrace{\hat{U}(t,t_0)|\psi(t_0)\rangle}_{|\psi(t)\rangle} = \hat{H}(t)\underbrace{\hat{U}(t,t_0)|\psi(t_0)\rangle}_{|\psi(t)\rangle} \quad \text{valid } \forall\ |\psi(t_0)\rangle$$

$\Rightarrow$ Operator equation $\boxed{i\hbar\frac{\partial}{\partial t}\hat{U}(t,t_0) = \hat{H}(t)\,\hat{U}(t,t_0)}$ with $\boxed{\hat{U}(t_0,t_0) = \hat{I}}$

▶ Iterative solution: $\underbrace{\hat{U}(t,t_0)}_{*} = \hat{I} - \frac{i}{\hbar}\int\limits_{t_0}^{t}\hat{H}(t_1)\underbrace{\hat{U}(t_1,t_0)}_{*=\ldots}\,dt_1 = \ldots$

$\Rightarrow$ **Dyson series**

$$\hat{U}(t,t_0) = \hat{I} + \left(-\tfrac{i}{\hbar}\right)^1\int\limits_{t_0}^{t}\hat{H}(t_1)\,dt_1 + \left(-\tfrac{i}{\hbar}\right)^2\int\limits_{t_0}^{t}\int\limits_{t_0}^{t_1}\hat{H}(t_1)\hat{H}(t_2)\,dt_2dt_1 + \ldots$$
$$+\left(-\tfrac{i}{\hbar}\right)^n\int\limits_{t_0}^{t}\int\limits_{t_0}^{t_1}\ldots\int\limits_{t_0}^{t_{n-1}}\hat{H}(t_1)\hat{H}(t_2)\ldots\hat{H}(t_n)\,dt_n\ldots dt_2dt_1 + \ldots$$

In general, the Dyson series can be summed up to a compact form only in case (a) of the Hamiltonian time dependence. In case (b), which is much more generic, the evolution operator can only be expressed in the infinite-series form.

▶ **Case (a)**: $[\hat{H}(t),\hat{H}(t')] = 0$

$$\int\limits_{t_0}^{t}\int\limits_{t_0}^{t_1}\hat{H}(t_1)\hat{H}(t_2)\,dt_2dt_1 = \frac{1}{2}\int\limits_{t_0}^{t}\int\limits_{t_0}^{t}\hat{H}(t_1)\hat{H}(t_2)\,dt_2dt_1 = \frac{1}{2}\left[\int\limits_{t_0}^{t}\hat{H}(t_1)\,dt_1\right]^2$$

$$\int\limits_{t_0}^{t}\int\limits_{t_0}^{t_1}\ldots\int\limits_{t_0}^{t_{n-1}}\hat{H}(t_1)\hat{H}(t_2)\ldots\hat{H}(t_n)\,dt_n\ldots dt_2dt_1 = \frac{1}{n!}\left[\int\limits_{t_0}^{t}\hat{H}(t_1)\,dt_1\right]^n$$

Compact expression of the evolution operator: $\boxed{\hat{U}(t,t_0) = e^{-\frac{i}{\hbar}\int_{t_0}^{t}\hat{H}(t_1)\,dt_1}}$

▶ **Case (b)**: $[\hat{H}(t),\hat{H}(t')] \neq 0$

Note that $\int\limits_{t_0}^{t}\int\limits_{t_0}^{t_1}\ldots\int\limits_{t_0}^{t_{n-1}}\hat{H}(t_1)\hat{H}(t_2)\ldots\hat{H}(t_n)\,dt_n\ldots dt_2dt_1 =$

$$\frac{1}{n!}\int\limits_{t_0}^{t}\int\limits_{t_0}^{t}\ldots\int\limits_{t_0}^{t}\underbrace{\mathfrak{T}\left[\hat{H}(t_1)\hat{H}(t_2)\ldots\hat{H}(t_n)\right]}_{\hat{H}(t_{i_1})\hat{H}(t_{i_2})\ldots\hat{H}(t_{i_n})}\,dt_n\ldots dt_2dt_1$$

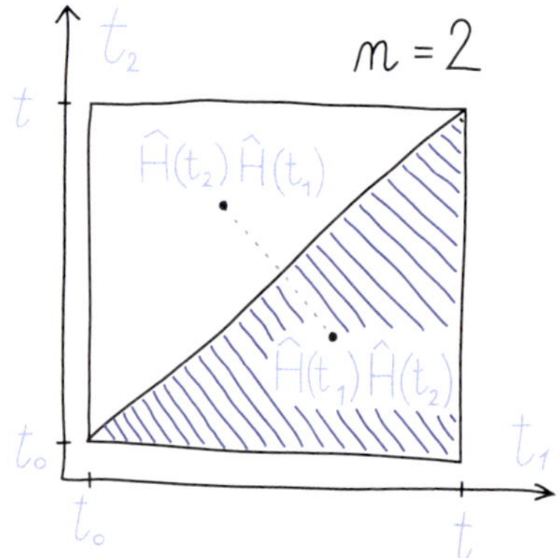

$\hat{H}(t_{i_1})\hat{H}(t_{i_2})\ldots\hat{H}(t_{i_n})$ **time ordering**
$(t_1,t_2\ldots t_n)\mapsto(t_{i_1}\geq t_{i_2}\geq\cdots\geq t_{i_n})$

In each term of Dyson series do the following:
(1) change the subintegral operator function to the t-ordered product: $[\ldots]\mapsto\mathfrak{T}[\ldots]$
(2) extend integ. domain $\Rightarrow$ all upper limits $= t$
(3) reduce the integral by factor $\frac{1}{n!}$

The resulting series looks like exponent. expansion and can be *abbreviated* by the symbolic expression: $\boxed{\hat{U}(t,t_0) = \mathfrak{T}e^{-\frac{i}{\hbar}\int_{t_0}^{t}\hat{H}(t_1)\,dt_1}}$

Alternative descriptions of time evolution

So far we practiced an approach in which the vectors corresponding to physical states vary in time while the operators associated with observables mostly stay constant. This is indeed the most common description of time evolution, but not the only one. All equivalent descriptions can be split into 3 groups.

▶ **3 equivalent ways** to express action of any unitary transformation $\hat{U}$

(1) $\begin{array}{lcl} |\psi\rangle & \mapsto & \hat{U}|\psi\rangle \\ \hat{A} & \mapsto & \hat{A} \end{array}$ **(2)** $\begin{array}{lcl} |\psi\rangle & \mapsto & |\psi\rangle \\ \hat{A} & \mapsto & \hat{U}^{-1}\hat{A}\hat{U} \end{array}$ **(3)** $\begin{array}{lcl} |\psi\rangle & \mapsto & \hat{U}_1|\psi\rangle \\ \hat{A} & \mapsto & \hat{U}_0^{-1}\hat{A}\hat{U}_0 \end{array}$

for any factorization $\hat{U} = \hat{U}_0\hat{U}_1$

In all cases, matrix elements $\langle\psi'|\hat{A}|\psi\rangle$ are the same $\Rightarrow$ equivalent descriptions

These possibilities constitute 3 equivalent types of description of quantum evolution with unitary operator $\boxed{\hat{U}(t) = e^{-i\frac{\hat{H}t}{\hbar}}}$

▶ **(1) Schrödinger picture**

$$\boxed{\begin{array}{l} |\psi(t)\rangle_{\rm S} = \hat{U}(t)|\psi(0)\rangle_{\rm S} \\ \hat{A}_{\rm S} \equiv \text{const.} \end{array}} \Rightarrow \begin{cases} \text{usual time evolution of state vectors} \\ \qquad \boxed{i\hbar\frac{d}{dt}|\psi(t)\rangle_{\rm S} = \hat{H}_{\rm S}|\psi(t)\rangle_{\rm S}} \\ \text{time independent operators} \end{cases}$$

▶ **(2) Heisenberg picture**

$$\boxed{\begin{array}{l} |\psi(t)\rangle_{\rm H} = |\psi\rangle_{\rm H} \equiv \text{const.} \\ \hat{A}_{\rm H}(t) = \hat{U}^{\dagger}(t)\,\hat{A}_{\rm S}\,\hat{U}(t) \end{array}} \Rightarrow \begin{cases} \text{time independent state vectors} \\ \qquad |\psi\rangle_{\rm H} = \hat{U}^{\dagger}(t)|\psi(t)\rangle_{\rm S} \\ \text{time dependent operators} \end{cases}$$

$\Rightarrow$ Hamiltonian $\hat{H}_{\rm H} = \hat{H}_{\rm S} \equiv \hat{H}$

$\Rightarrow$ General observable evolution equation: $\boxed{i\hbar\frac{d}{dt}\hat{A}_{\rm H}(t) = [\hat{A}_{\rm H}(t), \hat{H}]}$

▶ **(3) Dirac picture** (intermediate between Schrödinger and Heisenberg)

Hamiltonian splitting $\boxed{\hat{H} = \hat{H}_0 + \hat{H}'}$

In general, $[\hat{H}_0, \hat{H}'] \neq 0 \Rightarrow$ the factorization $\boxed{\hat{U}(t) = \overbrace{\hat{U}_0(t)}^{=e^{-i\frac{\hat{H}_0 t}{\hbar}}}\ \overbrace{\hat{U}_1(t)}^{\neq e^{-i\frac{\hat{H}' t}{\hbar}}}}$ is not trivial

Operators evolve by $\hat{U}_0(t) \Rightarrow \boxed{\hat{A}_{\rm D}(t) = \hat{U}_0^{\dagger}(t)\,\hat{A}_{\rm S}\,\hat{U}_0(t)} \Rightarrow \hat{H}_{0\rm D}{=}\hat{H}_{0\rm S}{\equiv}\hat{H}_0$

$\Rightarrow$ they satisfy differential eq.: $\boxed{i\hbar\frac{d}{dt}\hat{A}_{\rm D}(t) = [\hat{A}_{\rm D}(t), \hat{H}_0]}$

State vectors evolve by $\hat{U}_1(t) \Rightarrow \boxed{|\psi(t)\rangle_{\rm D} = \hat{U}_0^{\dagger}(t)|\psi(t)\rangle_{\rm S}}$

$$i\hbar\frac{d}{dt}|\psi(t)\rangle_{\rm D} = -\hat{H}_0|\psi(t)\rangle_{\rm D} + \hat{U}_0^{\dagger}(t)\underbrace{\left(i\hbar\frac{d}{dt}|\psi(t)\rangle_{\rm S}\right)}_{(\hat{H}_0+\hat{H}')\hat{U}_0(t)|\psi(t)\rangle_{\rm D}} = \underbrace{\hat{U}_0^{\dagger}(t)\hat{H}'\hat{U}_0(t)}_{\hat{H}'_{\rm D}(t)}|\psi(t)\rangle_{\rm D}$$

$$\boxed{i\hbar\frac{d}{dt}|\psi(t)\rangle_{\rm D} = \hat{H}'_{\rm D}(t)|\psi(t)\rangle_{\rm D}}$$

Schwinger-Tomonaga equation

[Schrödinger eq. with $\hat{H} \mapsto \hat{H}'_{\rm D}(t)$]

The evolution according to this equation can be represented by **state evolution operator** $\hat{U}(t,t_0)_{\rm D}$, which is expressed via the **Dyson series** with $\hat{H}(t) \equiv \hat{H}'_{\rm D}(t)$. In this case, due to the assumed "smallness" of $\hat{H}'_{\rm D}$, the series can be used in a **perturbative way**, i.e., neglecting higher-order terms (see Sec. 5.3).

◀ **Historical remark**
1925-6: W. Heisenberg & E. Schrödinger use the two descriptions of QM dynamics
1930: Paul Dirac connects these descriptions in a unified picture
1934: Julian Schwinger (S.-I. Tomonaga in 1940's) introduce the interaction picture
1949: Freeman Dyson uses the expansion of the evolution operator in QED

■ Green operator

We briefly outline an approach to evolution which becomes very useful later, in the context of relativistic quantum theory. It is based on the old idea of Green's function, known from the general theory of differential equations, and leads to a very enlightening view of quantum dynamics.

▶ **"Retarded" Green operator** for nonstationary Schrödinger equation

Defined as evolution oper. for $t \geq t_0$:

$$\hat{G}(t,t_0) = \underbrace{\Theta(t-t_0)}_{=\begin{cases}1 \text{ for } t\geq t_0\\ 0 \text{ for } t<t_0\end{cases}} \hat{U}(t,t_0)$$

Satisfies Green-like operator equation:

$$\left[i\hbar\tfrac{\partial}{\partial t} - \hat{H}(t)\right]\hat{G}(t,t_0) = i\hbar\delta(t-t_0)$$

Note: The meaning of "retarded" should be understood here as evolving the system from past t_0 to future t. Similarly, "advanced" Green operator is defined by $\hat{G}_-(t,t_0) = [1-\Theta(t-t_0)]\hat{U}(t,t_0)$ and satisfies the same Green-like equation.

▶ **Transition from known to unknown Green operator**

Splitting of Hamiltonian into "free" & "interaction" parts: $\hat{H}(t) = \hat{H}_0(t) + \hat{H}'(t)$

Assume we know Green operator $\hat{G}_0$ for the "free" part:

$$\left[i\hbar\tfrac{\partial}{\partial t} - \hat{H}_0(t)\right]\hat{G}_0(t,t_0) = i\hbar\delta(t-t_0)$$

⇒ The full Green operator satisfies the following integral equation:

$$\hat{G}(t,t_0) = \hat{G}_0(t,t_0) - \tfrac{i}{\hbar}\int\limits_{-\infty}^{+\infty} \hat{G}_0(t,t_1)\hat{H}'(t_1)\hat{G}(t_1,t_0)\,dt_1$$

Proof: application of $\left[i\hbar\frac{\partial}{\partial t} - \hat{H}_0\right]$ to the first term and inside the integral yields the defining eq. of $\hat{G}$:

$$\left[i\hbar\tfrac{\partial}{\partial t} - \hat{H}_0(t)\right]\hat{G}(t,t_0) = i\hbar\delta(t-t_0) + \overbrace{\int\limits_{-\infty}^{+\infty} \delta(t-t_1)\hat{H}'(t_1)\hat{G}(t_1,t_0)\,dt_1}^{\hat{H}'(t)\hat{G}(t,t_0)}$$

▶ **Iterative solution** of the integral equation for $\hat{G}$

$$\hat{G}(t,t_0) = \hat{G}_0(t,t_0) - \tfrac{i}{\hbar}\int_{-\infty}^{+\infty} \hat{G}_0(t,t_1)\hat{H}'(t_1)\hat{G}_0(t_1,t_0)\,dt_1 +$$

$$\dots + \left(-\tfrac{i}{\hbar}\right)^n \underbrace{\int_{-\infty}^{+\infty}\dots\int_{-\infty}^{+\infty}}_{n\times} \hat{G}_0(t,t_n)\hat{H}'(t_n)\hat{G}_0(t_n,t_{n-1})\dots\hat{G}_0(t_2,t_1)\hat{H}'(t_1)\hat{G}_0(t_1,t_0)\,dt_n..dt_1 + \dots\dots$$

This series is analogous to the Dyson series [except (a) the const. term $=\hat{G}_0 \neq \hat{I}$, (b) all integrals have the same limits, and (c) alternating operators $\hat{G}_0$ & $\hat{H}'$ inside the integral]. If $\hat{H}'$ is "small" compared to $\hat{H}_0$, the series can again be used in the perturbative way, i.e., neglecting the terms of higher order. The meaning of this expansion will become clear in the following.

▶ **Propagator**

Coordinate representation of single-particle Green operator

$$\langle\vec{x}|\psi(t)\rangle = \langle\vec{x}|\hat{G}(t,t_0)|\psi(t_0)\rangle = \int \underbrace{\langle\vec{x}|\hat{G}(t,t_0)|\vec{x}_0\rangle}_{G(\vec{x}t|\vec{x}_0t_0)}\langle\vec{x}_0|\psi(t_0)\rangle\,d\vec{x}_0$$

$G(\vec{x}t|\vec{x}_0t_0) \equiv$ propagator

$\equiv$ wavefunction evolved from ideally localized init. state $\psi(\vec{x},t_0)=\delta(\vec{x}-\vec{x}_0)\equiv|\vec{x}_0\rangle$

Wavefunction evolved from a general initial state is the convolution:

$$\boxed{\psi(\vec{x},t) = \int G(\vec{x}t|\vec{x}_0t_0)\psi(\vec{x}_0,t_0)\,d\vec{x}_0}$$

Propagator satisfies the following eq.:

$$\left[i\hbar\tfrac{\partial}{\partial t} + \tfrac{\hbar^2}{2M}\Delta - V(\vec{x},t)\right] G(\vec{x}t|\vec{x}_0t_0) = i\hbar\delta(t-t_0)\delta(\vec{x}-\vec{x}_0)$$

Let $\boxed{V(\vec{x},t)=V_0(\vec{x},t)+V'(\vec{x},t)}$ and $G_0(\vec{x}t|\vec{x}_0t_0)$ be the solution for $V_0(\vec{x},t)$

The iterative solution reads as:

$$\boxed{\begin{aligned} G(\vec{x}t|\vec{x}_0t_0) \;=\; & G_0(\vec{x}t|\vec{x}_0t_0) + \dots + \left(-\tfrac{i}{\hbar}\right)^n \underbrace{\int\dots\int}_{2n\times} G_0(\vec{x}t|\vec{x}_nt_n)V'(\vec{x}_n,t_n)\dots \\ & \dots G_0(\vec{x}_2t_2|\vec{x}_1t_1)V'(\vec{x}_1,t_1)G_0(\vec{x}_1t_1|\vec{x}_0t_0)\,d\vec{x}_ndt_n\dots d\vec{x}_1dt_1 + \dots \end{aligned}}$$

This series has a visual interpretation:

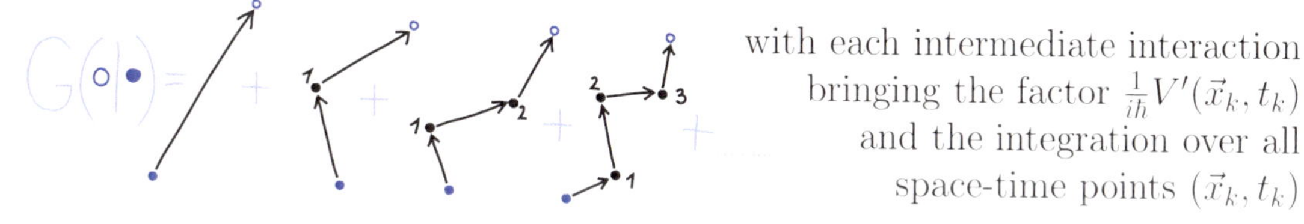

with each intermediate interaction bringing the factor $\frac{1}{i\hbar}V'(\vec{x}_k,t_k)$ and the integration over all space-time points $(\vec{x}_k,t_k)$

▶ Green operator for **time-independent Hamiltonian** $\hat{H}(t)\equiv\hat{H}$

Expansion in stationary states: $\hat{G}(t,t_0) = \Theta(t-t_0)\sum_{i,k} e^{-i\frac{E_i(t-t_0)}{\hbar}}|E_ik\rangle\langle E_ik| = \dots$

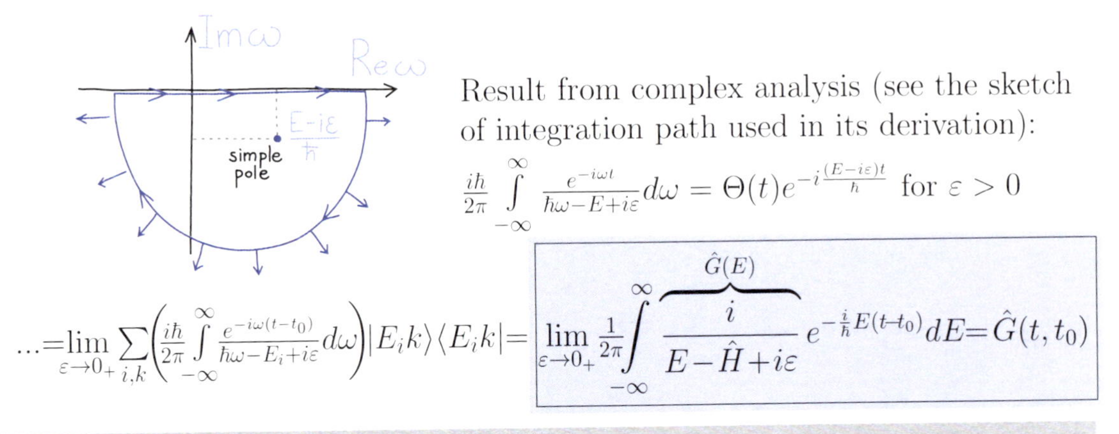

Result from complex analysis (see the sketch of integration path used in its derivation):

$$\frac{i\hbar}{2\pi}\int_{-\infty}^{\infty}\frac{e^{-i\omega t}}{\hbar\omega-E+i\varepsilon}d\omega=\Theta(t)e^{-i\frac{(E-i\varepsilon)t}{\hbar}}\quad\text{for }\varepsilon>0$$

$$\ldots=\lim_{\varepsilon\to 0_+}\sum_{i,k}\left(\frac{i\hbar}{2\pi}\int_{-\infty}^{\infty}\frac{e^{-i\omega(t-t_0)}}{\hbar\omega-E_i+i\varepsilon}d\omega\right)|E_ik\rangle\langle E_ik|=\boxed{\lim_{\varepsilon\to 0_+}\frac{1}{2\pi}\int_{-\infty}^{\infty}\overbrace{\frac{i}{E-\hat{H}+i\varepsilon}}^{\hat{G}(E)}e^{-\frac{i}{\hbar}E(t-t_0)}dE=\hat{G}(t,t_0)}$$

◀ **Historical remark**

1828: George Green applies math. analysis in electromagnetism ⇒ Green function
1949: Richard Feynman applies Green funcs. in QM+QED (later "propagator")

2.5 Examples of quantum evolution

Having digested all the general approaches to the description of quantum evolution, we need to see some concrete applications. A few examples discussed below represent just a personal selection—a multitude of other cases could serve the purpose as well.

■ Two-level system

Two-level systems yield periodic evolution. A lot of examples of such behavior exists in nature: from oscillation phenomena in particle physics to excitation-deexcitation cycles in quantum optics. Note that any system with Hilbert space of a finite dimension $n\geq 2$ exhibits in general a *quasiperiodic* evolution: it can be expressed via a finite number of periodic motions, like the function $f(t)=g(e^{i\omega_1 t},e^{i\omega_2 t},\ldots)$ where $\omega_1,\omega_2,\ldots$ represent partial frequencies.

▶ General Hamiltonian

$$\boxed{\hat{H}=\begin{pmatrix}\hbar\omega_0+\hbar\omega_3 & \hbar\omega_1-i\hbar\omega_2\\ \hbar\omega_1+i\hbar\omega_2 & \hbar\omega_0-\hbar\omega_3\end{pmatrix}=\hbar\left[\omega_0\hat{I}+\omega_1\hat{\sigma}_1+\omega_2\hat{\sigma}_2+\omega_3\hat{\sigma}_3\right]}\qquad\sqrt{\omega_1^2+\omega_2^2+\omega_3^2}\equiv\omega$$

Evolution operator calculated as the spinor transform. (see p. 61)

$$e^{-i\frac{\hat{H}t}{\hbar}}=e^{-i(\omega_0 t)\hat{I}}\underbrace{e^{-i(\vec{\omega}\cdot\hat{\vec{\sigma}})t}}$$

$$\boxed{\hat{U}(t)=e^{-i\omega_0 t}\left[(\cos\omega t)\,\hat{I}-i(\sin\omega t)\left(\frac{\vec{\omega}}{\omega}\cdot\hat{\vec{\sigma}}\right)\right]}$$

$$=e^{-i\omega_0 t}\begin{pmatrix}\cos\omega t-i\frac{\omega_3}{\omega}\sin\omega t & -\frac{\omega_2+i\omega_1}{\omega}\sin\omega t\\ \frac{\omega_2-i\omega_1}{\omega}\sin\omega t & \cos\omega t+i\frac{\omega_3}{\omega}\sin\omega t\end{pmatrix}$$

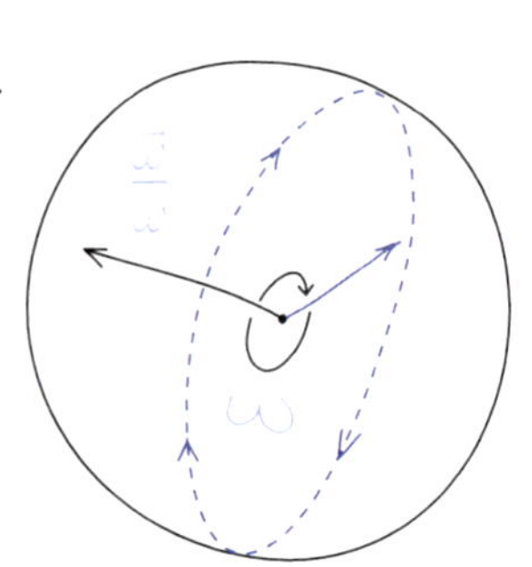

Quasiperiodic evolution with partial frequencies ω_0 and ω (but ω_0 not relevant, just a global phase). The motion is equivalent to a steady rotation.

▶ **Special case:** $\hat{H} = \begin{pmatrix} \hbar\omega_0 & \hbar\omega \\ \hbar\omega & \hbar\omega_0 \end{pmatrix} \Rightarrow \hat{U}(t) = e^{-i\omega_0 t}\begin{pmatrix} \cos\omega t & -i\sin\omega t \\ -i\sin\omega t & \cos\omega t \end{pmatrix}$

$|\psi(0)\rangle = \binom{1}{0} \xrightarrow{t} |\psi(t)\rangle = e^{-i\omega_0 t}\binom{\cos\omega t}{-i\sin\omega t}$

$$\Rightarrow \begin{Bmatrix} P_0(t) = \cos^2\omega t \\ P_1(t) = \sin^2\omega t \end{Bmatrix}$$ **oscillations** period $T = \frac{\pi}{\omega}$

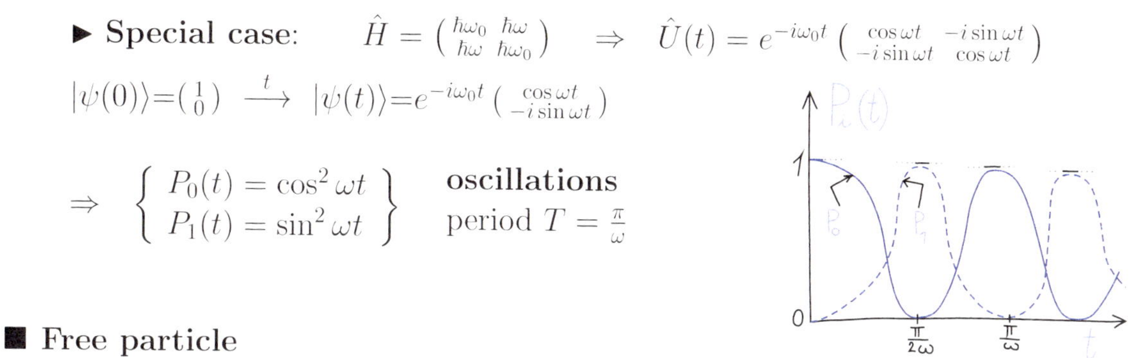

■ Free particle

Although particle moving in empty space (no fields) represents the most trivial example of evolution expressed in terms of an ordinary wavefunction, the particular calculation is a bit unpleasant. Nevertheless, it is worth of effort because of two benefits of general importance: discovery of the wavepacket spreading phenomenon and quantification of the limits of validity of nonrelativistic QM.

▶ Free-particle propagator

Green operator: $\hat{G}(t,t_0) = \Theta(t-t_0)\, e^{-i\frac{t-t_0}{2M\hbar}\hat{\vec{p}}^2}$ $\qquad t - t_0 \equiv \Delta t$

Propagator: $G(\vec{x}t|\vec{x}_0t_0) \equiv \langle\vec{x}|\hat{G}(t,t_0)|\vec{x}_0\rangle$ $\qquad \vec{x} - \vec{x}_0 \equiv \Delta\vec{x}$

$$= \Theta(\Delta t)\iint \underbrace{\langle\vec{x}|\vec{p}\rangle}_{\frac{1}{\sqrt{2\pi\hbar}^3}e^{+i\vec{p}\cdot\vec{x}/\hbar}} \underbrace{\langle\vec{p}|e^{-i\frac{\Delta t}{2M\hbar}\hat{\vec{p}}^2}|\vec{p}_0\rangle}_{e^{-i\frac{\Delta t}{2M\hbar}\vec{p}^2}\delta(\vec{p}-\vec{p}_0)} \underbrace{\langle\vec{p}_0|\vec{x}_0\rangle}_{\frac{1}{\sqrt{2\pi\hbar}^3}e^{-i\vec{p}_0\cdot\vec{x}_0/\hbar}} d\vec{p}\,d\vec{p}_0 = \frac{\Theta(\Delta t)}{(2\pi\hbar)^3}\int e^{\frac{i}{\hbar}\left[\vec{p}\cdot\Delta\vec{x} - \frac{\vec{p}^2}{2M}\Delta t\right]} d\vec{p}$$

$$= \frac{\Theta(\Delta t)}{(2\pi\hbar)^3} \underbrace{\int e^{a(\vec{p}-\vec{q})^2+b} d\vec{p}}_{\left(-\frac{\pi}{a}\right)^{3/2} e^b \text{ for } \mathrm{Re}\, a<0} \quad \text{with} \quad a = -i\frac{\Delta t}{2\hbar M} \quad b = i\frac{M(\Delta\vec{x})^2}{2\hbar\Delta t} \quad \vec{q} = \frac{M\Delta\vec{x}}{\Delta t}$$

To get $\mathrm{Re}\, a < 0$ assume: $\Delta t \longrightarrow \Delta t - i\varepsilon$ with $\varepsilon \to 0+$

$$\cdots = \lim_{\varepsilon\to 0+} \frac{\Theta(\Delta t)}{(2\pi\hbar)^3}\left(\frac{2\pi\hbar M}{\varepsilon + i\Delta t}\right)^{\frac{3}{2}} e^{i\frac{M(\Delta\vec{x})^2}{2\hbar\Delta t}} = \boxed{\Theta(\Delta t)\left(\frac{M}{2i\pi\hbar\Delta t}\right)^{\frac{3}{2}} e^{\frac{i}{\hbar}\frac{M}{2}\left(\frac{\Delta\vec{x}}{\Delta t}\right)^2\Delta t} = G(\Delta\vec{x},\Delta t)}$$

$|G(\Delta\vec{x},\Delta t)|^2 = \left(\frac{M}{2\pi\hbar\Delta t}\right)^3$ for $\Delta t > 0$ $\quad\Rightarrow$ immediate spread of the particle in the whole space $\Leftarrow$ nonrelativistic theory

▶ Evolution of Gaussian wavepackets

If the particle localization is imperfect, the spreading rate of its wavefunction should become finite.

$$\psi(\vec{x},t) = \frac{1}{(2\pi\hbar)^{3/2}}\int \underbrace{\tilde{\psi}(\vec{p})}_{\frac{e^{-(\vec{p}-\vec{p}_0)^2/4\sigma_p^2}}{(2\pi\sigma_p^2)^{3/4}}} e^{\frac{i}{\hbar}\left[\vec{p}\cdot\vec{x} - \frac{\vec{p}^2}{2M}t\right]} d\vec{p} = \frac{1}{(8\pi^3\hbar^2\sigma_p^2)^{3/4}}\int \underbrace{e^{\left(-\frac{1}{4\sigma_p^2} - \frac{it}{2\hbar M}\right)\vec{p}^2 + \left(\frac{\vec{p}_0}{2\sigma_p^2} + \frac{i\vec{x}}{\hbar}\right)\cdot\vec{p} - \frac{\vec{p}_0^2}{4\sigma_p^2}}}_{e^{a(\vec{p}-\vec{q})^2+b} \quad \mathrm{Re}\, a<0} =$$

$$= \frac{1}{(8\pi^3\hbar^2\sigma_p^2)^{3/4}}\left(-\frac{\pi}{a}\right)^{3/2} e^b \qquad a = -\frac{1}{4\sigma_p^2}\left(1 + i\frac{2\sigma_p^2 t}{\hbar M}\right) \quad \vec{q} = -\frac{1}{2a}\left(\frac{\vec{p}_0}{2\sigma_p^2} + i\frac{\vec{x}}{\hbar}\right) \quad b = -a\vec{q}^2 - \frac{\vec{p}_0^2}{4\sigma_p^2}$$

Probability density: $|\psi(\vec{x},t)|^2 = \left(\frac{1}{8\pi\hbar^2\sigma_p^2|a|^2}\right)^{\frac{3}{2}} e^{2\mathrm{Re}\, b}$

Define $\sigma_x^2(t) \equiv 4\hbar^2\sigma_p^2|a|^2 = \frac{\hbar^2}{4\sigma_p^2}\left[1+\frac{4\sigma_p^4}{\hbar^2M^2}t^2\right]$ and evaluate the exponent:

$$2\mathrm{Re}\, b = -\frac{1}{2\sigma_x(t)^2}\left[16\hbar^2\sigma_p^2|a|^2\mathrm{Re}(a\vec{q}^{\,2}) + 4\hbar^2|a|^2\vec{p}_0^{\,2}\right] = -\frac{\left(\vec{x}-\frac{\vec{p}_0}{M}t\right)^2}{2\sigma_x(t)^2}$$

$$|\psi(\vec{x},t)|^2 = \frac{1}{[2\pi\sigma_x(t)^2]^{3/2}}e^{-\frac{[\vec{x}-\vec{x}_0(t)]^2}{2\sigma_x(t)^2}}$$

$\vec{x}_0(t) = \frac{\vec{p}_0}{M}t$ **shift**

$\sigma_x(t) = \sigma_x(0)\sqrt{1+\left[\frac{\hbar}{2M\sigma_x(0)^2}\right]^2 t^2}$ **spreading**

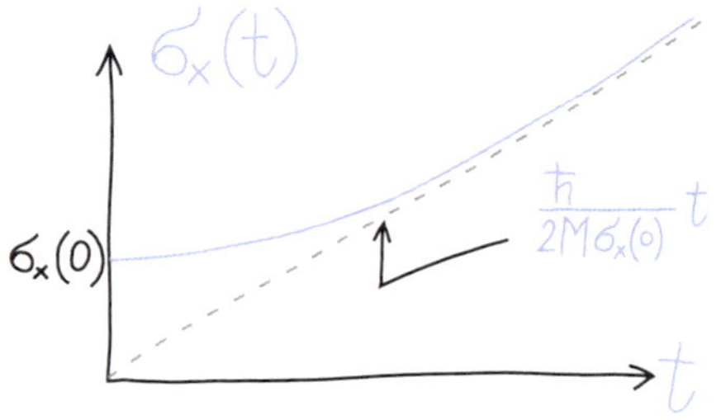

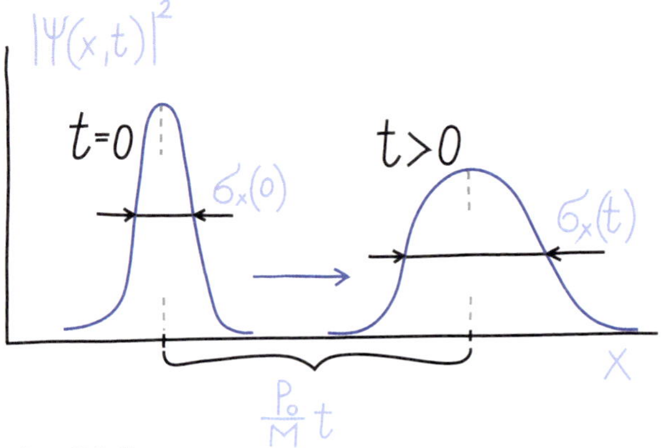

▶ **Validity limit of nonrelativistic QM**

Spreading speed of the wavepacket: $s \equiv \frac{1}{2}\frac{d}{dt}\sigma_x(t) = \frac{\frac{\sigma_x(0)}{2}\left[\frac{\hbar}{2M\sigma_x(0)^2}\right]^2 t}{\sqrt{1+\left[\frac{\hbar}{2M\sigma_x(0)^2}\right]^2 t^2}} \xrightarrow{\text{large}\, t} \frac{\hbar}{4M\sigma_x(0)}$

Nonrelativistic QM becomes *invalid* for $s \gtrsim c \quad\Rightarrow$

$\Rightarrow$ for the initial particle localization $\sigma_x(0) \lesssim \lambda_C$

$$\sigma_x(0) \lesssim \frac{\hbar}{4Mc} = \frac{1}{8\pi}\underbrace{\frac{h}{Mc}}_{\lambda_C}$$

Schrödiger equation applicable *iff* $\sigma_x(0) \gg \lambda_C$

$\lambda_C \equiv$ **Compton wavelength** (for electron $\lambda_C \doteq 2.4\cdot 10^{-12}\,$m)

▶ **Phase & group velocities**

(a) Monochromatic planar wave $\quad \psi(\vec{x},t) = e^{i\overbrace{[\vec{k}\cdot\vec{x}-\omega(\vec{k})t]}^{\phi(\vec{k},\vec{x},t)}}$

Phase velocity $\vec{v}_{\rm ph}$ given by condition of a constant phase:

$\phi(\vec{k},\, \underbrace{\vec{v}_{\rm ph}t}_{\vec{x}},\, t) = \mathrm{const} \quad\Rightarrow\quad \vec{v}_{\rm ph} = \frac{\omega(\vec{k})}{k^2}\vec{k}$

$\Rightarrow$ in QM: $\vec{k} = \frac{\vec{p}}{\hbar},\ \hbar\omega(\vec{k}) = \frac{(\hbar\vec{k})^2}{2M} \quad\Rightarrow\quad \boxed{\vec{v}_{\rm ph} = \frac{\vec{p}}{2M} = \frac{1}{2}\vec{v}_{\rm clas}}$

(b) Superposition of planar waves $\quad \psi(\vec{x},t) = \int a(\vec{k})\, e^{i[\vec{k}\cdot\vec{x}-\omega(\vec{k})t]}\, d\vec{k}$

with the amplitude function $a(\vec{k})$ having a sharp maximum at $\vec{k}_0$

Group velocity $\vec{v}_{\rm gr}$ represents motion of the $\psi(\vec{x},t)$ maximum; it is given by a stationary point of the phase:

$\vec{\nabla}_{\vec{k}}\,\phi(\vec{k},\, \underbrace{\vec{v}_{\rm gr}t}_{\vec{x}},\, t)|_{\vec{k}=\vec{k}_0} = 0 \quad\Rightarrow\quad \vec{v}_{\rm gr} = \vec{\nabla}_{\vec{k}}\,\omega(\vec{k})|_{\vec{k}=\vec{k}_0} \quad\Rightarrow$ in QM: $\boxed{\vec{v}_{\rm gr} = \frac{\vec{p}_0}{M} = \vec{v}_{0\rm clas}}$

■ Coherent states in harmonic oscillator

The harmonic oscillator potential has the magic power to prevent Gaussian wavepackets from spreading. It provides the simplest specimen from the large family of coherent states. These states generalized to more complex situations represent an important tool to construct the classical limit of a quantum system (cf. Chapter 3). For the sake of simplicity we will stay now in 1D space.

▶ Algebraic solution of harmonic oscillator

1D oscillator Hamiltonian $\hat{H} = \frac{1}{2M}\hat{p}^2 + \frac{M\omega^2}{2}\hat{x}^2$ can be expressed through ladder operators

$$\begin{aligned}\hat{b}^\dagger &= \sqrt{\tfrac{M\omega}{2\hbar}}\left(\hat{x} - i\tfrac{1}{M\omega}\hat{p}\right) \\ \hat{b} &= \sqrt{\tfrac{M\omega}{2\hbar}}\left(\hat{x} + i\tfrac{1}{M\omega}\hat{p}\right)\end{aligned} \quad\Leftrightarrow\quad \begin{aligned}\hat{x} &= \sqrt{\tfrac{\hbar}{2M\omega}}\left(\hat{b}^\dagger + \hat{b}\right) \\ \hat{p} &= i\sqrt{\tfrac{M\hbar\omega}{2}}\left(\hat{b}^\dagger - \hat{b}\right)\end{aligned} \quad\Leftrightarrow\quad \boxed{\hat{H} = \hbar\omega\left(\hat{b}^\dagger\hat{b} + \tfrac{1}{2}\right)}$$

$\boxed{[\hat{b}, \hat{b}^\dagger] = \hat{I}}$ **boson commutation** relation (see Sec. 7.1)

Commutation relations of $\hat{b}^\dagger, \hat{b}$ with $\hat{H}$ are those of a translation operator (Sec. 2.4) $\Rightarrow$ ladder operators make jumps between individual eigenstates

$$\begin{aligned}\left[\hat{H}, \hat{b}^\dagger\right] &= +\hbar\omega\hat{b}^\dagger \\ \left[\hat{H}, \hat{b}\right] &= -\hbar\omega\hat{b}\end{aligned} \quad\Rightarrow\quad \boxed{\begin{aligned}\hat{b}^\dagger|E_n\rangle &= \sqrt{n+1}|E_{n+1}\rangle \\ \hat{b}|E_n\rangle &= \sqrt{n}|E_{n-1}\rangle\end{aligned}} \quad\Rightarrow\quad \begin{aligned}&\hat{b}^\dagger \equiv \textbf{raising operator} \\ &\hat{b} \equiv \textbf{lowering operator}\end{aligned}$$

Normalization factors are okay: $\left\{\begin{aligned}\langle E_{n-1}|E_{n-1}\rangle &= \tfrac{1}{n}\overbrace{\langle E_n|\hat{b}^\dagger\hat{b}|E_n\rangle}^{n} = 1 \\ \langle E_{n+1}|E_{n+1}\rangle &= \tfrac{1}{n+1}\langle E_n|\underbrace{\hat{b}\hat{b}^\dagger}_{1+\hat{b}^\dagger\hat{b}}|E_n\rangle = 1\end{aligned}\right.$

Operators $\hat{b}^\dagger$ / $\hat{b}$ are thought to create/annihilate quanta of vibrations—effective particles of bosonic nature, so called **phonons**

▶ Coherent states in the energy eigenbasis

$$\boxed{|\psi_z\rangle = e^{-\frac{|z|^2}{2}}\sum_{n=0}^{\infty}\frac{z^n}{\sqrt{n!}}|E_n\rangle} \qquad z \in \mathbb{C}$$

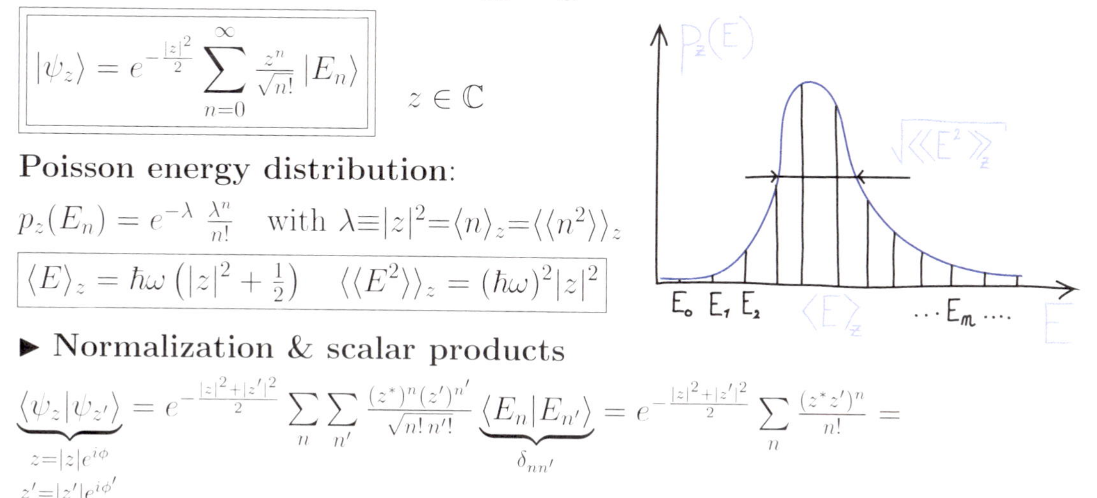

Poisson energy distribution:

$p_z(E_n) = e^{-\lambda}\frac{\lambda^n}{n!}$ with $\lambda \equiv |z|^2 = \langle n\rangle_z = \langle\langle n^2\rangle\rangle_z$

$$\boxed{\langle E\rangle_z = \hbar\omega\left(|z|^2 + \tfrac{1}{2}\right) \quad \langle\langle E^2\rangle\rangle_z = (\hbar\omega)^2|z|^2}$$

▶ Normalization & scalar products

$$\underbrace{\langle\psi_z|\psi_{z'}\rangle}_{\substack{z=|z|e^{i\phi} \\ z'=|z'|e^{i\phi'}}} = e^{-\frac{|z|^2+|z'|^2}{2}}\sum_n\sum_{n'}\frac{(z^*)^n(z')^{n'}}{\sqrt{n!\,n'!}}\underbrace{\langle E_n|E_{n'}\rangle}_{\delta_{nn'}} = e^{-\frac{|z|^2+|z'|^2}{2}}\sum_n\frac{(z^*z')^n}{n!} =$$

$$\cdots = e^{-\frac{|z|^2+|z'|^2}{2}+|z||z'|[\cos(\phi'-\phi)+i\sin(\phi'-\phi)]} = \boxed{e^{-\frac{|z'-z|^2}{2}+i|z||z'|\sin(\phi'-\phi)} = \langle\psi_z|\psi_{z'}\rangle}$$

Coherent states $\{|\psi_z\rangle\}_{z\in\mathbb{C}}$ form an **overcomplete set** in $\mathcal{H}$: $\begin{cases}\langle\psi_z|\psi_z\rangle=1\\ \langle\psi_z|\psi_{z'}\rangle\neq 0 \text{ for } z\neq z'\end{cases}$

▶ Coherent states as **eigenstates of lowering operator** $\boxed{\hat{b}|\psi_z\rangle = z|\psi_z\rangle}$

Proof: $\hat{b}|\psi_z\rangle = e^{-\frac{|z|^2}{2}}\sum_{n=0}^{\infty}\frac{z^n}{\sqrt{n!}}\underbrace{\hat{b}|E_n\rangle}_{\sqrt{n}|E_{n-1}\rangle} = z\,\underbrace{e^{-\frac{|z|^2}{2}}\sum_{n=1}^{\infty}\frac{z^{n-1}}{\sqrt{(n-1)!}}|E_{n-1}\rangle}_{|\psi_z\rangle}$

Note: There exists **no** eigenstate of $\hat{b}^\dagger$ (think!)

▶ **Coordinate & momentum** averages

$$\langle\psi_z|\hat{x}|\psi_z\rangle = \sqrt{\tfrac{\hbar}{2M\omega}}\overbrace{\langle\psi_z|(\hat{b}^\dagger+\hat{b})|\psi_z\rangle}^{z^*+z} = \boxed{\sqrt{\tfrac{2\hbar}{M\omega}}\,\mathrm{Re}\,z = \langle x\rangle_z}$$

$$\langle\psi_z|\hat{p}|\psi_z\rangle = i\sqrt{\tfrac{M\hbar\omega}{2}}\underbrace{\langle\psi_z|(\hat{b}^\dagger-\hat{b})|\psi_z\rangle}_{z^*-z} = \boxed{\sqrt{2M\hbar\omega}\,\mathrm{Im}\,z = \langle p\rangle_z}$$

▶ Coordinate representation $\psi_z(x) = \langle x|\psi_z\rangle$

$$\cdots = e^{-\frac{|z|^2}{2}}\sum_{n=0}^{\infty}\frac{z^n}{\sqrt{n!}}\underbrace{\langle x|E_n\rangle}_{\left(\frac{M\omega}{\pi\hbar}\right)^{1/4}\frac{1}{\sqrt{2^n n!}}e^{-\frac{M\omega}{2\hbar}x^2}H_n\left(\sqrt{\frac{M\omega}{\hbar}}x\right)} = \left(\tfrac{M\omega}{\pi\hbar}\right)^{\frac14}e^{-\frac{|z|^2}{2}-\frac{M\omega}{2\hbar}x^2}\underbrace{\sum_{n=0}^{\infty}\frac{\left(\frac{z}{\sqrt{2}}\right)^n}{n!}H_n\left(\sqrt{\tfrac{M\omega}{\hbar}}x\right)}_{e^{\frac{M\omega}{\hbar}x^2-\left(\sqrt{\frac{M\omega}{\hbar}}x-\frac{z}{\sqrt{2}}\right)^2}}$$

$$H_n(\xi) \equiv \frac{d^n}{d\eta^n}e^{\xi^2-(\xi-\eta)^2}\Big|_{\eta=0} \quad\Rightarrow\quad e^{\xi^2-(\xi-\eta)^2} = \sum_n H_n(\xi)\frac{\eta^n}{n!}$$

$$\cdots = \left(\tfrac{M\omega}{\pi\hbar}\right)^{\frac14}e^{-\frac{|z|^2}{2}+\frac{M\omega}{2\hbar}x^2}e^{-\left(\sqrt{\frac{M\omega}{\hbar}}x-\frac{z}{\sqrt{2}}\right)^2} = \left(\tfrac{M\omega}{\pi\hbar}\right)^{\frac14}e^{-\frac{M\omega}{2\hbar}x^2+2z\sqrt{\frac{M\omega}{2\hbar}}x-z\mathrm{Re}\,z}$$

$$\boxed{|\psi_z(x)|^2 = \left(\tfrac{M\omega}{\pi\hbar}\right)^{\frac12}e^{-\frac{M\omega}{\hbar}\left(x-\langle x\rangle_z\right)^2}}$$

Gaussian distribution with $\sigma_x^2 = \frac{\hbar}{2M\omega}$

▶ **Time evolution**

$$e^{-i\frac{\hat{H}t}{\hbar}}|\psi_z\rangle = e^{-\frac{|z|^2}{2}}\sum_{n=0}^{\infty}\frac{z^n}{\sqrt{n!}}e^{-i\left(n+\frac12\right)\omega t}|E_n\rangle = e^{-i\frac{\omega t}{2}}\underbrace{e^{-\frac{|z|^2}{2}}\sum_{n=0}^{\infty}\frac{\overbrace{(ze^{-i\omega t})^n}^{(z')^n}}{\sqrt{n!}}|E_n\rangle}_{|\psi_{z'}\rangle}$$

$$\boxed{\hat{U}(t)|\psi_{z(0)}\rangle = e^{-i\frac{\omega t}{2}}|\psi_{z(t)}\rangle \qquad z(t) = z(0)e^{-i\omega t}}$$

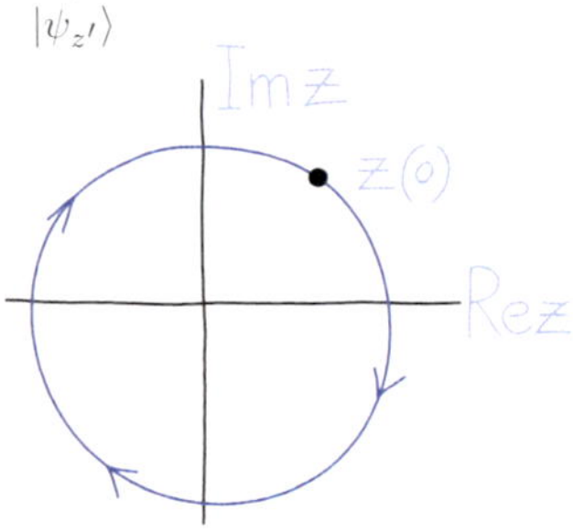

▶ Evolution of coordinate & momentum averages

$$\langle x\rangle_t = \underbrace{\sqrt{\tfrac{2\hbar}{M\omega}}\Big[\mathrm{Re}z(0)}_{\langle x\rangle_0}\cos(\omega t) + \mathrm{Im}z(0)\sin(\omega t)\Big]$$

$$\langle p\rangle_t = \underbrace{-\sqrt{2M\hbar\omega}\Big[\mathrm{Im}z(0)}_{\langle p\rangle_0}\cos(\omega t) - \mathrm{Re}z(0)\sin(\omega t)\Big]$$

The averages satisfy the following equation of an **ellipse**:

$$\frac{1}{2M}\langle p\rangle_t^2 + \frac{M\omega^2}{2}\langle x\rangle_t^2 = \underbrace{\hbar\omega|z(0)|^2}_{\langle E\rangle_{z(0)} - \frac{\hbar\omega}{2}}$$

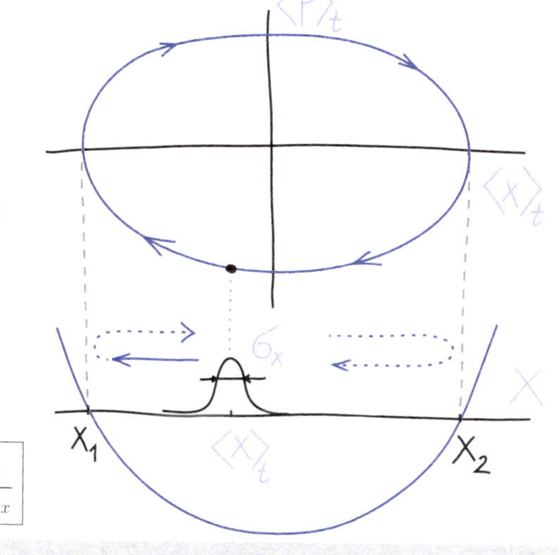

Coherent state imitates approximately the **classical trajectory** in phase space: $\frac{1}{2M}p^2 + \frac{M\omega 2}{2}x^2 = E$

Since $\langle E\rangle_z \gg \frac{\hbar\omega}{2}$ for $|z|^2 \gg 1$, the approximation gets improved with increasing $|z| \leftrightarrow \langle E\rangle_z$

Stationary widths: $\sigma_x = \sqrt{\frac{\hbar}{2M\omega}} \quad \sigma_p = \frac{\hbar}{2\sigma_x}$

◀ **Historical remark**

1925: Erwin Schrödinger discovers oscillator coherent states (he wrongly anticipates that such states will make the notion of pointlike particles irrelevant)

1950-60's: J. Schwinger and J.Klauder use coherent states in the field-theory context

1963: Roy Glauber shows the key importance of coherent states in quantum optics

■ Spin in rotating magnetic field

The following example of quantum evolution is based on a time dependent external field. Although the time dependence of the Hamiltonian is of the nontrivial type [case (b) of Sec. 1.5], the solution can be found analytically—not in the form of Dyson series. This is rather exceptional! Moreover, the example captures the physics of so-called nuclear magnetic resonance, which respresents a rather important tool to "engineer" the evolution of a quantum system (nuclear spin) with a number of brilliant applications.

▶ **Nuclear magnetic resonance** (NMR) situation: a particle with magnetic moment (operator $\hat{\mu}$) is placed in a combined stationary (homogeneous) + variable (rotating) magnetic field. Hamiltonian reads as:

$$\hat{H}(t) = \overbrace{-\hat{\mu}_z B_0}^{\text{stationary field}} \overbrace{-\hat{\vec{\mu}}\cdot\underbrace{\vec{B}_1(t)}_{B_1\vec{n}(t)}}^{\text{varying field}}$$

Magnetic dipole operator: $\hat{\vec{\mu}} = g\mu_N \frac{1}{\hbar}\hat{\vec{S}}$

$$\hat{H}(t) = -\underbrace{g\mu_N B_0}_{\hbar\omega_0}\frac{1}{\hbar}\hat{S}_3 - \underbrace{g\mu_N B_1}_{\hbar\omega_1}\left(\vec{n}(t)\cdot\frac{1}{\hbar}\hat{\vec{S}}\right)$$

$$\vec{n}(t) = \begin{pmatrix} \sin\vartheta\cos\omega t \\ -\sin\vartheta\sin\omega t \\ \cos\vartheta \end{pmatrix}$$ **rotating field**

Z $\vec{B}_0$ $\vec{B}_1(t)$ x y

In the NMR case, the frequency ω of rotating field B_1 ($\ll B_0$) is tuned to the Larmor frequency ω_0 of the spin precession in the stationary filed B_0, and is applied in the form of pulses of certain duration. These pulses are used to prepare the spin in a desired state. Most commonly $\vec{B}_1(t)\perp\vec{B}_0$

▶ Hamiltonians at various time instants *do not commute*:

$$[\hat{H}(t),\hat{H}(t')]=\omega_1^2\Big[\big(\vec{n}(t)\cdot\hat{\vec{S}}\big),\big(\vec{n}(t')\cdot\hat{\vec{S}}\big)\Big]+\omega_0\omega_1\Big[\hat{S}_3,\big(\vec{n}(t')\cdot\hat{\vec{S}}\big)\Big]+\omega_1\omega_0\Big[\big(\vec{n}(t)\cdot\hat{\vec{S}}\big),\hat{S}_3\Big]=$$

$$=i\hbar\omega_1\left(\omega_1[\vec{n}(t)\times\vec{n}(t')]\cdot\hat{\vec{S}}+\omega_0[\vec{n}(t')\times\hat{\vec{S}}]_3-\omega_0[\vec{n}(t)\times\hat{\vec{S}}]_3\right)\neq 0$$

▶ **Separation of the time dependence**

$$\hat{H}(t)=-(\omega_0+\omega_1\cos\vartheta)\hat{S}_3-\omega_1\sin\vartheta\underbrace{\Big[(\cos\omega t)\hat{S}_1-(\sin\omega t)\hat{S}_2\Big]}_{e^{+\frac{i}{\hbar}\omega t\hat{S}_3}\hat{S}_1e^{-\frac{i}{\hbar}\omega t\hat{S}_3}}$$

BCH formula: $e^{+\hat{A}}\hat{B}e^{-\hat{A}}=\sum_k\frac{1}{k!}[\hat{A},[\hat{A},\ldots[\hat{A},\hat{B}]\ldots]_k$

$$e^{\frac{i\varphi}{\hbar}\hat{S}_3}\hat{S}_1e^{-\frac{i\varphi}{\hbar}\hat{S}_3}=\hat{S}_1+\tfrac{1}{1!}\left(\tfrac{i\varphi}{\hbar}\right)^1\underbrace{[\hat{S}_3,\hat{S}_1]}_{i\hbar\hat{S}_2}+\tfrac{1}{2!}\left(\tfrac{i\varphi}{\hbar}\right)^2\underbrace{[\hat{S}_3,i\hbar\hat{S}_2]}_{\hbar^2\hat{S}_1}+\tfrac{1}{3!}\left(\tfrac{i\varphi}{\hbar}\right)^3\underbrace{[\hat{S}_3,\hbar^2\hat{S}_1]}_{i\hbar^3\hat{S}_2}+\cdots=$$

$$=\underbrace{\left(1-\tfrac{\phi^2}{2!}+\ldots\right)}_{\cos\varphi}\hat{S}_1-\underbrace{\left(\tfrac{\phi}{1!}-\tfrac{\phi^3}{3!}+\ldots\right)}_{\sin\varphi}\hat{S}_2$$

$$\hat{H}(t)=e^{+\frac{i}{\hbar}\omega t\hat{S}_3}\underbrace{\Big[-(\omega_0+\omega_1\cos\vartheta)\hat{S}_3-(\omega_1\sin\vartheta)\hat{S}_1\Big]}_{\hat{H}(0)}e^{-\frac{i}{\hbar}\omega t\hat{S}_3}$$

"rotating" Hamiltonian

The Hamiltonian time dependence was separated to the overall rotation. This enables one to solve the dynamics explicitly, using the rotating frame:

▶ **Transformation to rotating frame**: $|\psi(t)\rangle\mapsto|\psi'(t)\rangle\equiv e^{-\frac{i}{\hbar}\omega t\hat{S}_3}|\psi(t)\rangle$

$$i\hbar\tfrac{d}{dt}|\psi'(t)\rangle=\omega\hat{S}_3\underbrace{e^{-\frac{i}{\hbar}\omega t\hat{S}_3}|\psi(t)\rangle}_{|\psi'(t)\rangle}+\underbrace{e^{-\frac{i}{\hbar}\omega t\hat{S}_3}\hat{H}(t)e^{+\frac{i}{\hbar}\omega t\hat{S}_3}}_{\hat{H}(0)}\underbrace{e^{-\frac{i}{\hbar}\omega t\hat{S}_3}|\psi(t)\rangle}_{|\psi'(t)\rangle}$$

Schrödinger equation in rotating frame:

$$i\hbar\tfrac{d}{dt}|\psi'(t)\rangle=\underbrace{\Big[\hat{H}(0)+\omega\hat{S}_3\Big]}_{\hat{H}_{\rm eff}}|\psi'(t)\rangle$$

$$\hat{H}_{\rm eff}=(\omega-\omega_0-\omega_1\cos\vartheta)\hat{S}_3-(\omega_1\sin\vartheta)\hat{S}_1$$

▶ The evolution induced by the effective Hamiltonian in the rotating frame can be written analytically (just sas an appropriate rotation). Finally, to get solution in the lab. frame, one concludes with inverse of the above transform:

Solution: $|\psi(t)\rangle=e^{+\frac{i}{\hbar}\omega t\hat{S}_3}e^{-\frac{i}{\hbar}\hat{H}_{\rm eff}t}|\psi(0)\rangle$ (assuming $|\psi(0)\rangle\equiv|\psi'(0)\rangle$)

Expression in terms of **rotations**:

$$\hat{U}(t)=e^{+\frac{i}{\hbar}\omega t\hat{S}_3}e^{-\frac{i}{\hbar}\Omega t(\vec{n}_\Omega\cdot\hat{\vec{S}})}$$

$$\Omega=\sqrt{(\omega-\omega_0)^2-2(\omega-\omega_0)\omega_1\cos\vartheta+\omega_1^2}$$

$$\vec{n}_\Omega=\frac{1}{\Omega}\begin{pmatrix}-\omega_1\sin\vartheta\\0\\\omega-\omega_0-\omega_1\cos\vartheta\end{pmatrix}$$

Resonant & $[\vec{B}_1(t)\perp\vec{B}_0]$ case:

$$\left.\begin{matrix}\omega=\omega_0\\\vartheta=\frac{\pi}{2}\end{matrix}\right\}\Rightarrow\left\{\begin{matrix}\Omega=\omega_1\\\vec{n}_\Omega=\begin{pmatrix}1\\0\\0\end{pmatrix}\end{matrix}\right.$$

◄ **Historical remark**
1938: I. Rabi proposes NMR as a method to measure nuclear magnetic moments
1946: F. Bloch & E.M. Purcell extends NMR to solids & liquids, F. Bloch provides theoretical description of NMR and, in general, of evolution of 2-state systems

1.6 Quantum measurement

Besides spontaneous evolution, described by the nonstationary Schrödinger equation, quantum mechanics assumes also another type of dynamics—a sudden change of the state vector induced by a measurement performed on the system. In contrast to classical physics, where measurements just specify states of the system without essentially disturbing them (in an ideal case, the influence of measurement can be reduced to zero), quantum physics needs a special treatment of measurements. Their impact on the system is irreducible and rather dramatic! This "sector" of QM has quite unusual consequences and is a permanent subject of a vivid debate.

■ State vector reduction

The spontaneous quantum evolution is smooth and deterministic (in the sense of uniqueness of the evolved state vector in $\mathcal{H}$). We may call this motion "**process U**" (from its unitary character). In contrast, the evolution induced by quantum measurement—at least in the form assumed by the present-day QM—is abrupt and indeterministic. We will call it "**process R**" ("reduction of state vector").¶ The real nature of this process is still unknown.

► **Why we need process R ?**

Correlation of **repeated measurements** on the same system: conditional probability to measure eigenvalue a_j of $\hat{A}$ at time $t = t_0 + \Delta t$ given the result of the same measurement at t_0 was a_i: $\boxed{p(a_j t|a_i t_0) = \langle\bar{\psi}|\hat{U}^{\dagger}(\Delta t)\hat{P}_{a_j}\hat{U}(\Delta t)|\bar{\psi}\rangle}$

where $|\bar{\psi}\rangle \equiv \left\{\begin{array}{l}\text{state vector immediately}\\ \text{after the first measurement}\end{array}\right.$ For $\Delta t \to 0$ the 2nd measurement must yield the same outcome as the 1st one: $\lim_{\Delta t\to 0} p(a_j t|a_i t_0) = \delta_{ij} \quad \Leftrightarrow \quad \boxed{|\bar{\psi}\rangle = |a_i\rangle}$

Example:

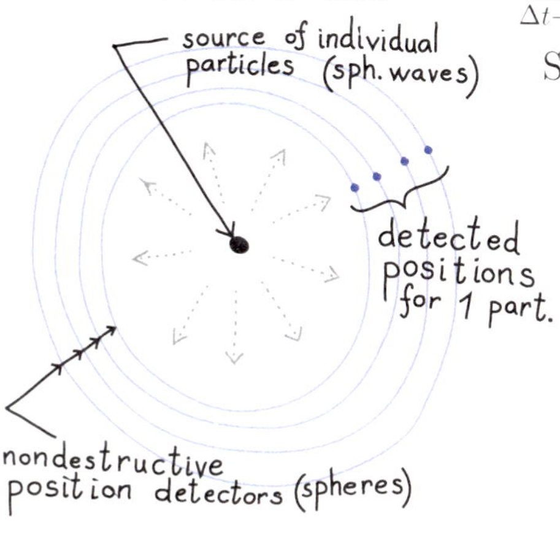

Sketch of the "U" & "R" evolutions:

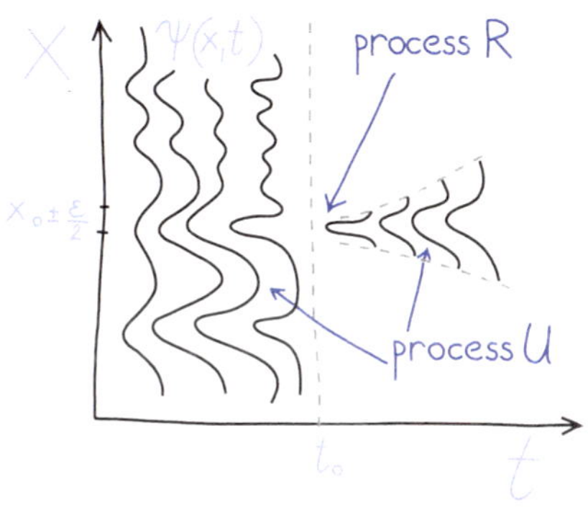

¶This terminology is due to R. Penrose, whose way of thinking on quantum measurement is partly exploited here.

▶ Measurement postulate

The instantenous evolution induced by a measurement of observable A:

$$|\psi\rangle \xrightarrow{\text{measurement of } A} |\bar{\psi}\rangle \equiv \hat{R}_A|\psi\rangle = \begin{cases} |a_1\rangle & \text{iff } a_1 \text{ measured, prob.} = \langle\psi|\hat{P}_{a_1}|\psi\rangle \\ |a_2\rangle & \text{iff } a_2 \text{ measured, prob.} = \langle\psi|\hat{P}_{a_2}|\psi\rangle \\ \vdots & \end{cases}$$

$$\boxed{\hat{R}_A|\psi\rangle \stackrel{a_i}{=} \frac{1}{\sqrt{\langle\psi|\hat{P}_{a_i}|\psi\rangle}}\hat{P}_{a_i}|\psi\rangle}$$

where $\stackrel{a_i}{=}$ means "conditional equality": it holds *iff* the outcome of measurement is a_i

Terminology: (a) gentle: "state vector **reduction**"
(b) dramatic: "**collapse** of wavefunction"

Note: The "collapse" does not mean the "end of wavefunction" :-) After the measurement, the wavefunction (instantenously localized in the respective space) continues its evolution according to ordinary Schrödinger equation.

▶ Example: photon polarization measurement

Polarization = manifestation of **photon spin** ($s = 1$)

Linear polarization basis: $\{|x\rangle, |y\rangle\}$ ... both $\vec{n}_x \perp \vec{n}_y$ directions $\perp \vec{n}_c = \frac{\vec{c}}{c}$

Rotated linear polar. basis: $\boxed{\begin{pmatrix} |x'\rangle \\ |y'\rangle \end{pmatrix} = \begin{pmatrix} \cos\vartheta & \sin\vartheta \\ -\sin\vartheta & \cos\vartheta \end{pmatrix} \begin{pmatrix} |x\rangle \\ |y\rangle \end{pmatrix}}$

Circular polarization basis: $\boxed{\begin{pmatrix} |L\rangle \\ |R\rangle \end{pmatrix} = \frac{1}{\sqrt{2}} \begin{pmatrix} 1 & -i \\ 1 & +i \end{pmatrix} \begin{pmatrix} |x\rangle \\ |y\rangle \end{pmatrix}}$

Spin-1 projection states in the flight direction:

$$\begin{pmatrix} |L\rangle \\ |R\rangle \end{pmatrix} \equiv \begin{pmatrix} |s{=}1, m_s{=}{+}1\rangle \\ |s{=}1, m_s{=}{-}1\rangle \end{pmatrix}_{\vec{n}_c}$$

Note: state $|s{=}1, m_s{=}0\rangle_{\vec{n}_c}$ does *not* exist for massless ($v{=}c$) particles

General polarization state: $\boxed{|\psi\rangle = \begin{cases} \alpha|x\rangle + \beta|y\rangle \\ \alpha'|x'\rangle + \beta'|y'\rangle \\ \lambda|L\rangle + \rho|R\rangle \end{cases}}$ with $\begin{Bmatrix} \alpha, \beta \\ \alpha', \beta' \\ \lambda, \rho \end{Bmatrix} \in \mathbb{C}$ $\quad \begin{matrix} |\alpha|^2 + |\beta|^2 = 1 \\ |\alpha'|^2 + |\beta'|^2 = 1 \\ |\lambda|^2 + |\rho|^2 = 1 \end{matrix}$

Consider measurement of linear polarization realized by passage of photon through a birefringent crystal (transmission/reflection):

Observable $\hat{\Theta} = 0|x\rangle\langle x| + \frac{\pi}{2}|y\rangle\langle y|$
$\equiv$ deviation angle on the crystal

$$\Rightarrow \hat{R}_\Theta|\psi\rangle = \begin{cases} |x\rangle & \text{for } \theta{=}0, \text{ prob.} = |\alpha|^2 \\ |y\rangle & \text{for } \theta{=}\frac{\pi}{2}, \text{ prob.} = |\beta|^2 \end{cases}$$

$\Rightarrow$ after the measurement the photon gets localized along the respective path from the crystal

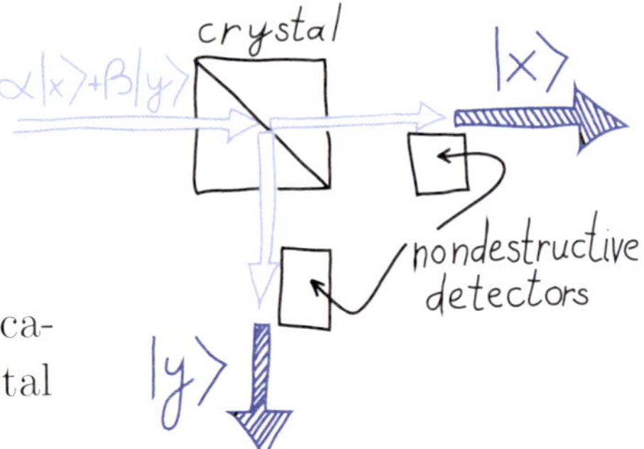

▶ **Properties** of "operator" $\hat{R}_A$

non-deterministic: one knows only probabilities of possible outputs

non-linear: $\left.\begin{matrix}\hat{R}_A|a_1\rangle=|a_1\rangle \\ \hat{R}_A|a_2\rangle=|a_2\rangle\end{matrix}\right\}$ but $\hat{R}_A\big(\alpha|a_1\rangle+\beta|a_2\rangle\big)=\left\{\begin{matrix}|a_1\rangle \\ |a_2\rangle\end{matrix}\right\}\neq\alpha\hat{R}_A|a_1\rangle+\beta\hat{R}_A|a_2\rangle$

non-unitary: $\left.\begin{matrix}|\psi\rangle=\alpha|a_1\rangle+\beta|a_2\rangle \\ |\psi'\rangle=\alpha'|a_1\rangle+\beta'|a_2\rangle\end{matrix}\right\}\xrightarrow{\hat{R}_A}\left\{\begin{matrix}\langle\begin{matrix}|a_1\rangle \\ |a_2\rangle\end{matrix} \\ \langle\begin{matrix}|a_1\rangle \\ |a_2\rangle\end{matrix}\end{matrix}\right. \Rightarrow \underbrace{\langle\psi|\psi'\rangle}_{\text{arbitrary}}\xrightarrow{\hat{R}_A}\underbrace{\langle\bar{\psi}|\bar{\psi}'\rangle}_{0\text{ or }1}$

"non-local",**"acausal"**: $\psi(\vec{x},t)$ collapses simultaneously in the whole space ?
(at least in the present unspecified form of the measurement postulate)

◀ **Historical remark**
1927: Werner Heisenberg first explicitly considers the wavefunction collapse
1932: John von Neumann includes the reduction postulate into the mathematical formulation of QM and discusses its properties

■ Consequences for measurement sequences

The reduction postulate has an immediate dynamical consequence for incompatible observables: A joint statistical distribution of measurement outcomes of quantities A and B for a given initial state $|\psi\rangle$ depends on whether the quantities are measured in succession (A,B) or (B,A).

▶ Measurement sequences (A,B) and (B,A) performed at times t_0 and $t_0{+}\Delta$ with $\Delta t\to 0$ on an initial state $|\psi(t_0)\rangle\equiv|\psi\rangle$

Joint probabilities of results $A{=}a$ and $B{=}b$: $\overbrace{p_\psi(a,b)}^{\text{joint}}=\overbrace{p_\psi(b|a)}^{\text{conditional}}\,p_\psi(a)$

(i) $p_\psi^{(AB)}(a,b)=\langle\bar{\psi}|\hat{P}_b|\bar{\psi}\rangle\langle\psi|\hat{P}_a|\psi\rangle=\frac{\langle\psi|\hat{P}_a\hat{P}_b\hat{P}_a|\psi\rangle}{\langle\psi|\hat{P}_a|\psi\rangle}\langle\psi|\hat{P}_a|\psi\rangle=\langle\psi|\hat{P}_a\hat{P}_b\hat{P}_a|\psi\rangle$

(ii) $p_\psi^{(BA)}(b,a)=\langle\psi|\hat{P}_b\hat{P}_a\hat{P}_b|\psi\rangle$

▶ **Compatible vs. incompatible observables**

$[\hat{A},\hat{B}]{=}0{=}[\hat{P}_a,\hat{P}_b] \quad\Rightarrow\quad \boxed{p_\psi^{(AB)}(a,b)=p_\psi^{(BA)}(b,a)}$ independent of succession

$[\hat{A},\hat{B}]{\neq}0{\neq}[\hat{P}_a,\hat{P}_b] \quad\Rightarrow\quad \boxed{p_\psi^{(AB)}(a,b)\neq p_\psi^{(BA)}(b,a)}$ dependent on succession

▶ **Statistical dependence of results**

The reduction postulate ⇒ results of subsequent A and B measurements are in general statistically dependent, **correlated**
(for both compatible & incompatible cases)

$$\boxed{\begin{matrix}p_\psi(a|b)\neq p_\psi(a) \\ p_\psi(b|a)\neq p_\psi(b) \\ p_\psi(a,b)\neq p_\psi(a)p_\psi(b)\end{matrix}}$$

■ Measurements on entangled states: EPR situation

A real puzzle arises when we start thinking about the effects of quantum measurements on coupled systems. If such a system is in an entangled state, any local measurement on one of the subsystems can alter the potential outcomes of local measurements on the second subsystem. This is independent of how large is the spatial separation of both subsystems. The acronym "EPR" stands for Einstein, Podolsky, and Rosen, who first noticed the phenomenon in 1935.

▶ Local measurements on a coupled system

Hilbert space: $\mathcal{H} = \mathcal{H}_1 \otimes \mathcal{H}_2$

"Local" observables defined on both subsystems: $\begin{cases} \hat{A} \equiv \hat{A}_1 \otimes \hat{I}_2 \\ \hat{B} \equiv \hat{I}_1 \otimes \hat{B}_2 \end{cases}$

$[\hat{A}, \hat{B}] = 0 \quad \Rightarrow$ **compatible observables**

In a coupled system, the statistical dependence (see above) of the results of subsequent *local* measurements appears only for *entangled* states. It generates a possibility to influence subsystem 2 by a local action on 1 and vice versa:

▶ Effect of $\left\{ \begin{matrix} A \\ B \end{matrix} \right\}$ measurements in $\mathcal{H}$:
$$\boxed{\hat{R}_k \propto \begin{cases} (\hat{R}_A)_1 \otimes \hat{I}_2 & k=1 \\ \hat{I}_1 \otimes (\hat{R}_B)_2 & k=2 \end{cases}}$$

(a) Separable state: $\boxed{|\psi\rangle = |\psi_1\rangle_1 |\psi_2\rangle_2}$ $\quad \Rightarrow \hat{R}_k|\psi\rangle = \begin{cases} |a\rangle_1 |\psi_2\rangle_2 & k=1 \\ |\psi_1\rangle_1 |b\rangle_2 & k=2 \end{cases}$

$\Rightarrow$ measurement on subsystem 1 has no consequence on 2 and vice versa (statistical independence of results)

(b) Entangled state: $\boxed{|\psi\rangle = \sum_{i,j} \gamma_{ij} |\phi_i\rangle_1 |\phi_j\rangle_2} \Rightarrow \hat{R}_k|\psi\rangle = \begin{cases} \mathcal{N}_1 \sum_{ij} \gamma_{ij} \langle a|\phi_i\rangle_1 |a\rangle_1 |\phi_j\rangle_2 & k=1 \\ \mathcal{N}_2 \sum_{ij} \gamma_{ij} \langle b|\phi_j\rangle_2 |\phi_i\rangle_1 |b\rangle_2 & k=2 \end{cases}$

$\mathcal{N}_1 = (\sum_{ii'j} \gamma^*_{ij} \gamma_{i'j} \langle \phi_i | \hat{P}_a | \phi_{i'} \rangle_1)^{-1/2}$ and $\mathcal{N}_2 = \cdots$ are normalization factors

$\Rightarrow$ both measurements change the state from entangled to separable

$\Rightarrow$ measurement on subsystem 1 generally alters probabilities of measurement outcomes for subsystem 2 and vice versa:

Before: $p_{|\psi\rangle}(b) = \langle \psi | \hat{I} \otimes \hat{P}_b | \psi \rangle = \sum_{ijj'} \gamma^*_{ij} \gamma_{ij'} \langle \phi_j | \hat{P}_b | \phi_{j'} \rangle_2$

After: $p_{|\hat{R}_1\psi\rangle}(b) = \langle \hat{R}_1 \psi | \hat{I} \otimes \hat{P}_b | \hat{R}_1 \psi \rangle = \mathcal{N}_1^2 \sum_{ii'jj'} \gamma^*_{ij} \gamma_{i'j'} \langle \phi_i | \hat{P}_a | \phi_{i'} \rangle_1 \langle \phi_j | \hat{P}_b | \phi_{j'} \rangle_2 \neq p_{|\psi\rangle}(b)$

$\Rightarrow$ local measurements on entangled states have **non-local effects!** However, the nature of these effects must prevent any possibility of causality violation.

▶ EPR example

To be specific, we consider an entangled **spin state** of two spin-$\frac{1}{2}$ particles. Essentially the same results can be obtained for analogous entangled states, like polarization states of two photons...

$$\boxed{|\psi_{\rm EPR}\rangle = \frac{1}{\sqrt{2}} \left(|\uparrow\rangle_1 |\downarrow\rangle_2 - |\downarrow\rangle_1 |\uparrow\rangle_2 \right)}$$

Remark: $|\psi_{\rm EPR}\rangle$ invariant under rotations $\hat{R}_{\vec{n}\phi}=\hat{U}\otimes\hat{U}$ with $\hat{U}=\begin{pmatrix}\alpha & -\beta^* \\ \beta & \alpha^*\end{pmatrix}\equiv\mathbf{S}_{\vec{n}\phi}$

$\hat{R}_{\vec{n}\phi}|\psi_{\rm EPR}\rangle =$

$$\tfrac{1}{\sqrt{2}}\Big[\big(\alpha|\!\uparrow\rangle_1+\beta|\!\downarrow\rangle_1\big)\big(-\beta^*|\!\uparrow\rangle_2+\alpha^*|\!\downarrow\rangle_2\big)-\big(-\beta^*|\!\uparrow\rangle_1+\alpha^*|\!\downarrow\rangle_1\big)\big(\alpha|\!\uparrow\rangle_2+\beta|\!\downarrow\rangle_2\big)\Big]=\overbrace{(|\alpha|^2+|\beta|^2)}^{|\alpha|^2+|\beta|^2=1}|\psi_{\rm EPR}\rangle$$

$|\psi_{\rm EPR}\rangle$ may originate from decay of spin-0 object to two spin-$\frac{1}{2}$ particles with orbital ang. momentum $=0$.

Note: in Sec.4.1 we derive that $\frac{1}{\sqrt{2}}(|\!\uparrow\rangle_1|\!\downarrow\rangle_2-|\!\downarrow\rangle_1|\!\uparrow\rangle_2)$ results from addition of two $s=\frac{1}{2}$ spins to total $s=0$ (spin-singlet state).

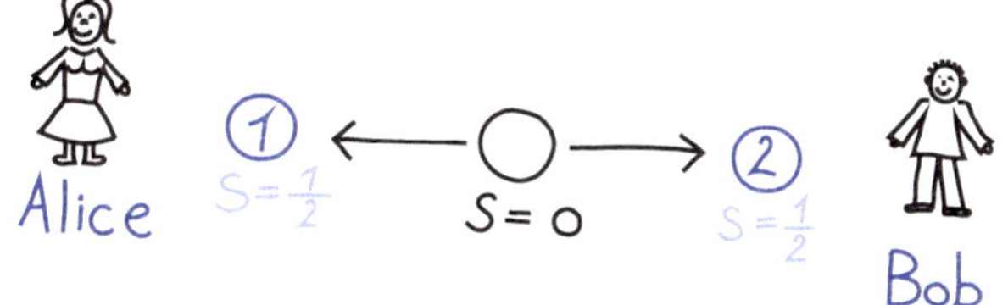

Outcomes & probabilities of local spin measurements

Due to the invariance of $|\psi_{\rm EPR}\rangle$ under rotations we have:

$p(\uparrow_1)=p(\uparrow_2)=p(\downarrow_1)=p(\downarrow_2)=p(\nearrow_1)=p(\swarrow_1)=p(\nearrow_2)=p(\swarrow_2)=\cdots\cdots=\frac{1}{2}$

(A) **Alice** makes measurement on **spin 1** in the basis $\{|\!\uparrow\rangle_1,|\!\downarrow\rangle_1\}$:

$$|\psi_{\rm EPR}\rangle\xrightarrow{\text{Alice}}\hat{R}_1|\psi_{\rm EPR}\rangle=\begin{cases}|\!\uparrow\rangle_1|\!\downarrow\rangle_2 & \text{iff } \uparrow_1 \text{ measured} \quad \ldots\text{case (a)}\\ |\!\downarrow\rangle_1|\!\uparrow\rangle_2 & \text{iff } \downarrow_1 \text{ measured} \quad \ldots\text{case (b)}\end{cases}$$

(B) **Bob** makes measurement (after Alice) on **spin 2** in the basis $\{|\!\uparrow\rangle_2,|\!\downarrow\rangle_2\}$:

$$\begin{bmatrix}p(\uparrow_2)\\p(\downarrow_2)\end{bmatrix}=\begin{bmatrix}0\\1\end{bmatrix}\text{ in case (a)},\ \begin{bmatrix}1\\0\end{bmatrix}\text{ in case (b)}$$

$\neq\frac{1}{2}=$ probability before Alice's measurement

◀ **Historical remark**

1935: Albert Einstein, Boris Podolsky, Nathan Rosen publish the EPR paper, questioning "completeness" of the quantum description

1951: David Bohm reformulates the "EPR paradox" to the spin language

■ Interpretation problems

The results of the previous paragraph invoke some questions concerning locality, which is believed to be an untouchable ingredient of an ultimate physical theory. Although quantum mechanics—even with entangled states and the reduction postulate—remains local on the operational level, there is a shadow of doubt: Do we really understand the nature of the measurement process? Probably not.

▶ Problem of **superluminal communication**

Question: Bob's measurement may be far off the light cone of Alice's measurement. Does QM break the general assumption of finite-speed propagation of all physical impulses?

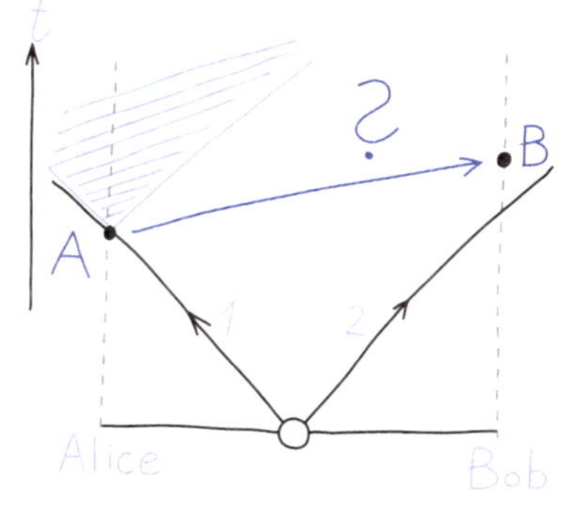

Note: this question seems inappropriate for non-relativistic QM, but the EPR problem is not modified by the crossover to relativistic theory.

Answer: The EPR correlation does *not* enable real superluminal communication. Assume the following scheme for such communication: The state $|\psi_{\rm EPR}\rangle$ is repeatedly produced, particles 1 and 2 being always sent to Alice and Bob, respectively. Alice may encode a binary message for Bob into an altering sequence of directions of particle-1 spin measurements (direction $\left\{\begin{smallmatrix}\vec{n}_z\equiv 0\\ \vec{n}'_z\equiv 1\end{smallmatrix}\right\}$ bit value). The measurement direction actually used is imprinted in the resulting state of particle 2. However, the spin state vector of particle 2 cannot be determined by Bob having only one specimen of this state.
No-cloning theorem: it is not possible to copy the state vector to more carriers since the ideal "cloning" transformation $|\psi\rangle_1|\bullet\rangle_2 \mapsto |\psi\rangle_1|\psi\rangle_2\ \forall|\psi\rangle$ violates linearity:

$$\left.\begin{array}{l}|\psi_a\rangle_1|\phi\rangle_2 \mapsto |\psi_a\rangle_1|\psi_a\rangle_2\\ |\psi_b\rangle_1|\phi\rangle_2 \mapsto |\psi_b\rangle_1|\psi_b\rangle_2\end{array}\right\} \Rightarrow \underbrace{(\alpha|\psi_a\rangle_1+\beta|\psi_b\rangle_1)}_{|\psi\rangle_1}|\phi\rangle_2 \mapsto \underbrace{\alpha|\psi_a\rangle_1|\psi_a\rangle_2+\beta|\psi_b\rangle_1|\psi_b\rangle_2}_{\neq|\psi\rangle_1|\psi\rangle_2}$$

► **Causality problem**

Question: Time order of two events which are off the relative light cone can be reversed by a Lorentz transformation ⇒ in the new frame, Bob makes the measurement (and the state reduction) first. Which picture is true?
Answer: Both pictures yield the same probabilities of measurable outcomes. This follows from mutual compatibility of local measurements on subsystems 1 & 2, which implies independence of the joint probabilities on the succession of measurements (see p. 85).

► What is the **nature of process R**?

The final answer is still unknown, but so far the following possibilities proposed:

(a) *Classical answer*: R is an unavoidable and irreducible consequence of interaction between a quantum system and a "classical apparatus". This early-day answer is not considered satisfactory as everything is made of quantum constituents: Where ends the quantum domain and starts the classical one?

(b) *Metaphysical answers*: R "happens" on the interface between the quantum world and (human?) consciousness. The hard form of this idea (consciousness having an impact on physical reality) seems inadmissible, but a softer form looks okay: the state vector is not the "reality" itself but just a maximal (ultimate?) "information on reality". R captures a sudden change of this information and thus does not have to conform with "materialistic" forms of causality. Another answer of this type was given by the many-worlds interpretation in which consciousness is a part of quantum description.

(c) *Logical answer*: R is to be eliminated in the proper formulation of QM. Example: formulation in terms of the path integral or quantum histories (the notion of a state vector, hence also its reduction, is completely avoided from the formalism; theory considered as a "machinery" to compute observable results)

(d) *Physical answers* (to be elaborated): R results from a so far unknown, but completely natural process, which happens spontaneously as soon as the "amount of matter" involved in unitary quantum evolution becomes "macroscopic". Examples: spontaneous-localization hypothesis, extended decoherence theory, hypothesis of gravitationally-induced collapse

◂ **Historical remark**

1926-9: N. Bohr & W. Heisenberg put cornerstones of "Copenhagen interpretation"
1930's: J. von Neumann & E. Wigner consider consciousness-induced collapse
1957: H. Everett proposes the "many-worlds" interpretation
1980's-90's: attempts to introduce R as a new process (G.C. Ghirardi-A. Rimini-T. Weber, R. Penrose) and to explain R from the decoherence theory (W. Zurek)
1990's: attempts to formulate collapse-free QM (R.B. Griffiths, M. Gell-Mann)

2.6 Implications & applications of quantum measurement

We may hope that the reduction postulate in the present minimal form will be—in an unspecified future—replaced by a more sophisticated and physically transparent formulation. Nevertheless, already at the present stage of knowledge we can discuss several implications. Some of them are rather interesting for the theory itself, some others have an appreciable potential for practical applications.

■ Paradoxes of quantum measurement

What is a paradox? In the following, we adopt the view of paradox as a counterintuitive, surprising, unexpected kind of behavior. Quantum measurement is responsible for several paradoxes.

▸ **Three polarizers paradox**

◂ 1930: noticed by Dirac

2 polarization filters with $\varphi=0°$ & $90°$ stop every individual photon. The insertion of a 3rd filter with $\varphi=45°$ between the two enables some photons to pass.

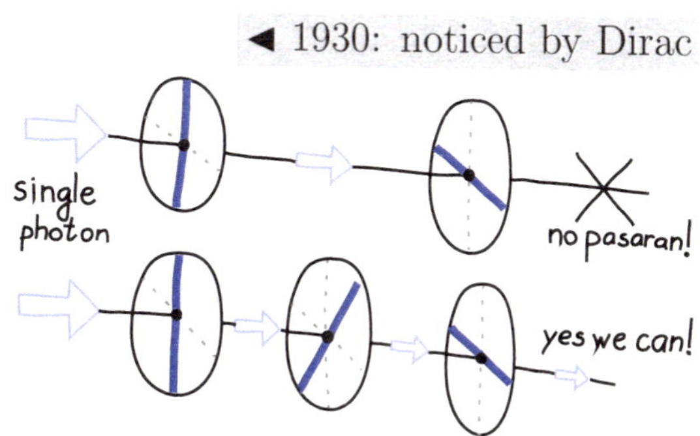

▸ **EPR paradox**

◂ 1935: formulated by Einstein, Podolsky, Rosen

Process R on entangled states ⇒ "spooky action at distance" (see p. 86)

▸ **Schrödinger cat**

◂ 1935, 67: discussed by Schrödinger and Wigner

Quantum superpositions can be extended to macroscopic objects.

For example, consider poor cat whose life and death depends on a quantum process such that it happens to be in a superposition state

$$\boxed{|\psi_{\rm cat}\rangle = \alpha|\psi_{\rm dead}\rangle + \beta|\psi_{\rm alive}\rangle}.$$

On which level the process R takes place? Or: who kills the cat?

▶ **Quantum logic** ◀ 1936: introduced by Birkhoff & von Neumann

Which-path & interference setups of the two-slit experiment indicate that $(A \vee B) \wedge S \neq (A \wedge S) \vee (B \wedge S)$ (see p. 9)

▶ **Quantum Zeno paradox** ◀ 1977: discovered by Sudarshan & Misra

Repeated measurements slow down or even stop (in the limiting case of infinite frequency) the decay process.

Survival probability of a decaying system for evolution without measurement:

$P_0(t) = |\langle\psi(0)|\psi(t)\rangle|^2 \approx 1 - \left(\frac{t}{\tau}\right)^2$ (see p. 70)

Periodic measurement of $\hat{A} \equiv |\psi(0)\rangle\langle\psi(0)|$ with interval $\Delta t = \frac{t}{n} \to 0$:

$$P_0'(t) = \left[P_0\left(\tfrac{t}{n}\right)\right]^n \approx \left[1 - \left(\tfrac{t}{n\tau}\right)^2\right]^n = \underbrace{\left(1 - \tfrac{t}{n\tau}\right)^n}_{\to e^{-t/\tau}} \underbrace{\left(1 + \tfrac{t}{n\tau}\right)^n}_{\to e^{+t/\tau}} \xrightarrow{n\to\infty} 1$$

Note: for exponential decay no effect: $P_0'(t) = [e^{-\lambda\frac{t}{n}}]^n = e^{-\lambda t} = P_0(t)$

▶ **"Bomb-testing" paradox** ◀ 1993: discovered by Elitzur & Vaidman

General name "**interaction-free measurement**": Measurement at one of the paths in a double-slit-type experiment destroys the interference behavior. Detection of the particle in a forbidden direction indicates that the measurement was done—it verifies functionality of the measuring device without necessarily locating the particle on the path where the device is placed.

Example: photon in Mach-Zehnder interferometer with arms I, II. Symbolic expression of the photon state evolution before the 2nd beam splitter (b.s.):

$$|1\rangle \xrightarrow{1^{\rm st}\text{b.s.}} \tfrac{1}{\sqrt{2}}(|{\rm I}\rangle + i|{\rm II}\rangle) \xrightarrow{\text{mirrors}} \tfrac{1}{\sqrt{2}}(i|{\rm I}\rangle - |{\rm II}\rangle) \equiv |\psi\rangle$$

Since $\left.\begin{array}{l} |{\rm I}\rangle \xrightarrow{2^{\rm nd}\text{b.s.}} \frac{1}{\sqrt{2}}(|2\rangle + i|1\rangle) \\ |{\rm II}\rangle \xrightarrow{2^{\rm nd}\text{b.s.}} \frac{1}{\sqrt{2}}(|1\rangle + i|2\rangle) \end{array}\right\}$ with $\left.\begin{array}{l}|1\rangle \\ |2\rangle\end{array}\right\} \equiv$ two exit directions,

interference is observed: $|\psi\rangle \xrightarrow{2^{\rm nd}\text{b.s.}} -|1\rangle \equiv |1\rangle$.

A bomb with *single-photon sensitive* trigger, placed e.g. in arm II, acts as a which-path measurement device: $\frac{1}{\sqrt{2}}(|{\rm I}\rangle + i|{\rm II}\rangle) \xrightarrow{R} \left\{\begin{array}{l} |{\rm I}\rangle \;\; 50\,\% \\ |{\rm II}\rangle \;\; 50\,\% \end{array}\right.$. In both cases, the photon can exit in states $\left\{\begin{smallmatrix}|1\rangle\\|2\rangle\end{smallmatrix}\right.$. Sequence $|1\rangle \to .. \xrightarrow{R} |\mathrm{I}\rangle \xrightarrow{2^{\mathrm{nd}}\mathrm{b.s.}} |2\rangle$ (with probability 25 %) indicates functionality of the bomb *without causing its explosion*! More sophisticated setups have been described which enable one to increase the efficiency of the "bomb detection" arbitrarily close to 100 %

■ Applications of quantum measurement

Present-day people are no more impressed by mere paradoxes. They seek for *practical applications*! So here are some.

▶ Quantum cryptography

The measurement-induced collapse of wavefuction can, in principle, disclose any hidden measurement performed on the system. This can be used to detect eavesdropping in secret communications: Alice sends a binary sequence by individual photons in linear polarization states $\left.\begin{smallmatrix}|x\rangle\\|x'\rangle\end{smallmatrix}\right\} \equiv 0$ and $\left.\begin{smallmatrix}|y\rangle\\|y'\rangle\end{smallmatrix}\right\} \equiv 1$, selecting between 2 rotated *polarization frames* S & S′. Bob measures photon polarizations using *independent selection* of the *same* frames S or S′. The photons for which Alice's & Bob's frames coincide must yield the same initial & final polarizations. Any violation of this rule (detected on a released sample of photons) indicates that the photon state was distorted during transmission (Eve's measurement). If no eavesdropping detected, the states of the remaining photons (for which Alice's & Bob's frames equal) may be used as a private key.

▶ Quantum teleportation

Teleportation means transfer of a physical state of a given object to another carrier. The simplest quantum realization is for a 2-state object, e.g., spin $\frac{1}{2}$.

Setup: $\xrightarrow[\text{spin 1}]{|\psi\rangle}$ Alice $\left\{ \begin{matrix} \rightsquigarrow\rightsquigarrow\rightsquigarrow & \rightsquigarrow\rightsquigarrow\rightsquigarrow & \rightsquigarrow\rightsquigarrow\rightsquigarrow \\ \xleftarrow[\text{spin 2}]{} & \text{EPR source} & \xrightarrow[\text{spin 3}]{} \end{matrix} \right\}$ Bob $\xrightarrow[\text{spin 3}]{|\psi\rangle}$

Unknown state $|\psi\rangle = \alpha|\uparrow\rangle + \beta|\downarrow\rangle$ of spin 1 is reconstructed on spin 3, which is a part of the entangled pair in state $|\psi_{\mathrm{EPR}}\rangle_{23}$, using results of Alice's measurement on spins 1+2 communicated to Bob by a classical channel $\rightsquigarrow\rightsquigarrow\rightsquigarrow$.

Alice measures in entang. basis: $\left\{\begin{matrix} |\phi^{\mathrm{A}}\rangle_{12} = \frac{1}{\sqrt{2}}(|\uparrow\rangle_1|\downarrow\rangle_2 - |\downarrow\rangle_1|\uparrow\rangle_2),\ |\phi^{\mathrm{B}}\rangle_{12} = \frac{1}{\sqrt{2}}(|\uparrow\rangle_1|\downarrow\rangle_2 + |\downarrow\rangle_1|\uparrow\rangle_2), \\ |\phi^{\mathrm{C}}\rangle_{12} = \frac{1}{\sqrt{2}}(|\uparrow\rangle_1|\uparrow\rangle_2 - |\downarrow\rangle_1|\downarrow\rangle_2),\ |\phi^{\mathrm{D}}\rangle_{12} = \frac{1}{\sqrt{2}}(|\uparrow\rangle_1|\uparrow\rangle_2 + |\downarrow\rangle_1|\downarrow\rangle_2). \end{matrix}\right\}$

$$\underbrace{(\alpha|\uparrow\rangle_1 + \beta|\downarrow\rangle_1)}_{|\psi\rangle_1} \underbrace{\tfrac{1}{\sqrt{2}}(|\uparrow\rangle_2|\downarrow\rangle_3 - |\downarrow\rangle_2|\uparrow\rangle_3)}_{|\psi_{\mathrm{EPR}}\rangle_{23}} =$$

$$\frac{1}{\sqrt{4}}\Big[|\phi^{\mathrm{A}}\rangle_{12}\overbrace{(-\alpha|\uparrow\rangle_3 - \beta|\downarrow\rangle_3)}^{|\psi^{\mathrm{A}}\rangle_3} + |\phi^{\mathrm{B}}\rangle_{12}\overbrace{(-\alpha|\uparrow\rangle_3 + \beta|\downarrow\rangle_3)}^{|\psi^{\mathrm{B}}\rangle_3} + |\phi^{\mathrm{C}}\rangle_{12}\overbrace{(\alpha|\downarrow\rangle_3 + \beta|\uparrow\rangle_3)}^{|\psi^{\mathrm{C}}\rangle_3} + |\phi^{\mathrm{D}}\rangle_{12}\overbrace{(\alpha|\downarrow\rangle_3 - \beta|\uparrow\rangle_3)}^{|\psi^{\mathrm{D}}\rangle_3}\Big]$$

Correlated with $\{|\phi^{\mathrm{A}}\rangle_{12}, |\phi^{\mathrm{B}}\rangle_{12}, |\phi^{\mathrm{C}}\rangle_{12}, |\phi^{\mathrm{D}}\rangle_{12}\}$ (results of Alice's measurement), Bob receives states $\{|\psi^{\mathrm{A}}\rangle_3, |\psi^{\mathrm{B}}\rangle_3, |\psi^{\mathrm{C}}\rangle_3, |\psi^{\mathrm{D}}\rangle_3\}$, each of them allowing specific unitary transformation $\hat{U}^{\bullet}|\psi^{\bullet}\rangle_3 = |\psi\rangle_3$ ($\bullet$=A,B,C, or D) to the desired state $|\psi\rangle$.

▶ Quantum computation

Quantum generalization of classical bit $b = \{{0 \atop 1}\}$: **qubit** $\boxed{|\psi\rangle = \alpha|0\rangle + \beta|1\rangle}$

Replacing classical bits by qubits can essentially *speed up some computations*! Instantaneous state of an n-bit classical computer $\equiv (b_0, b_1, \ldots, b_{n-1})$ encodes a single number $x = \sum_{i=0}^{n-1} b_i 2^i$. A state of n-qubit **quantum computer** corresponds to a general superposition of numbers: $\boxed{|\Psi\rangle = \sum_{x=0}^{2^n-1} \alpha_x |x\rangle}$ $\begin{matrix} \alpha_x \in \mathbb{C} \\ \sum_x |\alpha_x|^2 = 1 \end{matrix}$

$$\{|x\rangle\}_{x=0}^{2^n-1} \equiv \{|b_0 b_1 \ldots b_{n-1}\rangle\}_{b_i=0,1} \equiv \text{separable basis in } \mathcal{H}=\mathcal{H}_0\otimes\mathcal{H}_1\otimes\ldots\otimes\mathcal{H}_{n-1}$$

Quantum computation: controlled sequence of elementary unitary operations on a system of qubits (only 1- and 2-qubit operations allowed) concluded by a specific quantum measurement. The same sequence repeated N times to yield a sufficiently large statistical sample of outputs.

Possible task: **probing of unknown function** $f(x)$ $\begin{cases} n \text{ qubits} \equiv x \\ m \text{ qubits} \equiv f(x) \end{cases}$

Examples: period determination, distinction of constant/nonconst. functions...

General computation scheme:

$$|0\rangle_n|0\rangle_m \xrightarrow{\hat{U}_1\otimes\hat{I}} \frac{1}{\sqrt{2^n}}\sum_x |x\rangle_n|0\rangle_m \xrightarrow{\hat{U}_2} \frac{1}{\sqrt{2^n}}\sum_x |x\rangle_n|f(x)\rangle_m \xrightarrow{\hat{U}_3\otimes\hat{I}} \frac{1}{\sqrt{2^n}}\sum_{x,y} \alpha_{xy}|y\rangle_n|f(x)\rangle_m$$

Measurement of $y \Rightarrow$ output probabilities $p(y)$ contain information on $f(x)$

In general, quantum computation uses both $\begin{cases} \text{superpositions} \Rightarrow \text{parallelism} \\ \text{entanglement} \Rightarrow \text{link } x \leftrightarrow f(x) \end{cases}$

◀ Historical remark

1982: R. Feynman comments on potential use of quantum systems for computation
1984: C.H. Bennet & G. Brassard propose a scheme for quantum cryptography
1993: C.H. Bennett *et al.* discover quantum teleportation
1994: P. Shor formulates an efficient quantum algorithm for factorization problem
till now: multiple experimental attempts in all these areas

■ Bell inequalities

Let us return to the EPR situation. Above, we presented the perfect correlation (anticorrelation) of Alice's & Bob's spin measurements as a paradox. But is it really a paradox? Given that both spins have a common origin and both observers use the same (pre-agreed!) orientations of measuring devices, who can be surprised by the correlation of results?[||] But what would happen if Alice & Bob select orientations of their respective spin measurements independently?

[||]The correlation *is* surprising if one insists on the *reality* of wavefunction. If the wavefunction represents an element of the world "out there" (and not just our information on it), Alice's measurement indeed *acts* out of its light cone!

Would the correlation of results remain stronger than might be expected from classical considerations? The positive answer to this question, as elaborated below, is the real puzzle of quantum theory. Moreover, it has changed the debate on possible crossover to a "more complete" (hidden-variable) description from endless discussions to experimental efforts. Today it is an *experimental fact* that quantum theory cannot be replaced by a classical-like theory.

▶ **Generalization of the EPR situation**

Both observers perform spin measurements in *different coordinate frames* ⇒ rotation angles Φ_A & Φ_B of Alice's & Bob's instruments selected independently

▶ **Description in terms of a hidden-variable theory**

Consider a **classical-like**, but **probabilistic** description of the EPR situation:

Probabilities of Alice's & Bob's outputs $\boxed{a,b\in\{\overbrace{+1}^{\text{spin up}},\overbrace{-1}^{\text{down}}\}}$ controlled by: (1) instrument angles Φ_A,Φ_B, (2) so far unknown physical parameters of the entire system, so-called **hidden variables**. We sort these variables to 3 groups: $\{\alpha_1\ldots\}$, $\{\beta_1\ldots\}$ related to individual particles 1 & 2, respectively, and the corresponding measurements, $\{\gamma_1\ldots\}$ related to the emitted pair as a whole.

Scheme:

$$\left.{}^{+1}_{-1}\right\}=a\ \underbrace{\overbrace{\boxed{\text{Alice}}}^{\text{angle }\Phi_A}\xleftarrow[\text{particle 1}]{}}_{\text{hidden variables }\alpha}\ \underbrace{\boxed{\text{EPR source}}}_{\text{hidden variables }\gamma}\ \underbrace{\xrightarrow[\text{particle 2}]{}\overbrace{\boxed{\text{Bob}}}^{\text{angle }\Phi_B}}_{\text{hidden variables }\beta}\ b=\left\{{}^{+1}_{-1}\right.$$

▶ **Strict locality required!**

Probabilities of outputs & h.variables:

$$\boxed{\begin{array}{ll}\left.\begin{array}{ll}P_{\Phi_A}(a|\alpha\gamma) & P_{\Phi_B}(b|\beta\gamma)\\ P_{\Phi_A}(\alpha|\gamma) & P_{\Phi_B}(\beta|\gamma)\end{array}\right\} & \text{conditional}\\ \qquad P(\gamma) & \text{apriori}\end{array}}$$

Locality condition: joint probability $\boxed{P_{\Phi_A\Phi_B}(ab|\alpha\beta\gamma)=P_{\Phi_A}(a|\alpha\gamma)P_{\Phi_B}(b|\beta\gamma)}$

⇒ for *fixed* γ,Φ_A,Φ_B the average $\langle ab\rangle$ factorizes: $\boxed{\langle ab\rangle_{\Phi_A\Phi_B\gamma}=\langle a\rangle_{\Phi_A\gamma}\langle b\rangle_{\Phi_B\gamma}}$

Since variable γ is out of control, we calculate $\langle ab\rangle_{\Phi_A\Phi_B}=\int\langle ab\rangle_{\Phi_A\Phi_B\gamma}P(\gamma)d\gamma$

▶ Define the following 4-angle quantity:

$$\overbrace{\boxed{\langle ab\rangle_{\Phi_A\Phi_B}+\langle ab\rangle_{\Phi_A\Phi'_B}+\langle ab\rangle_{\Phi'_A\Phi_B}-\langle ab\rangle_{\Phi'_A\Phi'_B}}}^{\mathcal{B}(\Phi_A,\Phi'_A,\Phi_B,\Phi'_B)}=$$

$$\int\left[\langle ab\rangle_{\Phi_A\Phi_B\gamma}+\langle ab\rangle_{\Phi_A\Phi'_B\gamma}+\langle ab\rangle_{\Phi'_A\Phi_B\gamma}-\langle ab\rangle_{\Phi'_A\Phi'_B\gamma}\right]P(\gamma)d\gamma=$$

$$\int\underbrace{\left[\langle a\rangle_{\Phi_A\gamma}\langle b\rangle_{\Phi_B\gamma}+\langle a\rangle_{\Phi_A\gamma}\langle b\rangle_{\Phi'_B\gamma}+\langle a\rangle_{\Phi'_A\gamma}\langle b\rangle_{\Phi_B\gamma}-\langle a\rangle_{\Phi'_A\gamma}\langle b\rangle_{\Phi'_B\gamma}\right]}_{\in[-2,+2]\ \Leftarrow\ \langle a\rangle,\langle b\rangle\in[-1,+1]}P(\gamma)d\gamma\quad\in[-2,+2]$$

Locality conditions restrict the domain of $\mathcal{B}(\Phi_A,\Phi'_A,\Phi_B,\Phi'_B)$ to interval $[-2,+2]$

▶ **Bell inequalities**

Conditions necessarily satisfied by *any* classical-like theory (deterministic/non-deterministic) describing the EPR situation: $-2 \leq \mathcal{B}(\Phi_A, \Phi'_A, \Phi_B, \Phi'_B) \leq +2$

▶ **Quantum calculation** of $\langle ab \rangle_{\Phi_A \Phi_B}$

Spinor transformation between measuring frames (y-axis rotation by $\Phi_\bullet$):
$$\begin{pmatrix} |\!\uparrow\rangle_k \\ |\!\downarrow\rangle_k \end{pmatrix} = \begin{pmatrix} \cos\frac{\Phi_\bullet}{2} & \sin\frac{\Phi_\bullet}{2} \\ -\sin\frac{\Phi_\bullet}{2} & \cos\frac{\Phi_\bullet}{2} \end{pmatrix} \begin{pmatrix} |\nearrow_{\Phi_\bullet}\rangle_k \\ |\swarrow_{\Phi_\bullet}\rangle_k \end{pmatrix} \quad \begin{matrix} k=1, \bullet=A \\ k=2, \bullet=B \end{matrix}$$

$$\underbrace{\tfrac{1}{\sqrt{2}}(|\!\uparrow\rangle_1|\!\downarrow\rangle_2 - |\!\downarrow\rangle_1|\!\uparrow\rangle_2)}_{|\psi_{\rm EPR}\rangle} = \underbrace{\frac{\sin\frac{\Phi_A}{2}\cos\frac{\Phi_B}{2} - \cos\frac{\Phi_A}{2}\sin\frac{\Phi_B}{2}}{\sqrt{2}}}_{A_{++}(\Phi_A,\Phi_B)} \underbrace{|\nearrow_{\Phi_A}\rangle_1|\nearrow_{\Phi_B}\rangle_2}_{ab=+1} + \underbrace{\frac{\sin\frac{\Phi_A}{2}\sin\frac{\Phi_B}{2} + \cos\frac{\Phi_A}{2}\cos\frac{\Phi_B}{2}}{\sqrt{2}}}_{A_{+-}(\Phi_A,\Phi_B)}$$
$$\times \underbrace{|\nearrow_{\Phi_A}\rangle_1|\swarrow_{\Phi_B}\rangle_2}_{ab=-1} + \underbrace{\frac{-\sin\frac{\Phi_A}{2}\sin\frac{\Phi_B}{2} - \cos\frac{\Phi_A}{2}\cos\frac{\Phi_B}{2}}{\sqrt{2}}}_{A_{-+}(\Phi_A,\Phi_B)} \underbrace{|\swarrow_{\Phi_A}\rangle_1|\nearrow_{\Phi_B}\rangle_2}_{ab=-1} + \underbrace{\frac{\sin\frac{\Phi_A}{2}\cos\frac{\Phi_B}{2} - \cos\frac{\Phi_A}{2}\sin\frac{\Phi_B}{2}}{\sqrt{2}}}_{A_{--}(\Phi_A,\Phi_B)} \underbrace{|\swarrow_{\Phi_A}\rangle_1|\swarrow_{\Phi_B}\rangle_2}_{ab=+1}$$

$$\langle ab \rangle_{\Phi_A\Phi_B} = |A_{++}(\Phi_A,\Phi_B)|^2 - |A_{+-}(\Phi_A,\Phi_B)|^2 - |A_{-+}(\Phi_A,\Phi_B)|^2 + |A_{--}(\Phi_A,\Phi_B)|^2 = -\cos(\Phi_A - \Phi_B)$$

▶ **Quantum inequalities**

$$\mathcal{B} = -\cos(\Phi_A - \Phi_B) - \cos(\Phi_A - \Phi'_B) - \cos(\Phi'_A - \Phi_B) + \cos(\Phi'_A - \Phi'_B)$$

⇒ **Bell inequalities violated!** $-2\sqrt{2} \leq \mathcal{B}(\Phi_A, \Phi'_A, \Phi_B, \Phi'_B) \leq +2\sqrt{2}$

⇒ Predictions of QM differ from those of a general local hidden-variable theory

Conclusion: "**Quantum nonlocality**" does *not* exist in the sense of an *exploitable* superluminal communication. Nevertheless, a trace of nonlocality lies in correlations between Alice's & Bob's results in the generalized EPR situation. These correlations are stronger than possible classical ones if locality is required in the classical description. ⇒ The following soft form of nonlocality is valid: *Quantum mechanics cannot be replaced by any classical-like local theory.*

◀ **Historical remark**
1964, 70: John Bell derives various versions of his inequalities
1981: A. Aspect *et al.* provide the first reliable experimental confirmation of the Bell inequalities violation, many more (increasingly "loophole-free") tests up to now

1.7 Quantum statistical physics

Physics would have much less power if there is no statistical physics. This important branch of physics deals with situations, rather generic for all complex systems, when the initial state cannot be precisely determined. Instead, one has some knowledge on the probability distribution characterizing a multitude of possible states in which the system may occur. In classical statistical physics, a single realization of the given system at a point $(\vec{p}_0, \vec{q}_0)$ of a multidimensional phase space is replaced by a *statistical ensemble* of replicas of the system at different points. This means that $\delta(\vec{p}-\vec{p}_0, \vec{q}-\vec{q}_0)$ changes into a delocalized distribution $\rho(\vec{p}, \vec{q})$ (satisfying $\iint \rho(\vec{p}, \vec{q})\, d\vec{p}\, d\vec{q} = 1$). We are ready now to apply this kind of statistical description to quantum physics.

Pure and mixed states: density operator

Statistical description implies **statistical uncertainty**. However, quantum physics already contains **quantum uncertainty**. It is useful to unify both these types of uncertainty in a generalized notion of quantum state. It is expressed by a positive-definite Hermitian operator in $\mathcal{H}$, called density operator.

▶ Quantum statistical ensemble

In analogy to classical statistical ensemble, we want to introduce a quantum ensemble. Assume that the state vector describing the system is a random selection from the set: $\left\{\begin{array}{l}|\psi_1\rangle \ldots \text{probability } p_1 \\ |\psi_2\rangle \ldots \text{probability } p_2 \\ \vdots\end{array}\right\}$ with $\left\{\begin{array}{ll}\langle\psi_k|\psi_k\rangle = 1 & \sum_k p_k = 1 \\ \underset{\text{for } k\neq l}{\langle\psi_k|\psi_l\rangle \neq 0} & \text{(in general)}\end{array}\right.$

Density operator/matrix: $\boxed{\hat{\rho} \equiv \sum_k p_k |\psi_k\rangle\langle\psi_k|}$ $\quad \langle i|\hat{\rho}|j\rangle = \sum_k p_k \langle i|\psi_k\rangle\langle\psi_k|j\rangle$

Operator $\hat{\rho}$ generates **probabability distribution in the *entire* Hilbert space $\mathcal{H}$**: probability to find $|\psi\rangle$ given by $\sum_k p_k \underbrace{|\langle\psi|\psi_k\rangle|^2}_{p_{\psi_k}(\psi)} = \boxed{\langle\psi|\hat{\rho}|\psi\rangle = p_{\hat{\rho}}(\psi)}$

▶ Generalization of states in QM:

pure state	$\hat{\rho} = \|\psi\rangle\langle\psi\|$	$\Leftrightarrow$	$\|\psi\rangle$	$\exists$ state vector
mixed state	$\hat{\rho} = \sum_k p_k \|\psi_k\rangle\langle\psi_k\|$	$\Leftrightarrow$	$\times$	$\nexists$ state vector

▶ Properties of density operator

(a) Hermiticity $\boxed{\hat{\rho} = \hat{\rho}^\dagger}$

(b) $\mathrm{Tr}\left[\sum_k p_k|\psi_k\rangle\langle\psi_k|\right] = \sum_i \sum_k p_k \langle i|\psi_k\rangle\langle\psi_k|i\rangle = \sum_k p_k \overbrace{\sum_i \langle\psi_k|i\rangle\langle i|\psi_k\rangle}^{\langle\psi_k|\psi_k\rangle = 1} = \boxed{1 = \mathrm{Tr}\hat{\rho}}$

(c) $\langle\psi|\hat{\rho}|\psi\rangle \equiv p_\psi(\rho) \in [0,1] \quad \forall\,|\psi\rangle \quad \Rightarrow$ eigenvalues $\boxed{\rho_i \in [0,1]}$

(d) Diagonalized density matrix:

$$\boxed{\hat{\rho} = \sum_i \rho_i |\phi_i\rangle\langle\phi_i|} \equiv \begin{pmatrix} \rho_1 & 0 & 0 & \cdots \\ 0 & \rho_2 & 0 & \\ 0 & 0 & \rho_3 & \\ \vdots & & & \ddots \end{pmatrix} \qquad \begin{array}{l} \sum_i \rho_i = 1 \quad \rho_i \ldots \text{probability to find } |\phi_i\rangle \\ \sum_i \rho_i^2 \leq 1 \end{array}$$

Criterion to distinguish pure & mixed states: $\boxed{\mathrm{Tr}\hat{\rho}^2 \begin{cases} =1 & \textbf{pure state } (\rho_i = \delta_{ij}) \\ <1 & \textbf{mixed state} \end{cases}}$

▶ Ambiguity in the expansion of $\hat{\rho}$

The same diagonalized form $\hat{\rho} = \sum_i \rho_i |\phi_i\rangle\langle\phi_i|$ (with $\{|\phi_i\rangle\}$ orthonormal) can be produced by different expressions $\hat{\rho} = \sum_k p_k |\psi_k\rangle\langle\psi_k|$ $\left(\begin{array}{c}\text{with } \{|\psi_k\rangle\} \text{ normalized} \\ \text{but otherwise arbitrary}\end{array}\right)$.

▶ **Statistical properties of observables**

$\langle A\rangle_\rho \equiv$ **average** of quantity $\hat{A}$ in state $\hat{\rho} \equiv \sum_k p_k|\psi_k\rangle\langle\psi_k|$

$$\langle A\rangle_\rho = \int a \overbrace{\sum_k p_k\langle\psi_k|\hat{P}_a|\psi_k\rangle}^{p_\rho(a)}\, da = \sum_k p_k\langle\psi_k|\hat{A}|\psi_k\rangle = \sum_{ij}\sum_k p_k\langle\psi_k|i\rangle\langle i|\hat{A}|j\rangle\langle j|\psi_k\rangle =$$

$$= \sum_{ij}\underbrace{\sum_k p_k\langle j|\psi_k\rangle\langle\psi_k|i\rangle}_{\langle j|\hat{\rho}|i\rangle}\langle i|\hat{A}|j\rangle = \boxed{\mathrm{Tr}(\hat{\rho}\hat{A}) = \mathrm{Tr}(\hat{A}\hat{\rho}) = \langle A\rangle_\rho}$$

For a pure state: $\langle A\rangle_\psi = \sum_i\langle i|\psi\rangle\langle\psi|\hat{A}|i\rangle = \langle\psi|\hat{A}|\psi\rangle$

Dispersion: $\langle\langle A^2\rangle\rangle_\rho = \langle A^2\rangle_\rho - \langle A\rangle_\rho^2 = \boxed{\mathrm{Tr}(\hat{A}^2\hat{\rho}) - \mathrm{Tr}^2(\hat{A}\hat{\rho}) = \langle\langle A^2\rangle\rangle_\rho}$

Probability distribution $p_\rho(a) = \sum_k p_k\langle\psi_k|\hat{P}_a|\psi_k\rangle = \mathrm{Tr}(\hat{P}_a\,\hat{\rho})$

◀ **Historical remark**

1927: J. von Neumann introduces density operator to describe a general quantum state (simultaneous work by L. Landau and F. Bloch)

■ Entropy and canonical ensemble

The concept of entropy plays an important role in thermodynamics as well as in mathematical information theory. Statistical physics is a bridge between both these seemingly distant coasts. States with null entropy are the pure states of ordinary QM. In contrast, states whose entropy is maximal—within given constraints upon some physical averages—represent equilibrated systems in contact with a thermal bath.

▶ **Shannon information entropy**

General probability distribution for a finite set of events:

event $i \in \{1, 2, \ldots n\}$ $\leftrightarrow$ probability $\{p_i\} \equiv \{p_1, p_2, \ldots, p_n\}$

Information entropy is a functional on the space of probability distributions: $\boxed{S[\{p_i\}] = -\sum_{i=1}^{n} p_i \ln p_i}$

▶ **Properties**

Maximum	$S = \ln n$ for $p_i = \mathrm{const} = \frac{1}{n}$	maximal uncertainty
Minimum	$S = 0$ for $p_i = \delta_{ij}$	minimal uncertainty

Additivity: 2 sets of **independent events** $\left\{\begin{smallmatrix} i \leftrightarrow p_i \\ j \leftrightarrow p_j\end{smallmatrix}\right\} \Rightarrow$ entropy $\left\{\begin{smallmatrix} S_1 \\ S_2\end{smallmatrix}\right\}$

Joint distribution $(i \wedge j) \leftrightarrow p_{ij}{=}p_i p_j$ $\Rightarrow$ entropy $\boxed{S_{12} = S_1 + S_2}$

However, for correlated events: $S_{12} = S_1 + S_2 + \Delta S$ with $\Delta S \gtrless 0$

▶ **Von Neumann entropy**

$$\boxed{S_\rho = -k\sum_i \rho_i \ln \rho_i = -k\mathrm{Tr}\left[\hat{\rho}\ln\hat{\rho}\right]}$$

$\equiv$ **thermodynamic entropy**

$k{=}8.6\cdot 10^{-5}\mathrm{eV/K}$ Boltzmann const.

$S_\rho = 0$ for pure state
$S_\rho > 0$ for mixed state ($S_\rho = S_{\max} = \ln n$ for "maximally mixed" state)

▶ **Equilibrium state** of a quantum system which exchanges energy with the surrounding environment (**thermal bath**):

$\hat\rho$ diagonal in energy eigenbasis: $\hat\rho = \sum_i \rho_i |E_i\rangle\langle E_i|$

$$\Rightarrow \textbf{stationary}\text{ state: } \hat\rho(t) = \sum_i \rho_i e^{-i\frac{E_i t}{\hbar}} |E_i\rangle\langle E_i| e^{+i\frac{E_i t}{\hbar}} = \hat\rho(0)$$

Probabilities ρ_i determined from the **maximal entropy prinicple**:
S_ρ=max. for a fixed **energy average** $\langle E\rangle_\rho$ $\Rightarrow$ method of Lagrange multipliers

$$f = -\sum_i \rho_i \ln \rho_i + (\alpha+1)\sum_i \rho_i - \beta \sum_i \rho_i E_i$$

$$\frac{\partial f}{\partial \rho_i} = -\ln\rho_i - \rho_i \frac{1}{\rho_i} + (\alpha+1) - \beta E_i = 0 \quad \Rightarrow \quad \boxed{\rho_i = e^{\alpha - \beta E_i}} = \overbrace{e^{\alpha}}^{\mathcal{N}} e^{-\beta E_i}$$

Constants α, β determined from normalization $\mathrm{Tr}\,\hat\rho = 1$ & fixed average $\langle E\rangle_\rho$

▶ **Canonical density operator** (ensemble)

$$\boxed{\hat\rho_\beta = \frac{1}{Z(\beta)} e^{-\beta \hat H}} \quad \text{with} \quad \boxed{\beta = \frac{1}{kT}}$$

inverse temperature (the only parameter of the canon. state)

$$\boxed{Z(\beta) = \sum_i e^{-\beta E_i} = \mathrm{Tr}\, e^{-\beta \hat H}}$$

partition function
normalization $\mathcal{N} = e^{\alpha} \equiv \frac{1}{Z(\beta)}$

▶ Function $Z(\beta)$ contains complete information on thermal energy distribution:

$$\frac{d}{d\beta} Z(\beta) = \frac{d}{d\beta} \mathrm{Tr}\, e^{-\beta \hat H} = -\mathrm{Tr}\big[\hat H \underbrace{e^{-\beta \hat H}}_{Z(\beta)\hat\rho_\beta}\big] = -Z(\beta) \underbrace{\mathrm{Tr}[\hat H \hat\rho_\beta]}_{\langle E\rangle_\beta}$$

$$\boxed{\langle E\rangle_\beta = -\frac{1}{Z(\beta)} \frac{d}{d\beta} Z(\beta) = -\frac{d}{d\beta} \ln Z(\beta)}$$

energy average

$$-\frac{d}{d\beta}\langle E\rangle_\beta = kT^2 \underbrace{\frac{d}{dT}\langle E\rangle_T}_{c_V(T)}$$ specific heat at temperature T

$$-\frac{d}{d\beta}\langle E\rangle_\beta = \frac{d^2}{d\beta^2} \ln Z(\beta) = \overbrace{\frac{1}{Z(\beta)} \frac{d^2 Z(\beta)}{d\beta^2}}^{\frac{1}{Z(\beta)}\mathrm{Tr}[\hat H^2 e^{-\beta \hat H}]} - \overbrace{\frac{1}{Z(\beta)^2} \left[\frac{dZ(\beta)}{d\beta}\right]^2}^{\langle E\rangle_\beta^2} = \langle E^2\rangle_\beta - \langle E\rangle_\beta^2 = \langle\langle E^2\rangle\rangle_\beta$$

$$\boxed{\langle\langle E^2\rangle\rangle_\beta = \frac{1}{k\beta^2} c_V(\beta) = \frac{d^2}{d\beta^2} \ln Z(\beta)}$$

energy dispersion & specific heat

▶ Function $Z(\beta)$ also contains complete information on the whole energy spectrum $\{E_i\}$ and, equivalently, on the **level density** $\boxed{\varrho(E) = \sum_i \delta(E - E_i)}$

which can be obtained from inverse Laplace transform of $Z(\beta) = \int \varrho(E) e^{-\beta E} dE$

Thermal distribution of energy $w_\beta(E)$ (probabability density for finding the system at energy E if temperature is T) is expressed via the level density $\varrho(E)$:

$$\boxed{w_\beta(E) = \tfrac{1}{Z(\beta)}\varrho(E)e^{-\beta E}}$$

Usually the (increasing × decreasing) function product yields a peak at a certain value $[E]_\beta$ close to $\langle E\rangle_\beta$

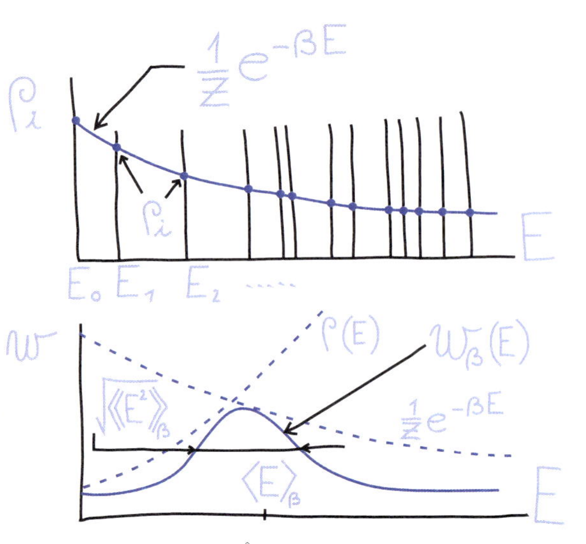

▶ Useful observation: evolution operator $\hat{U}(t) = e^{-i\frac{\hat{H}t}{\hbar}}$ for **imaginary time** $t = -i\hbar\beta \Rightarrow$ canonical density operator $e^{-\beta\hat{H}} = \boxed{Z(\beta)\,\hat{\rho}_\beta = \hat{U}(-i\hbar\beta)}$

This is used in some advanced calculations of thermal & dynamical properties.

▶ Similar procedure (using maximal entropy principle) is applicable also for systems with variable numbers N_i of particles (types $i = 1, 2, \ldots n$) but fixed averages $\langle N_i\rangle$ (particles exchanged with the bath) ⇒ grand-canonical ensemble characterized by inverse temperature β and chemical potentials μ_i (see Sec. 7.2).

◀ **Historical remark**

1878: J.W. Gibbs presents the concept of canonical ensemble & entropy formula
1927: J. von Neumann introduces the density operator & entropy in QM
1948: C. Shannon applies entropy in the information theory

■ Wigner distribution function

As the density-operator formalism merges statistical and quantal fluctuations into a unified picture, it may raise hopes of formulating quantum mechanics in a purely statistical language—via some appropriate statistical distributions in the classical phase space. Although it turns out that such a formulation is not possible, the product of this effort is useful by itself.

▶ $\left.\begin{array}{l}\text{Coordinate}\\ \text{Momentum}\end{array}\right\}$ representation of density operator: $\boxed{\begin{array}{l}\langle\vec{x}\,'|\hat{\rho}\,|\vec{x}\,\rangle \equiv \rho(\vec{x}\,',\vec{x}\,)\\ \langle\vec{p}\,'|\hat{\rho}\,|\vec{p}\rangle \equiv \rho(\vec{p}\,',\vec{p}\,)\end{array}}$

Could we get $\rho(\vec{x},\vec{p}\,) \equiv$ an analog of the classical phase-space distribution?

▶ Any probability distribution ρ is equivalently expressed through its so-called **characteristic function** ≡ Fourier transform of ρ. For a distribution $\rho(\vec{x},\vec{p}\,)$ in the classical phase space it reads as:

$$\chi_\rho(\vec{\xi},\vec{\eta}\,) = \int \rho(\vec{x},\vec{p}\,)e^{\frac{i}{s}(\vec{\eta}\cdot\vec{x}-\vec{\xi}\cdot\vec{p})}d\vec{x}\,d\vec{p} \;\leftrightarrow\; \rho(\vec{x},\vec{p}\,) = \tfrac{1}{(2\pi s)^{2n}}\int \chi_\rho(\vec{\xi},\vec{\eta}\,)e^{-\frac{i}{s}(\vec{\eta}\cdot\vec{x}-\vec{\xi}\cdot\vec{p})}d\vec{\xi}\,d\vec{\eta}$$

$\left\{\begin{smallmatrix}\vec{\xi}\\ \vec{\eta}\end{smallmatrix}\right\}$ n-dim variables in the same units as $\left\{\begin{smallmatrix}\vec{x}\\ \vec{p}\end{smallmatrix}\right\}$, $s \equiv$ constant in units of xp

$\Rightarrow$ characteristic function expressed as the average: $\boxed{\chi_\rho(\vec{\xi},\vec{\eta}) = \left\langle e^{\frac{i}{s}(\vec{\eta}\cdot\vec{x}-\vec{\xi}\cdot\vec{p})}\right\rangle_\rho}$

► The last expression makes it possible to find a **quantum analog** of characteristic function: $\boxed{C_\rho(\vec{\xi},\vec{\eta}) \equiv \mathrm{Tr}\left[e^{\frac{i}{\hbar}(\vec{\eta}\cdot\hat{\vec{x}}-\vec{\xi}\cdot\hat{\vec{p}})}\hat{\rho}\right]}$

Fourier inverse of $C_\rho(\vec{\xi},\vec{\eta})$ should be the quantum distribution in phase space:

$$\boxed{W_\rho(\vec{x},\vec{p}) \equiv \frac{1}{(2\pi\hbar)^{2n}} \int C_\rho(\vec{\xi},\vec{\eta})\, e^{-\frac{i}{\hbar}(\vec{\eta}\cdot\vec{x}-\vec{\xi}\cdot\vec{p})}\, d\vec{\xi}\, d\vec{\eta}}$$ **Wigner distribution**

► Characteristic function and Wigner distribution in dimension $\boldsymbol{n{=}1}$

$$C_\rho(\xi,\eta) = \int \langle x|\hat{\rho}\, e^{\frac{i}{\hbar}(\eta\hat{x}-\xi\hat{p})}|x\rangle\, dx \overset{\text{BCH}}{=} e^{i\frac{\eta\xi}{2\hbar}} \int \langle x|\hat{\rho}\, e^{\frac{i}{\hbar}\eta\hat{x}} e^{-\frac{i}{\hbar}\xi\hat{p}}|x\rangle\, dx = \ldots$$

Special BCH formula for $[\hat{A},\hat{B}]=\hat{C}$, $[\hat{A},\hat{C}]=[\hat{B},\hat{C}]=0$: $e^{\hat{A}+\hat{B}} = e^{\hat{A}}e^{\hat{B}}e^{\frac{1}{2}\hat{C}}$

$$\cdots = e^{i\frac{\eta\xi}{2\hbar}} \iint \underbrace{\langle x|\hat{\rho}e^{\frac{i}{\hbar}\eta\hat{x}}|x'\rangle}_{e^{\frac{i}{\hbar}\eta x'}\langle x|\hat{\rho}|x'\rangle} \underbrace{\langle x'|e^{-\frac{i}{\hbar}\xi\hat{p}}|x\rangle}_{\langle x'|x-\xi\rangle}\, dx\, dx' = e^{-i\frac{\eta\xi}{2\hbar}} \int \underbrace{\langle x|\hat{\rho}|x-\xi\rangle}_{\rho(x,x-\xi)}\, e^{\frac{i}{\hbar}\eta x}\, dx$$

$$\Rightarrow \boxed{C_\rho(\xi,\eta) = \int \rho(x'{+}\tfrac{\xi}{2}, x'{-}\tfrac{\xi}{2})\, e^{\frac{i}{\hbar}\eta x'}\, dx'}$$

$$W_\rho(x,p) = \frac{1}{(2\pi\hbar)^2} \iint \left[\int \rho(x'{+}\tfrac{\xi}{2}, x'{-}\tfrac{\xi}{2})\, e^{\frac{i}{\hbar}\eta x'}\, dx'\right] e^{-\frac{i}{\hbar}(\eta x-\xi p)}\, d\eta\, d\xi =$$
$$= \frac{1}{(2\pi\hbar)^2} \iint \rho(x'{+}\tfrac{\xi}{2}, x'{-}\tfrac{\xi}{2}) \underbrace{\left[\int e^{-\frac{i}{\hbar}\eta(x-x')}\, d\eta\right]}_{2\pi\hbar\delta(x-x')} e^{+\frac{i}{\hbar}\xi p}\, dx'\, d\xi$$

$$\boxed{W_\rho(x,p) = \frac{1}{2\pi\hbar} \int_{-\infty}^{+\infty} \rho\left(x{+}\tfrac{\xi}{2}, x{-}\tfrac{\xi}{2}\right)\, e^{+\frac{i}{\hbar}\xi p}\, d\xi}$$

where $\rho(x'{+}\frac{\xi}{2}, x'{-}\frac{\xi}{2}) = \langle x'{+}\frac{\xi}{2}|\hat{\rho}|x'{-}\frac{\xi}{2}\rangle$

This is the desired quantum analog of phase-space distribution. Indeed, this function is

real :	$W_\rho(x,p) = W_\rho(x,p)^*$
normalized :	$\int W_\rho(x,p)\, dx\, dp = 1$

However, it is *not* semi-positive, which indicates that $W_\rho(x,p)$ does *not* have the meaning of ordinary probability density. Moral: quantum oddity is unremovable!

◄ **Historical remark**
1927: H. Weyl derives a mapping of Hermitian operators to phase-space functions
1932: E. Wigner introduces quasiprobability distribution related to density operators
1940's-present: developments in the phase-space formulation of QM

■ Density operator for open systems

The way we introduced the density operator might invoke a picture of somebody drawing balls (quantum states) from a wheel of fortune. The states are prepared there, one just does not know what he will get. We may think of an accelerator delivering individual particles in different polarization states. However, there is another—and probably more important—use of the density-matrix formalism. It deals with coupled (open) systems: the systems that interact with other systems, environment, or internal degrees of freedom. Such composite objects generically occur in entangled quantum states and the density operator is the only entity that allows one to extract states of individual subsystems.

► Two coupled systems: $\boxed{\mathcal{H} = \mathcal{H}_1 \otimes \mathcal{H}_2}$ $\begin{cases} 1 \equiv \text{open quantum system} \\ 2 \equiv \text{environment} \begin{cases} \text{another system or} \\ \text{internal degs. of freedom} \end{cases} \end{cases}$

General pure state of 1+2: $\boxed{|\Psi\rangle = \sum_{ij} \alpha_{ij} |\phi_{1i}\rangle |\phi_{2j}\rangle}$ $\Big\{ |\phi_{kl}\rangle \Big\}_l \equiv \text{basis} \in \mathcal{H}_k$

► Reduced density operator

Information on the state of **subsystem 1** available only in the form of reduced density operator obtained by the technique of **partial trace**:

$$|\Psi\rangle \quad \longmapsto \quad |\Psi\rangle\langle\Psi| \equiv \hat{\rho}_{12} \quad \longmapsto \quad \boxed{\hat{\rho}_1 \equiv \mathrm{Tr}_2\, \hat{\rho}_{12} \equiv \sum_l \langle \phi_{2l} | \hat{\rho}_{12} | \phi_{2l} \rangle}$$

For the above state $|\Psi\rangle$: $\hat{\rho}_1 = \sum_l \sum_{ij} \sum_{i'j'} \alpha_{ij} \alpha^*_{i'j'} \underbrace{\langle \phi_{2l} | \phi_{2j} \rangle}_{\delta_{jl}} |\phi_{1i}\rangle\langle\phi_{1i'}| \underbrace{\langle \phi_{2j'} | \phi_{2l} \rangle}_{\delta_{j'l}}$

$$\Rightarrow \boxed{\hat{\rho}_1 = \sum_{ii'} \underbrace{\left(\sum_j \alpha_{ij} \alpha^*_{i'j} \right)}_{\rho_{1ii'} = \rho^*_{1i'i}} |\phi_{1i}\rangle\langle\phi_{1i'}|}$$

This is an operator on $\mathcal{H}_1$ which has (as shown below) the properties of a density operator

► Properties of $\hat{\rho}_1$:

(a) $\hat{\rho}_1^\dagger = \sum_{ii'} \rho^*_{1ii'} |\phi_{1i'}\rangle\langle\phi_{1i}| = \hat{\rho}_1$ (b) $\mathrm{Tr}_1 \hat{\rho}_1 = \sum_{ij} |\alpha_{ij}|^2 = 1$

(c) $\langle \psi_1 | \hat{\rho}_1 | \psi_1 \rangle \geq 0 \quad \forall\ |\psi_1\rangle \equiv \sum_l \beta_l |\phi_{1l}\rangle \quad \Rightarrow$ eigenvalues ≥ 0

Proof: $\langle \psi_1 | \hat{\rho}_1 | \psi_1 \rangle = \sum_{ll'} \beta^*_{l'} \beta_l \sum_{ii'} \big(\sum_j \alpha_{ij} \alpha^*_{i'j} \big) \underbrace{\langle \phi_{1l'} | \phi_{1i} \rangle}_{\delta_{l'i}} \underbrace{\langle \phi_{1i'} | \phi_{1l} \rangle}_{\delta_{i'l}} = \sum_j \big| \sum_i \beta^*_i \alpha_{ij} \big|^2 \geq 0$

(d) $\mathrm{Tr}_1 \hat{\rho}_1^2 \leq 1 \quad \Leftarrow$(b),(c)

(e) Average of a local observable $\hat{A} \equiv \hat{A}_1 \otimes \hat{I}_2$

$$\langle \Psi | \hat{A} | \Psi \rangle = \sum_{ij} \sum_{i'j'} \alpha_{ij} \alpha^*_{i'j'} \langle \phi_{1i'} | \hat{A}_1 | \phi_{1i} \rangle \underbrace{\langle \phi_{2j'} | \phi_{2j} \rangle}_{\delta_{jj'}} = \overbrace{\sum_{ii'} \underbrace{\sum_j \alpha_{ij} \alpha^*_{i'j}}_{\rho_{1ii'}} \langle \phi_{1i'} | \hat{A}_1 | \phi_{1i} \rangle}^{\mathrm{Tr}(\hat{A}_1 \hat{\rho}_1)}$$

$$\Rightarrow \boxed{\langle A \rangle_\Psi = \mathrm{Tr}(\hat{A}_1 \hat{\rho}_1)}$$

$\hat{\rho}_1 \equiv$ density operator of subsystem 1

► Pure states of the subsystem

The subsystem is in a pure state *iff* the whole system is in a separable state: $\hat{\rho}_1 = \mathrm{Tr}_2\,\hat{\rho}_{12}$ is a pure state $|\psi_1\rangle \equiv \sum_l \beta_l |\phi_{1l}\rangle \quad \Leftrightarrow \quad |\Psi\rangle \equiv |\psi_1\rangle|\psi_2\rangle$ separable

$$\left.\begin{array}{lll} \hat{\rho}_1 = |\psi_1\rangle\langle\psi_1| & \Rightarrow & \rho_{1ii'} = \langle\phi_{1i}|\hat{\rho}_1|\phi_{1i'}\rangle = \beta_i\beta^*_{i'} \\ |\Psi\rangle = |\psi_1\rangle|\psi_2\rangle & \Rightarrow & \rho_{1ii'} = \sum_j \underbrace{\alpha_{ij}}_{\beta_i\gamma_j}\,\underbrace{\alpha^*_{i'j}}_{\beta^*_{i'}\gamma^*_j} = \beta_i\beta^*_{i'}\overbrace{\sum_j |\gamma_j|^2}^{1} \end{array}\right\}\text{ same expressions}$$

► Schmidt decomposition

Any entangled state of a given coupled system can be expressed in a "canonical form", with the aid of eigenvectors of the respective reduced density matrices:

Consider general state $|\Psi\rangle = \sum_{ij}\alpha_{ij}|\phi_{1i}\rangle|\phi_{2j}\rangle$

Subsystem 1:
$\hat{\rho}_1 = \mathrm{Tr}_2\,\hat{\rho}_{12} = \sum_{ii'}\big(\sum_j \alpha_{ij}\alpha^*_{i'j}\big)|\phi_{1i}\rangle\langle\phi_{1i'}|$

Subsystem 2:
$\hat{\rho}_2 = \mathrm{Tr}_1\,\hat{\rho}_{12} = \sum_{jj'}\big(\sum_i \alpha_{ij}\alpha^*_{ij'}\big)|\phi_{2j}\rangle\langle\phi_{2j'}|$

Suppose $\boxed{\alpha_{ij} = \sqrt{\rho_i}\,\delta_{ij}} \Rightarrow \left\{\begin{array}{l} \rho_{1ii'} = \sum_j \alpha_{ij}\alpha^*_{i'j} = \sum_j \sqrt{\rho_i}\,\delta_{ij}\sqrt{\rho_{i'}}\,\delta_{i'j} = \rho_i\delta_{ii'} \\ \rho_{2jj'} = \sum_i \alpha_{ij}\alpha^*_{ij'} = \sum_i \sqrt{\rho_i}\,\delta_{ij}\sqrt{\rho_i}\,\delta_{ij'} = \rho_j\delta_{jj'} \end{array}\right\}$ diagonal

$\Rightarrow$ both $\left\{\begin{smallmatrix}\hat{\rho}_1\\ \hat{\rho}_2\end{smallmatrix}\right\}$ diagonalized with the same eigenvalues $\{\rho_i\}$

$\Rightarrow$ both subsystem's entropies equal: $\boxed{S_1 = S_2}$

In the eigenbases of $\hat{\rho}_1$ & $\hat{\rho}_2$ the state reads as: $\boxed{|\Psi\rangle = \sum_i \sqrt{\rho_i}\,|\chi_{1i}\rangle|\chi_{2i}\rangle}$

$\boxed{\text{number of terms} = \mathrm{Min}\{\dim\mathcal{H}_1, \dim\mathcal{H}_2\}}$

Example: $\mathcal{H}_1$ basis $\equiv \{|\uparrow\rangle, |\downarrow\rangle\}$ $\quad\mathcal{H}_2$ basis $\equiv \{|1\rangle, |2\rangle, |3\rangle\}$

$$|\psi\rangle = \tfrac{1}{\sqrt{6}}\Big[|\uparrow\rangle_1|1\rangle_2 + |\uparrow\rangle_1|2\rangle_2 + |\uparrow\rangle_1|3\rangle_2 + \sqrt{2}|\downarrow\rangle_1|1\rangle_2 - \tfrac{1}{\sqrt{2}}|\downarrow\rangle_1|2\rangle_2 - \tfrac{1}{\sqrt{2}}|\downarrow\rangle_1|3\rangle_2\Big]$$

$$= \underbrace{\sqrt{\tfrac{1}{2}}}_{\sqrt{\rho_1}}\,\underbrace{|\uparrow\rangle_1}_{|\chi_{11}\rangle}\,\underbrace{\tfrac{1}{\sqrt{3}}\big[|1\rangle_2 + |2\rangle_2 + |3\rangle_2\big]}_{|\chi_{21}\rangle} + \underbrace{\sqrt{\tfrac{1}{2}}}_{\sqrt{\rho_2}}\,\underbrace{|\downarrow\rangle_1}_{|\chi_{12}\rangle}\,\underbrace{\tfrac{1}{\sqrt{3}}\big[\sqrt{2}|1\rangle_2 - \tfrac{1}{\sqrt{2}}|2\rangle_2 - \tfrac{1}{\sqrt{2}}|3\rangle_2\big]}_{|\chi_{22}\rangle}$$

$$\left.\begin{array}{l} \langle\chi_{1i}|\chi_{1i'}\rangle = \delta_{ii'} \\ \langle\chi_{2j}|\chi_{2j'}\rangle = \delta_{jj'} \end{array}\right\} \Rightarrow \left\{\begin{array}{l} \hat{\rho}_1 = \tfrac{1}{2}|\chi_{11}\rangle\langle\chi_{11}| + \tfrac{1}{2}|\chi_{12}\rangle\langle\chi_{12}| \\ \hat{\rho}_2 = \tfrac{1}{2}|\chi_{21}\rangle\langle\chi_{21}| + \tfrac{1}{2}|\chi_{22}\rangle\langle\chi_{22}| + 0|\chi_\perp\rangle\langle\chi_\perp| \end{array}\right.$$

◄ Historical remark

1907: E. Schmidt formulates the decomposition theorem (in theory of integral eqs.)

■ Evolution of density operator

The density operator in general depends on time. The form of this dynamics can be easily deduced from the evolution of individual states in $\mathcal{H}$. However, we come to an essential point here: There is a fundamental difference between the evolutions of density operators for closed and open systems! The density operator of a *closed system* undergoes just a continuous unitary transformation by ordinary evolution operator. This implies a fully reversible picture of dynamics. In contrast, the evolution of a reduced density operator associated with an *open system* is more complicated. Since the environment in general interacts with the system, one cannot write its dynamical equation in an autonomous way (i.e., just in terms of the system's degrees of freedom). This is the place where irreversibility enters the physical description!

▶ **Evolution of a closed system**

Consider density operator in the form given by an initial set of state vectors. Evolution of the density operator determined by evolution of individual vectors:

$$\underset{\text{initial state}}{\hat\rho(0)=\sum_k p_k|\psi_k\rangle\langle\psi_k|}\quad\xrightarrow{t}\quad\underset{\text{evolved state}}{\hat\rho(t)=\sum_k p_k\hat U(t)|\psi_k\rangle\langle\psi_k|\hat U(t)^{-1}}$$

General evolution:

$$\boxed{\hat\rho(t)=\hat U(t)\hat\rho(0)\hat U(t)^{-1}}\quad\leftarrow\text{operator}\quad\text{differential}\rightarrow\quad\boxed{i\hbar\tfrac{d}{dt}\hat\rho(t)=[\hat H,\hat\rho(t)]}$$

forms

▶ Analogy with the classical **Liouville equation** for distribution $\rho(p,q,t)$ in phase space: $\frac{d}{dt}\rho(\vec p,\vec q,t)=\sum_i \frac{\partial\rho}{\partial p_i}\underbrace{\frac{dp_i}{dt}}_{-\frac{\partial H}{\partial q_i}}+\frac{\partial\rho}{\partial q_i}\underbrace{\frac{dq_i}{dt}}_{+\frac{\partial H}{\partial p_i}}+\frac{\partial\rho}{\partial t}=0\quad\Rightarrow\quad\boxed{\frac{\partial\rho}{\partial t}=-\{H,\rho\}}$

▶ Evolution of a closed system does not change traces and entropy

Unitary transformation $\hat\rho(t)=\hat U(t)\hat\rho(0)\hat U(t)^{-1}=\sum_k\rho_i\underbrace{\hat U(t)|\phi_i\rangle}_{|\phi_i(t)\rangle}\underbrace{\langle\phi_i|\hat U(t)^{-1}}_{\langle\phi_i(t)|}$

$$\mathrm{Tr}\,\hat\rho(t)=\mathrm{Tr}\,\hat\rho(0)\quad\Rightarrow\quad\text{normalization conserved}$$

$$\mathrm{Tr}\,\hat\rho(t)^2=\mathrm{Tr}\,\hat\rho(0)^2\quad\Rightarrow\quad\left.\begin{matrix}\text{pure}\\\text{mixed}\end{matrix}\right\}\text{remains}\left\{\begin{matrix}\text{pure}\\\text{mixed}\end{matrix}\right.$$

Eigenvalues ρ_i conserved $\Rightarrow$ **entropy** $\boxed{S_\rho(t)=-k\sum\rho_i\ln\rho_i=S_\rho(0)}$ $=$**const**

▶ **Evolution of open systems: non-interacting case**

Consider first the case when the system under study and its environment do *not* interact with each other. Below we verify that this effectively coincides with the isolated case, as may be immediately anticipated.

$$\boxed{\hat H=\hat H_1\otimes\hat I_2+\hat I_1\otimes\hat H_2}\quad\Rightarrow\quad\text{separable evolution}$$

$$\left.\begin{matrix}\hat U_1(t)=e^{-i\frac{\hat H_1t}{\hbar}}\\\hat U_2(t)=e^{-i\frac{\hat H_2t}{\hbar}}\\\hat U(t)=\hat U_1(t)\otimes\hat U_2(t)\end{matrix}\right\}\Rightarrow\left\{\begin{matrix}\hat\rho_1(t)=\hat U_1(t)\hat\rho_1(0)\hat U_1(t)^{-1}\quad i\hbar\frac{d}{dt}\hat\rho_1(t)=[\hat H_1,\hat\rho_1(t)]\\\hat\rho_2(t)=\hat U_2(t)\hat\rho_2(0)\hat U_2(t)^{-1}\quad i\hbar\frac{d}{dt}\hat\rho_2(t)=[\hat H_2,\hat\rho_2(t)]\\|\Psi(t)\rangle=\sum_i\sqrt{\rho_i}\,\underbrace{\hat U_1(t)|\chi_{1i}\rangle}_{|\chi_{1i}(t)\rangle}\underbrace{\hat U_2(t)|\chi_{2i}\rangle}_{|\chi_{2i}(t)\rangle}\end{matrix}\right.$$

Both **entropies** equal & conserved: $\boxed{S_1(t)=S_2(t)=-k\sum_i \rho_i \ln \rho_i}$ = **const**

Separable states remain separable

▶ Evolution of open systems: interacting case

If the system-environment interaction is turned on, the system's evolution becomes qualitatively different.

$\boxed{\hat{H}=\hat{H}_1\otimes\hat{I}_2+\hat{I}_1\otimes\hat{H}_2+\hat{V}_{12}}$ where $\hat{V}_{12}$ acts irreducibly on $\mathcal{H}\equiv\mathcal{H}_1\otimes\mathcal{H}_2 \Rightarrow$ non-separable evolution of the whole system

$$\hat{U}(t)\neq\hat{U}_1(t)\otimes\hat{U}_2(t) \;\Rightarrow |\Psi(t)\rangle=\sum_{ij}\alpha_{ij}(t)|\phi_{1i}\rangle|\phi_{2j}\rangle=\sum_i\sqrt{\rho_i(t)}\,|\chi_{1i}(t)\rangle|\chi_{2i}(t)\rangle$$

Eigenvalues $\rho_i(t)$ vary $\Rightarrow$ entropy $\boxed{S_1(t)=S_2(t)=-k\sum_i \rho_i(t)\ln\rho_i(t)}\neq$ const

$\Rightarrow$ **non-unitary evolution**

of partial density operators $\hat{\rho}_1(t)$ & $\hat{\rho}_2(t)$ corresponding to both subsystems

▶ Decoherence

Let the [system $\otimes$ environment] evolve from a *separable* [pure $\otimes$ general] initial state at $t{=}0$:

$$\underbrace{|\psi\rangle\langle\psi|_1}_{\hat{\rho}_1(0)}\otimes\underbrace{\hat{\rho}_2(0)}_{\text{may be a pure state }|\tilde{\psi}\rangle\langle\tilde{\psi}|_2}=\hat{\rho}_{12}(0)\xrightarrow{t}\underbrace{\hat{\rho}_{12}(t)\neq\hat{\rho}_1(t)\otimes\hat{\rho}_2(t)}_{\text{unfactorizable}}$$

For the non-separable evolution, $\hat{\rho}_1(t)\equiv\mathrm{Tr}_2\,\hat{\rho}_{12}(t)$ for $t>0$ is most probably a mixed state $\Rightarrow$ loss of the system's initial coherence (purity):

$$\boxed{\text{pure state }\hat{\rho}_1(0)\xrightarrow{t}\hat{\rho}_1(t)\text{ mixed state}}$$

Entropy relations:

$$\underbrace{\overbrace{S_1(0)}^{=0}+\overbrace{S_2(0)}^{\geq 0}+\overbrace{\Delta S(0)}^{=0}}_{S_{12}(0)}=\underbrace{\overbrace{S_1(t)}^{>0}+\overbrace{S_2(t)}^{>0}+\overbrace{\Delta S(t)}^{\neq 0}}_{S_{12}(t)}$$

where the correlation-induced term $\Delta S(t)$ compensates the change of $S_1(t){+}S_2(t)$

The decoherence process results from the system's entanglement with environment, which takes place due to their mutual interaction. An increase of the system's entropy can be interpreted as *spreading of information* (quantum correlations) from the system alone to the composite system + environment. Since mixed states often carry semiclassical properties, decoherence usually induces loss of quantum features and emergence of classical behavior (cf. Chap. 3).

Note: The canonical (micro-canonical, grand-canonical) density operators represent equilibrium states resulting from a "generic" and "long-enough" interaction of the system with a "large-enough" environment. The reason why nature prefers these states is their maximal (under given constraints) entropy.

◀ Historical remark

1970: H.Dieter Zeh introduces the concept of environmentally-induced decoherence
1980's-present: intense research of various aspects of decoherence (W. Zurek *et al.*)

2.7 Examples of statistical description

We will briefly present a few applications of the above-described ideas. It is worth emphasizing here that the density operator is not just a superfluous appendix of the quantum formalism, suitable only in some more or less exotic situations. Strictly speaking, no system of ordinary quantum theory is perfectly isolated. Therefore, the density operator represents the most fundamental language of QM.

■ Harmonic oscillator at nonzero temperature

Let us start with the most familiar system, harmonic oscillator. It will be immersed now into a heat bath with temperature $T > 0$. This example has a great historical importance as it indicates the correct quantum solution of a so-called **specific-heat paradox**—the fact (classically inexplicable) that the specific heat of solids gradually vanishes with the temperature going down to absolute zero. The same calculation, just in slightly different clothes, applies also to the well-known problem of thermal blackbody radiation.

▸ Partition function of a 3D oscillator

Energies: $E_{n_1 n_2 n_3} = \sum_{i=1}^{3} \hbar\omega_i \left(n_i + \tfrac{1}{2}\right) \qquad n_i = 0, 1, 2, \dots$

$$Z(\beta) = \sum_{\left\{\begin{smallmatrix} n_1 \\ n_2 \\ n_3 \end{smallmatrix}\right\}=0}^{\infty} e^{-\beta E_{n_1 n_2 n_3}} = \prod_{i=1}^{3} \Bigg[e^{-\beta \frac{\hbar\omega_i}{2}} \underbrace{\sum_{n_i=0}^{\infty} e^{-\beta\hbar\omega_i n_i}}_{\frac{1}{1-e^{-\beta\hbar\omega_i}}} \Bigg] = \prod_{i=1}^{3} \frac{e^{-\beta\frac{\hbar\omega_i}{2}}}{1-e^{-\beta\hbar\omega_i}} = \prod_{i=1}^{3} \underbrace{\left(e^{+\beta\frac{\hbar\omega_i}{2}} - e^{-\beta\frac{\hbar\omega_i}{2}}\right)^{-1}}_{\left[2\sinh\left(\beta\frac{\hbar\omega_i}{2}\right)\right]^{-1}}$$

$$\boxed{\ln Z(\beta) = -\sum_{i=1}^{3} \ln\left(e^{+\beta\frac{\hbar\omega_i}{2}} - e^{-\beta\frac{\hbar\omega_i}{2}}\right)}$$

▸ Specific heat

Energy average: $\langle E \rangle_\beta = -\frac{d}{d\beta} \ln Z(\beta) = \sum_{i=1}^{3} \frac{\hbar\omega_i}{2} \overbrace{\frac{e^{+\beta\frac{\hbar\omega_i}{2}} + e^{-\beta\frac{\hbar\omega_i}{2}}}{e^{+\beta\frac{\hbar\omega_i}{2}} - e^{-\beta\frac{\hbar\omega_i}{2}}}}^{\coth\frac{\beta\hbar\omega_i}{2}}$

Molar specific heat: $c_V^{\rm mol}(\beta) = N_A k \beta^2 \frac{d^2}{d\beta^2} \ln Z(\beta) = N_A k \sum_{i=1}^{3} \left(\frac{\beta\hbar\omega_i}{e^{+\beta\frac{\hbar\omega_i}{2}} - e^{-\beta\frac{\hbar\omega_i}{2}}} \right)^2$

High-$T \Rightarrow \beta \ll (\hbar\omega_i)^{-1}$

$$\boxed{c_V^{\rm mol}(T) \approx 3N_A k = \text{const}}$$

classical behavior

Low-$T \Rightarrow \beta \gg (\hbar\omega_i)^{-1}$

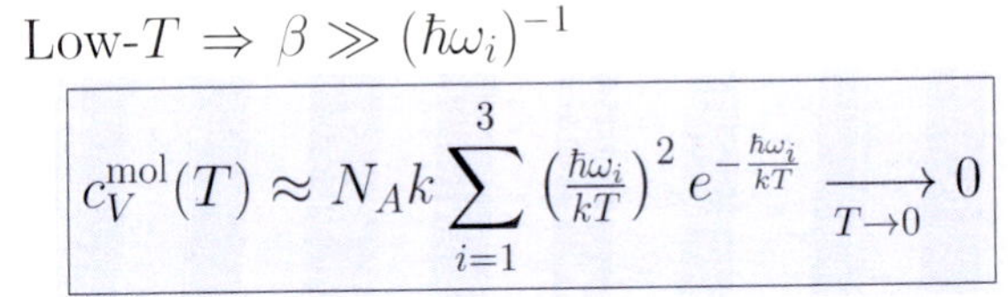

$$c_V^{\rm mol}(T) \approx N_A k \sum_{i=1}^{3} \left(\frac{\hbar\omega_i}{kT}\right)^2 e^{-\frac{\hbar\omega_i}{kT}} \xrightarrow[T\to 0]{} 0$$

quantum behavior

◀ **Historical remark**

1907: A. Einstein derives the specific heat formula for a quantized oscillator

■ Coherent superposition vs. statistical mixture

The following example attempts to clarify the difference between a coherent superposition, which is a pure state composed of some components with the respective *amplitudes*, and a statistical mixture, which is a mixed state containing the same components but just with the corresponding *probabilities*.

▶ **Density operator** of a pure state given by **wavefunction** $\psi(\vec{x}) \equiv \langle\vec{x}|\psi\rangle$

$\hat{\rho} = |\psi\rangle\langle\psi| \;\Rightarrow$ coordinate representation $\boxed{\langle\vec{x}|\hat{\rho}|\vec{x}\,'\rangle = \rho(\vec{x},\vec{x}\,') = \psi(\vec{x})\psi^*(\vec{x}\,')}$

▶ **Coherent superposition** of two wavefunctions

$$\boxed{|\psi\rangle \propto |\psi_{\rm I}\rangle + |\psi_{\rm II}\rangle} \equiv \psi_{\rm I}(\vec{x}) + \psi_{\rm II}(\vec{x})$$

$$\rho(\vec{x},\vec{x}\,') \propto [\psi_{\rm I}(\vec{x}) + \psi_{\rm II}(\vec{x})]\,[\psi_{\rm I}^*(\vec{x}\,') + \psi_{\rm II}^*(\vec{x}\,')]$$

Probability distribution: $\rho(\vec{x},\vec{x}) \propto |\psi_{\rm I}(\vec{x})|^2 + |\psi_{\rm II}(\vec{x})|^2 + \underbrace{2{\rm Re}\,[\psi_{\rm I}(\vec{x})\psi_{\rm II}^*(\vec{x})]}_{\textbf{interference}}$

▶ **Statistical mixture** of the same wavefunctions

$$\boxed{\hat{\rho} = \tfrac{1}{2}|\psi_{\rm I}\rangle\langle\psi_{\rm I}| + \tfrac{1}{2}|\psi_{\rm II}\rangle\langle\psi_{\rm II}|} \qquad \rho(\vec{x},\vec{x}\,') = \tfrac{1}{2}\psi_{\rm I}(\vec{x})\psi_{\rm I}^*(\vec{x}\,') + \tfrac{1}{2}\psi_{\rm II}(\vec{x})\psi_{\rm II}^*(\vec{x}\,')$$

Probability distribution: $\rho(\vec{x},\vec{x}) = \tfrac{1}{2}|\psi_{\rm I}(\vec{x})|^2 + \tfrac{1}{2}|\psi_{\rm II}(\vec{x})|^2 \Rightarrow$ **no interference**

▶ **1D example**

(a) Superposition: $\boxed{\psi(x) \propto \delta_\epsilon(x+a) + \delta_\epsilon(x-a)}$ with $\delta_\epsilon(x) = \frac{1}{(2\pi\epsilon^2)^{\frac{1}{4}}}e^{-\frac{x^2}{4\epsilon^2}}$ $\quad(\epsilon\to 0)$

$$\rho(x,x') \propto \delta_\epsilon(x+a)\delta_\epsilon(x'+a) + \delta_\epsilon(x-a)\delta_\epsilon(x'-a) + \delta_\epsilon(x+a)\delta_\epsilon(x'-a) + \delta_\epsilon(x-a)\delta_\epsilon(x'+a)$$

$$\rho(x,x) \propto \delta_\epsilon^2(x+a) + \delta_\epsilon^2(x-a) + \underbrace{\delta_\epsilon(x+a)\delta_\epsilon(x-a) + \delta_\epsilon(x-a)\delta_\epsilon(x+a)}_{\to 0 \text{ for } \epsilon\to 0}$$

$$W_\rho(x,p) \propto \int_{-\infty}^{+\infty} \rho(x+\tfrac{\xi}{2}, x-\tfrac{\xi}{2})e^{\frac{i}{\hbar}\xi p}d\xi \overset{\epsilon\to 0}{\approx} \delta(x+a) + \delta(x-a) + 2\cos\left(\tfrac{2ap}{\hbar}\right)\delta(x)$$

$\neq$ probability density ($W \gtreqless 0$)

(b) Mixture with the same spatial distribution:

$$\boxed{\begin{aligned}\rho(x,x') = \tfrac{1}{2}\,&\delta_\epsilon(x+a)\delta_\epsilon(x'+a)\\ &+\tfrac{1}{2}\,\delta_\epsilon(x-a)\delta_\epsilon(x'-a)\end{aligned}}$$

$$W_\rho(x,p) \propto \left[\delta_\epsilon^2(x+a) + \delta_\epsilon^2(x-a)\right] e^{-\frac{p^2}{2(\hbar/2\epsilon)^2}}$$

$\equiv$ classical-like probability density

($W \geq 0$)

■ Density operator and decoherence for a two-state system

The rest of this section is devoted to the familiar spin-$\frac{1}{2}$ system—a qubit. The density operator and its evolution can be clearly visualized in this system, yielding an understandable picture of the spin coherence & decoherence.

▶ Parametrization of 2D density matrix

$$\boxed{\hat{\rho} = \tfrac{1}{2}\left[\hat{I} + \vec{b}\cdot\hat{\vec{\sigma}}\right] = \tfrac{1}{2}\begin{pmatrix} 1+b_3 & b_1-ib_2 \\ b_1+ib_2 & 1-b_3 \end{pmatrix}}$$

$\vec{b} \equiv (b_1, b_2, b_3)$ is a vector of parameters $\Rightarrow$ Normalization $\mathrm{Tr}\,\hat{\rho} = 1$ satisfied

$$\mathrm{Tr}\,\hat{\rho}^2 = \tfrac{1}{4}\,\mathrm{Tr}\left[\hat{I} + 2(\vec{b}\cdot\hat{\vec{\sigma}}) + \underbrace{(\vec{b}\cdot\hat{\vec{\sigma}})^2}_{|\vec{b}|^2\hat{I}}\right] = \tfrac{1+|\vec{b}|^2}{2}\begin{cases} = 1 \\ < 1 \end{cases} \text{for}$$

$|\vec{b}|=1$ **pure state**

$|\vec{b}|<1$ **mixed state**

$$(\vec{b}\cdot\hat{\vec{\sigma}})^2 = \tfrac{1}{2}\sum_{ij} b_i b_j \underbrace{(\hat{\sigma}_i\hat{\sigma}_j + \hat{\sigma}_j\hat{\sigma}_i)}_{2\delta_{ij}\hat{I}} = |\vec{b}|^2\hat{I}$$

▶ Spin polarization

The average values of the 3 spin components:

$$\hat{\vec{S}} \equiv \tfrac{\hbar}{2}\hat{\vec{\sigma}} \quad\Rightarrow\quad \langle S_i\rangle_\rho = \mathrm{Tr}\,(\hat{S}_i\hat{\rho}) = \tfrac{\hbar}{4}\,\mathrm{Tr}\left[\hat{\sigma}_i + (\vec{b}\cdot\hat{\vec{\sigma}})\hat{\sigma}_i\right] = \tfrac{\hbar}{4}\sum_j b_j \overbrace{\mathrm{Tr}(\hat{\sigma}_j\hat{\sigma}_i)}^{2\delta_{ij}} = \tfrac{\hbar}{2}b_i$$

Geometric interpretation with the aid of the **Bloch sphere** of the vector $\vec{b} \propto \langle\vec{S}\rangle$ (cf. the visualization of spin pure states in Sec. 2.2)

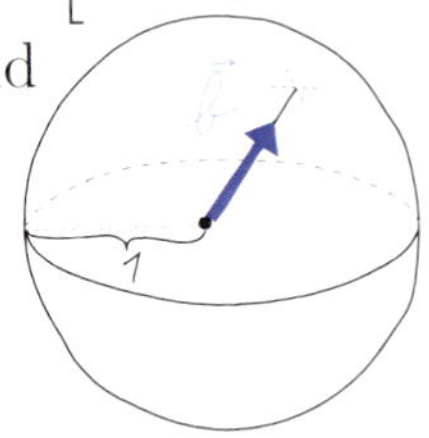

$$\boxed{\langle\vec{S}\,\rangle_\rho = \tfrac{\hbar}{2}\vec{b}}$$

polarization vector

▶ Thermal ensemble

Hamiltonian parametrization: $\boxed{\hat{H} = \hbar\omega_0\hat{I} + \hbar\vec{\omega}\cdot\hat{\vec{\sigma}}}$ with $|\vec{\omega}| \equiv \omega$ (cf. p. 76)

$$e^{-\beta\hat{H}} = e^{-\beta\hbar\omega_0}\sum_{k=0}^{\infty}\frac{(-\beta\hbar\omega)^k}{k!}\left(\frac{\vec{\omega}}{\omega}\cdot\hat{\vec{\sigma}}\right)^k = e^{-\beta\hbar\omega_0}\Bigg[\underbrace{\sum_{k=0,2,4\dots}\frac{(-\beta\hbar\omega)^k}{k!}}_{\frac{e^{+\beta\hbar\omega}+e^{-\beta\hbar\omega}}{2}}\hat{I} + \underbrace{\sum_{k=1,3,5\dots}\frac{(-\beta\hbar\omega)^k}{k!}}_{-\frac{e^{+\beta\hbar\omega}-e^{-\beta\hbar\omega}}{2}}\left(\frac{\vec{\omega}}{\omega}\cdot\hat{\vec{\sigma}}\right)\Bigg]$$

$$= e^{-\beta\hbar\omega_0}\left[\cosh(\beta\hbar\omega)\hat{I} - \sinh(\beta\hbar\omega)\left(\frac{\vec{\omega}}{\omega}\cdot\hat{\vec{\sigma}}\right)\right]$$

$$\mathrm{Tr}\,e^{-\beta\hat{H}} = \boxed{2e^{-\beta\hbar\omega}\cosh(\beta\hbar\omega) = Z(\beta)} \Rightarrow \hat{\rho}_\beta = \frac{1}{Z(\beta)}e^{-\beta\hat{H}} = \frac{1}{2}\left[\hat{I} - \tanh(\beta\hbar\omega)\left(\frac{\vec{\omega}}{\omega}\cdot\hat{\vec{\sigma}}\right)\right]$$

$$\boxed{\vec{b}_\beta = -\tanh(\beta\hbar\omega)\frac{\vec{\omega}}{\omega}} \Rightarrow \langle\vec{S}\rangle_\beta = -\frac{\hbar}{2}\tanh(\beta\hbar\omega)\frac{\vec{\omega}}{\omega}$$

The average spin polarization is oriented in the direction $\left(-\frac{\vec{\omega}}{\omega}\right)$ and increases with $T \to 0$

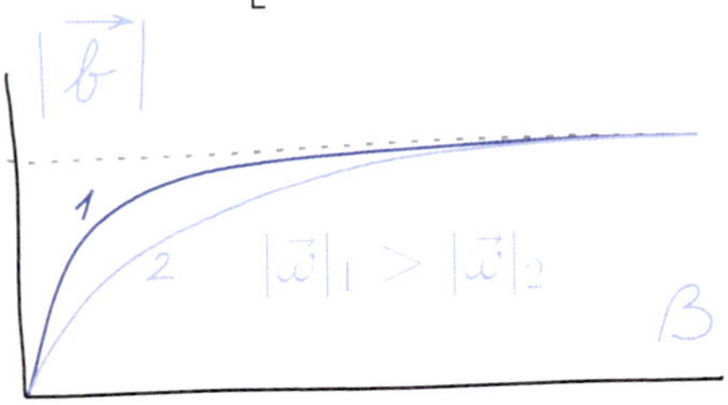

▶ Qubit coupled to environment

Bases in the spin & environment Hilbert spaces: $\overbrace{\{|\uparrow\rangle, |\downarrow\rangle\}}^{\mathcal{H}_1} \otimes \overbrace{\{|e_i\rangle\}_i}^{\mathcal{H}_2}$

Assume
$$\boxed{\begin{aligned} |\uparrow\rangle|e_i\rangle &\xrightarrow{t} |\uparrow\rangle|e_{i\uparrow}(t)\rangle \\ |\downarrow\rangle|e_i\rangle &\xrightarrow{t} |\downarrow\rangle|e_{i\downarrow}(t)\rangle \end{aligned}}$$

where $|e_{i\uparrow}(t)\rangle, |e_{i\downarrow}(t)\rangle \equiv$ some states $\in \mathcal{H}_2$. This defines a **special evolution** which does not affect the z component of spin.

Separable initial state:

$$\hat\rho_{12}(0)=\underbrace{|\psi\rangle\langle\psi|}_{\hat\rho_1(0)}\otimes\underbrace{\left(\sum_i w_i|e_i\rangle\langle e_i|\right)}_{\hat\rho_2(0)}$$

$$|\psi\rangle=\alpha|\uparrow\rangle+\beta|\downarrow\rangle \qquad \hat\rho_1(0)=\begin{pmatrix}|\alpha|^2 & \alpha\beta^* \\ \alpha^*\beta & |\beta|^2\end{pmatrix}$$

Evolution:

$$\hat\rho_{12}(t)=\sum_i w_i\big[|\alpha|^2|e_{i\uparrow}(t)\rangle|\uparrow\rangle\langle\uparrow|\langle e_{i\uparrow}(t)|+\alpha\beta^*|e_{i\uparrow}(t)\rangle|\uparrow\rangle\langle\downarrow|\langle e_{i\downarrow}(t)|+ \alpha^*\beta|e_{i\downarrow}(t)\rangle|\downarrow\rangle\langle\uparrow|\langle e_{i\uparrow}(t)|+|\beta|^2|e_{i\downarrow}(t)\rangle|\downarrow\rangle\langle\downarrow|\langle e_{i\downarrow}(t)|\big]$$

$$\hat\rho_1(t)=\mathrm{Tr}_2\hat\rho_{12}(t)=$$

$$|\alpha|^2|\uparrow\rangle\langle\uparrow|\underbrace{\Big[\sum_{ij}w_i\langle e_j|e_{i\uparrow}(t)\rangle\langle e_{i\uparrow}(t)|e_j\rangle\Big]}_{1}+\alpha\beta^*|\uparrow\rangle\langle\downarrow|\underbrace{\Big[\sum_{ij}w_i\langle e_j|e_{i\uparrow}(t)\rangle\langle e_{i\downarrow}(t)|e_j\rangle\Big]}_{\sum_i w_i\langle e_{i\downarrow}(t)|e_{i\uparrow}(t)\rangle\equiv\ D(t)}$$

$$+\alpha^*\beta|\downarrow\rangle\langle\uparrow|\underbrace{\Big[\sum_{ij}w_i\langle e_j|e_{i\downarrow}(t)\rangle\langle e_{i\uparrow}(t)|e_j\rangle\Big]}_{\sum_i w_i\langle e_{i\uparrow}(t)|e_{i\downarrow}(t)\rangle\equiv\ D(t)^*}+|\beta|^2|\downarrow\rangle\langle\downarrow|\underbrace{\Big[\sum_{ij}w_i\langle e_j|e_{i\downarrow}(t)\rangle\langle e_{i\downarrow}(t)|e_j\rangle\Big]}_{1}$$

$$\hat\rho_1(t)=\begin{pmatrix}|\alpha|^2 & \alpha\beta^*D(t)\\ \alpha^*\beta D(t)^* & |\beta|^2\end{pmatrix}$$

where $|D(t)|\leq\sum_i w_i\overbrace{|\langle e_{i\downarrow}(t)|e_{i\uparrow}(t)\rangle|}^{\leq 1}\leq 1$

► Spin decoherence

$$\mathrm{Tr}\,\hat\rho_1(t)^2=\mathrm{Tr}\begin{pmatrix}|\alpha|^4+|\alpha|^2|\beta|^2|D(t)|^2 & (|\alpha|^2+|\beta|^2)\alpha\beta^*D(t)\\ (|\alpha|^2+|\beta|^2)\alpha^*\beta D(t)^* & |\alpha|^2|\beta|^2|D(t)|^2+|\beta|^4\end{pmatrix}=|\alpha|^4+2|\alpha|^2|\beta|^2|D(t)|^2+|\beta|^4$$

$$=\underbrace{(|\alpha|^2+|\beta|^2)^2}_{1}-\underbrace{\big[1-|D(t)|^2\big]}_{\in[0,1]}\underbrace{2|\alpha|^2|\beta|^2}_{\in[0,1]}\begin{cases}=1 & \text{for } |D(t)|=1 \ \text{ or } \ \alpha\beta=0\\ <1 & \text{for } |D(t)|<1 \text{ and } \alpha\beta\neq 0\end{cases}$$

For a large environment, $|D(t)|$ is usually a very *quickly decreasing* function $\Rightarrow$

$$\left.\begin{matrix}\text{pure state}\\ |\psi\rangle=\alpha|\uparrow\rangle+\beta|\downarrow\rangle\end{matrix}\right\}\xrightarrow{t}\left\{\begin{matrix}\text{mixed state, for } t\to\infty:\\ \hat\rho_1=|\alpha|^2|\uparrow\rangle\langle\uparrow|+|\beta|^2|\downarrow\rangle\langle\downarrow|\end{matrix}\right.$$

qubit's decoherence

Parametrization: $D(t)=|D(t)|e^{i\chi(t)}$

Spin initially along direction $\vec n$: $\quad|\psi\rangle\equiv|s_{\vec n}=+\frac{\hbar}{2}\rangle\equiv\overbrace{e^{-i\varphi}\cos\frac{\vartheta}{2}}^{\alpha}|\uparrow\rangle+\overbrace{\sin\frac{\vartheta}{2}}^{\beta}|\downarrow\rangle$

Evolution of **polarization** vector:

$$\vec b(t)=\begin{pmatrix}|D(t)|\sin\vartheta\cos[\varphi-\chi(t)]\\ |D(t)|\sin\vartheta\sin[\varphi-\chi(t)]\\ \cos\vartheta\end{pmatrix}$$

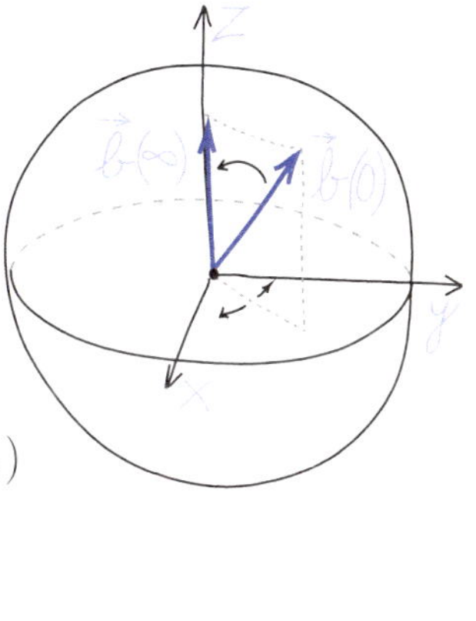

$$\Rightarrow\qquad \vec b(0)=(n_1,n_2,n_3)\ \stackrel{t\to\infty}{\longrightarrow}\ (0,0,n_3)=\vec b(\infty)$$

$\Rightarrow$ **dephasing** of the xy-projection of polarization (e.g., due to multiple Larmor freqs. in mag. field $\vec B\propto\vec e_z$)

3. QUANTUM-CLASSICAL CORRESPONDENCE

We may say that rough construction of the QM formalism (Sec. 1) and demonstration of its basic applications (Sec. 2) have been completed now. Before proceeding to more complex applications, we wish to explore the land on the border between quantum and classical physics. Quite surprisingly, one often finds here a rather inaccessible and hardly passable terrain. This is also the reason why the quantum-classical correspondence belongs to the most interesting topics in physics.

3.1 Classical limit of quantum mechanics

Once a new theory is formulated, an immediate task is to specify the circumstances under which the old theory is reproduced.

■ The singular limit $\hbar \to 0$

Physical theories—like various objects in the mathematical world—may be subject to a limiting procedure: variation of an essential constant of the theory to the limit in which another theory takes the reins. A well-known example is the limit $c \to \infty$ (or $\frac{v}{c} \to 0$), in which special relativity changes to classical mechanics, or $N \to \infty$, when statistical physics becomes thermodynamics. We are now interested in the limit $\hbar \to 0$ (or $\frac{\Delta S}{\hbar} \to \infty$). In this limit, quantum mechanics should peacefully crossover to classical mechanics. However, this process turns out to be rather tricky. The reason for difficulties is that quantum mechanics is apparently richer than classical mechanics, so a number of *emergent phenomena* appears on the quantum side of the border line.**

▶ Example I: harmonic-oscillator eigenstates

Classical motion $x(t) = \underbrace{x_{\max}(E)}_{\sqrt{\frac{2E}{M\omega^2}}} \sin\omega t$ with period $T = \frac{2\pi}{\omega}$

Probability to find the oscillator at position x in random time: $\frac{2}{T}\underbrace{\left|\frac{dt}{dx}\right|}_{1/|\dot{x}|}dx = \frac{1}{\pi}\frac{1}{x_{\max}(E)|\cos\omega t|}dx$

Probability density $\boxed{\rho_{\rm clas}(x)_E \equiv \frac{1}{\pi}\frac{1}{\sqrt{x_{\max}(E)^2 - x^2}}}$

What is the link to $\boxed{\rho_{\rm quant}(x)_{E_n} \equiv |\psi_n(x)|^2}$?

For $\hbar \to 0$ & $E_n = $const we get $n \to \infty$
$\Rightarrow$ infinitely dense oscillations of $\psi_n(x)$

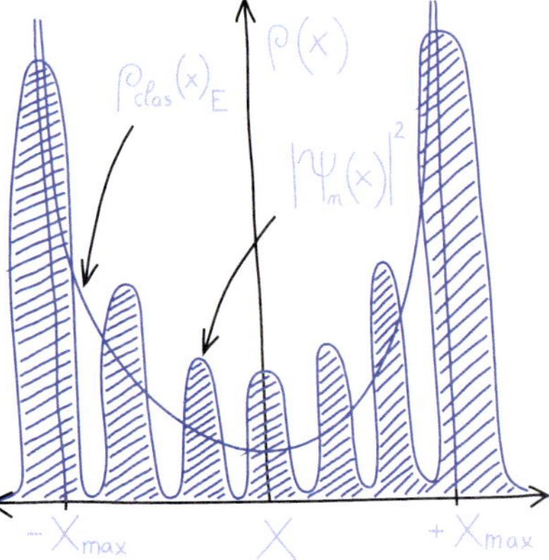

**This paragraph is formulated in the spirit of some of the writings by Michael Berry.

It turns out that
the smoothened distribution $\bar{\rho}_{\rm quant}(x)_{E_n} \equiv \frac{1}{dx}\int\limits_{x-\frac{dx}{2}}^{x+\frac{dx}{2}} |\psi_n(x')|^2 dx' \xrightarrow{n\to\infty} \rho_{\rm clas}(x)_E$

Therefore, the limit $\hbar \to 0$ reproduces the classical case only if the smoothening of $|\psi_n(x)|^2$ is performed along with the limiting procedure.

▶ **Example II: coefficient of transmission** through a potential barrier

Square barrier of width a and height V_0: parameter $\gamma \equiv a\sqrt{2MV_0}/\hbar$
Transmission coefficient for particle with energy $\boxed{E = \epsilon V_0}$

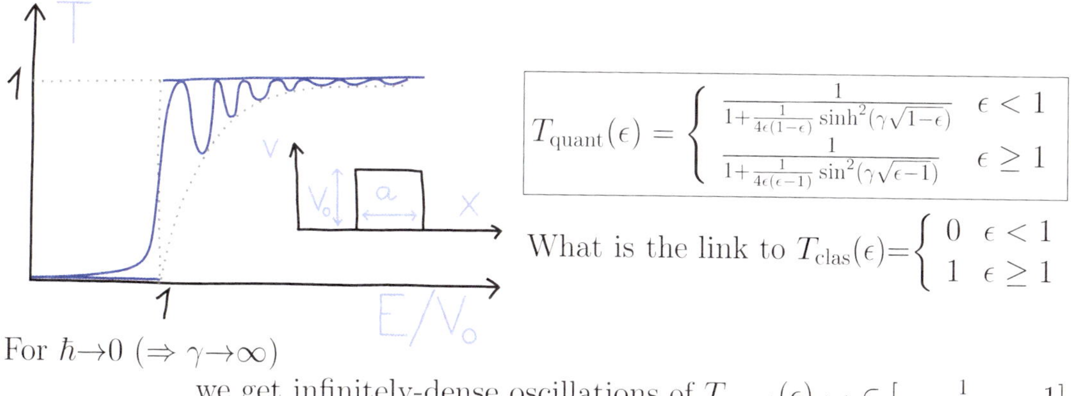

$$T_{\rm quant}(\epsilon) = \begin{cases} \frac{1}{1+\frac{1}{4\epsilon(1-\epsilon)}\sinh^2(\gamma\sqrt{1-\epsilon})} & \epsilon < 1 \\ \frac{1}{1+\frac{1}{4\epsilon(\epsilon-1)}\sin^2(\gamma\sqrt{\epsilon-1})} & \epsilon \geq 1 \end{cases}$$

What is the link to $T_{\rm clas}(\epsilon) = \begin{cases} 0 & \epsilon < 1 \\ 1 & \epsilon \geq 1 \end{cases}$

For $\hbar \to 0$ ($\Rightarrow \gamma \to \infty$)
we get infinitely-dense oscillations of $T_{\rm quant}(\epsilon)_{\epsilon\geq 1} \in [\frac{1}{1+1/4\epsilon(\epsilon-1)}, 1]$

Only a smoothened coefficient $\bar{T}_{\rm quant}(\epsilon) \equiv \frac{1}{d\epsilon}\int\limits_{\epsilon-\frac{d\epsilon}{2}}^{\epsilon+\frac{d\epsilon}{2}} T_{\rm quant}(\epsilon')d\epsilon' \xrightarrow{\hbar\to 0} T_{\rm clas}(\epsilon)$

◀ **Historical remark**

1913: N. Bohr discusses the quantum-classical correspondence within the “old QM”
1920’s-present: research of various aspects of quasiclassical quantum mechanics
1980’s-90’s: M. Berry points out the singularity of the $\hbar \to 0$ limit

■ Ehrenfest theorem

The question of quantum-classical correspondence was in the main focus of quantum theorists already in the early days of QM. An important result by Ehrenfest is often presented as the final answer to this question, although in reality it represents just a *beginning* of a still-unknown answer.

▶ **Derivation**

Consider a single spinless particle with Hamiltonian: $\hat{H} = \frac{1}{2M}\hat{\vec{p}}^2 + V(\hat{\vec{x}})$

Time **evolution of operators** in **Heisenberg representation**:

$$\left.\begin{array}{l} \frac{d}{dt}\hat{p}_i = \frac{1}{i\hbar}[\hat{p}_i, \hat{H}] = \frac{1}{i\hbar}[\hat{p}_i, V(\hat{\vec{x}})] = -\frac{\partial V}{\partial x_i}(\hat{\vec{x}}) \\ \frac{d}{dt}\hat{x}_i = \frac{1}{i\hbar}[\hat{x}_i, \hat{H}] = \frac{1}{i\hbar}[\hat{x}_i, \frac{1}{2M}\hat{p}_i^2] = \frac{1}{M}\hat{p}_i \end{array}\right\} \Rightarrow \left\{\boxed{\begin{array}{l} \frac{d}{dt}\hat{\vec{p}} = -\vec{\nabla}V(\hat{\vec{x}}) \\ \frac{d}{dt}\hat{\vec{x}} = \frac{1}{M}\hat{\vec{p}} \end{array}}\right.$$

$$\frac{d^2}{dt^2}\hat{x}_i = \frac{d}{dt}\left(\frac{1}{M}\hat{p}_i\right) = -\frac{1}{M}\frac{\partial V}{\partial x_i}(\hat{\vec{x}}) \Rightarrow \boxed{M\frac{d^2}{dt^2}\hat{\vec{x}} = -\vec{\nabla}V(\hat{\vec{x}})}$$ “**quantum Newton law**”

▶ Consequences

Consider an arbitrary state $|\psi_{\rm H}\rangle = |\psi_{\rm S}(t{=}0)\rangle$, e.g., a narrow wavepacket. Coordinate averages $\langle x_i(t)\rangle_\psi = \langle\psi_{\rm S}(t)|\hat{x}_{i{\rm S}}|\psi_{\rm S}(t)\rangle = \langle\psi_{\rm H}|\hat{x}_i(t)_{\rm H}|\psi_{\rm H}\rangle$ evolve in accord with an averaged Newton law. In particular: $\boxed{M\frac{d^2}{dt^2}\langle\vec{x}(t)\rangle_\psi = -\langle\vec{\nabla}V(\hat{\vec{x}})\rangle_\psi}$

⇒ **semiclassical behavior** can be obtained for convenient initial states (cf. coherent states in harmonic oscillator)

◀ **Historical remark**
1927: P. Ehrenfest formulates the relation between quantum and classical dynamics

▶ Limits of applicability

Spreading of wavepacket (in almost all potentials) ⇒ The semiclassical behavior terminated at a sufficiently long time $t \gtrsim \tau_{\rm Q}$ when the variation of the force across the wavepacket spread ≈ the force average: $\boxed{\sqrt{\langle\langle F_i^2\rangle\rangle_\psi} \approx \langle F_i\rangle_\psi}$

$$\Rightarrow \frac{\sqrt{\langle\langle F_i^2\rangle\rangle_\psi}}{\langle F_i\rangle_\psi} \approx \frac{{\rm Max}_j\left\langle\left|\frac{\partial^2 V}{\partial x_j x_i}\right|\right\rangle_\psi \Delta x_j}{\left\langle\left|\frac{\partial V}{\partial x_i}\right|\right\rangle_\psi} \quad \text{where } \Delta x_j \equiv \sigma_{x_j}(t) \text{ is spreading width along } x_j$$

"Quantum time" $\tau_{\rm Q}$ can be estimated from: ${\rm Max}_j\left\langle\left|\frac{\partial^2 V}{\partial x_j x_i}\right|\right\rangle \sigma_{x_j}(\tau_{\rm Q}) \approx \left\langle\left|\frac{\partial V}{\partial x_i}\right|\right\rangle$

Phase-space criterion for $\tau_{\rm Q}$: Consider $t{=}0$ state represented by Wigner function $W(\vec{x},\vec{p},0)$ in the form of a classical distribution $\rho(\vec{x},\vec{p},0)$ with the support $\mathcal{S}_\rho(0)$ being a simple compact phase-space domain of volume $\Omega_\rho(0)$. Semiclassical approximation holds if $W(\vec{x},\vec{p},t)$ evolved by the quantum dynamical equation coincides with $\rho(\vec{x},\vec{p},t)$ evolved by the classical Liouville equation.

The classical *volume* $\Omega_\rho(t){=}\Omega_\rho(0)$ is conserved but the *shape* of $\mathcal{S}_\rho(t)$ usually becomes more and more complicated [its maximal linear size grows typically as $L_\rho(t) \approx L_\rho(0)e^{t/\tau_{\rm chaos}}$, where $\tau_{\rm chaos}$ characterizes sensitivity of the system's evolution to initial conditions]. Semiclassical behavior is terminated when the **fine structures** of $\mathcal{S}_\rho(t)$ become of the size of "**elementary cells**" $\sim \hbar^f$ (f = number of degrees of freedom) deduced from the uncertainty principle. At about this time scale, $W(\vec{x},\vec{p},t)$ becomes partly negative ⇒ non-classical.

Schematic illustration: The classical phase-space domain $\mathcal{S}_\rho(t)$ becomes complicated and starts to interfere with cells $\sim \hbar^f$

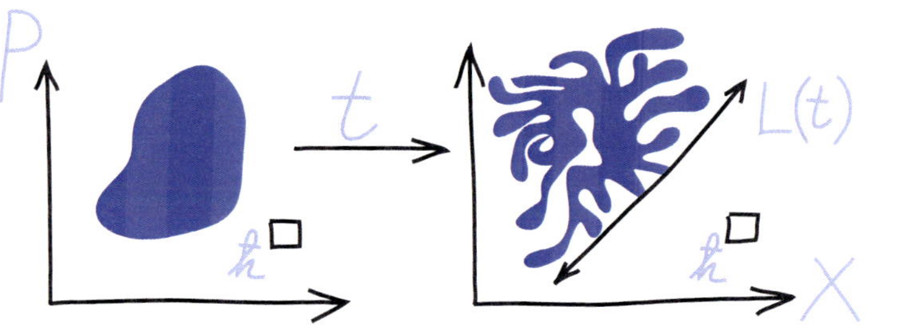

■ Role of decoherence

The process in which a quantum system loses its coherence due to an interaction with some "environment" is a hot candidate for the ultimate answer to the question of quantum-classical correspondence. Indeed, even if the system of interest is well isolated from the surrounding objects, its interaction with omnipresent matter (relict radiation, solar photons, dark matter etc.) or with some internal degrees of freedom is most likely out of control. Such interactions often make the system behave in accord with classical physics.

► **Classical behavior emerging due to interaction with environment**

The reduced density operator of the system interacting with "environment" evolves in a non-unitary way. In *generic case*, for $t \gtrsim t_{\rm decoh}$, a pure state $\hat\rho(0)$ becomes a mixed state $\hat\rho(t)$. This usually has the following consequences for the **Wigner phase-space distribution** function:

(a) *non-classical* $W(\vec{x},\vec{p},0) \lesseqgtr 0 \xrightarrow{t\gtrsim t_{\rm decoh}} W(\vec{x},\vec{p},t) \gtrsim 0$ *classical-like*
(b) *classical-like* $W(\vec{x},\vec{p},0) \geq 0 \xrightarrow{t\in[0,\infty)} W(\vec{x},\vec{p},t) \gtrsim 0$ *classical-like*

For classical-like initial states the permanent decoherence is likely to preserve the classical-quantum correspondence for $t\to\infty$ ($\Rightarrow$ no problem with $\tau_{\rm Q}$).

► **Quantum measurement as decoherence?** (tentative interpretation)

The decoherence process may be essential for a physical explanation of quantum measurement. Consider the following scheme based on the coupling:

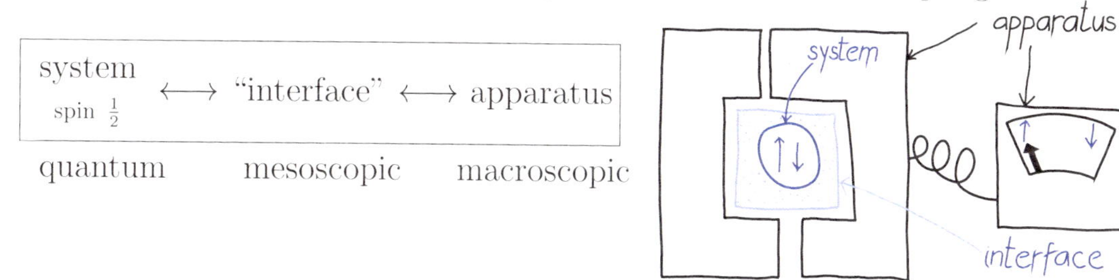

(a) A "pre-measurement" ≡ system-interface interaction

Interface, initially in state $|I_0\rangle$, responds to the spin states as follows: $\boxed{\begin{array}{l}|\!\uparrow\rangle|I_0\rangle\mapsto|\!\uparrow\rangle|I_\uparrow\rangle\\|\!\downarrow\rangle|I_0\rangle\mapsto|\!\downarrow\rangle|I_\downarrow\rangle\end{array}}$

where $|I_\uparrow\rangle$ and $|I_\downarrow\rangle$ are interface states that are *almost orthogonal* (because of the interface's high sensitivity to the system's spin states and a large dimension of the interface's Hilbert space): $\boxed{\langle I_\uparrow|I_\downarrow\rangle = \epsilon \approx 0}$

Unitary evolution of a general [system ⊗ interface] state:

$$|\Psi_0\rangle \equiv \big(\alpha|\!\uparrow\rangle + \beta|\!\downarrow\rangle\big)\otimes|I_0\rangle \xrightarrow{\rm (a)} \big(\alpha|\!\uparrow\rangle\otimes|I_\uparrow\rangle + \beta|\!\downarrow\rangle\otimes|I_\downarrow\rangle\big) \equiv |\Psi_1\rangle$$

Suppose that the evolution of the apparatus depends only on the interface, not on the spin itself, hence evaluate $\hat\rho_{{\rm int}\,i} = {\rm Tr}_{\rm sys}|\Psi_i\rangle\langle\Psi_i|$:

$$\hat\rho_{\rm int0} = |I_0\rangle\langle I_0| \xrightarrow{\rm (a)} |\alpha|^2|I_\uparrow\rangle\langle I_\uparrow| + |\beta|^2|I_\downarrow\rangle\langle I_\downarrow| = \hat\rho_{\rm int1}$$

$\hat\rho_{\rm int1}$ is the interface's mixed state, which effectively describes its *collapsed* wave-function *before* reading out the result $|I_\uparrow\rangle$ or $|I_\downarrow\rangle$ (with a precision determined by a residual overlap ϵ of the two interface states).

(b) The actual measurement ≡ interface-apparatus interaction
The apparatus, initially in a mixed state $\hat{\rho}_{\rm app0}$, responds to the relevant interface states as follows: $\boxed{\begin{matrix}|I_\uparrow\rangle\langle I_\uparrow|\otimes\hat{\rho}_{\rm app0}\mapsto|I_\uparrow\rangle\langle I_\uparrow|\otimes\hat{\rho}_{\rm app\uparrow}\\ |I_\downarrow\rangle\langle I_\downarrow|\otimes\hat{\rho}_{\rm app0}\mapsto|I_\downarrow\rangle\langle I_\downarrow|\otimes\hat{\rho}_{\rm app\downarrow}\end{matrix}}$ where $\hat{\rho}_{\rm app\uparrow}$ and $\hat{\rho}_{\rm app\downarrow}$ are macroscopic "pointer states" which are classical-like ($W\gtrsim 0$) and almost perfectly distinguishable: ${\rm Tr}\,\hat{\rho}_{\rm app\downarrow}\hat{\rho}_{\rm app\uparrow}\approx 0$ (to the extent in which $\langle I_\uparrow|I_\downarrow\rangle\approx 0$).
Unitary evolution of the [interface ⊗ apparatus] state from step (a):
$\left(|\alpha|^2|I_\uparrow\rangle\langle I_\uparrow|+|\beta|^2|I_\downarrow\rangle\langle I_\downarrow|\right)\otimes\hat{\rho}_{\rm app0}\xrightarrow{\rm (b)}|\alpha|^2|I_\uparrow\rangle\langle I_\uparrow|\otimes\hat{\rho}_{\rm app\uparrow}+|\beta|^2|I_\downarrow\rangle\langle I_\downarrow|\otimes\hat{\rho}_{\rm app\downarrow}$
The final reduced density operator of the apparatus $\boxed{\hat{\rho}_{\rm app1}\approx|\alpha|^2\hat{\rho}_{\rm app\uparrow}+|\beta|^2\hat{\rho}_{\rm app\downarrow}}$
This mixed state (mixture of two mixed states) describes two alternative distinguishable classical-like pointer states & their respective probabilities.

(c) The *role of observer* is to *select* the single alternative which "actually happens". This final reduction might be beyond the reach of physical description.

◀ **Historical remark**
1970-80's: H.D. Zeh and W. Zurek consider environmentally-induced decoherence as an effective mechanism for the wavefunction collapse
1990's-present: Examples of decoherence-based quantum-to-classical transitions

3.2 WKB approximation

Not only that the quantum-classical correspondence represents a problem of fundamental importance, its investigation also yields an effective approximation method. The acronym WKB associated with this method stands for Wentzel, Kramers, and Brillouin, who were among its first independent inventors.

■ Classical Hamilton-Jacobi theory

Classical mechanics can also be formulated in a wave form. The appearance of classical trajectories in this formulation is quite analogous to the way in which rays of light arise from wave optics. Before we derive the WKB approximation of QM, we have to outline this classical theory.

▶ Action as a function of coordinates and time

Action for a structureless particle ≡ functional on the space of trajectories $\vec{x}(t)$:

$$\boxed{S[\vec{x}(t)]_{t_0}^{t_1}=\int_{t_0}^{t_1}\mathcal{L}[\vec{x}(t),\dot{\vec{x}}(t)]dt}\qquad \mathcal{L}(\vec{x},\dot{\vec{x}})=\tfrac{M}{2}\dot{\vec{x}}^2-V(\vec{x})\equiv\text{Lagrangian}$$

For a fixed initial point $\vec{x}(t_0)=\vec{x}_0$ and a fixed final point $\vec{x}(t_1)=\vec{x}_1$ the classical equations of motion select the trajectory $\vec{x}_c(t)$ satisfying the variational principle $\delta S[\vec{x}_c(t)]_{t_0}^{t_1}=0$

Consider a bunch of classical trajectories $\{\vec{x}_c(t)\}$ (satisfying $\delta S = 0$) leading from a *fixed initial point* $(\vec{x}_0, t_0)$ to *variable final point* $(\vec{x}_1, t_1)$.

Action along these trajectories:

$$S(\vec{x}_1, t_1) = \int_{t_0}^{t_1} \mathcal{L}[\vec{x}_c(t), \dot{\vec{x}}_c(t)]dt$$

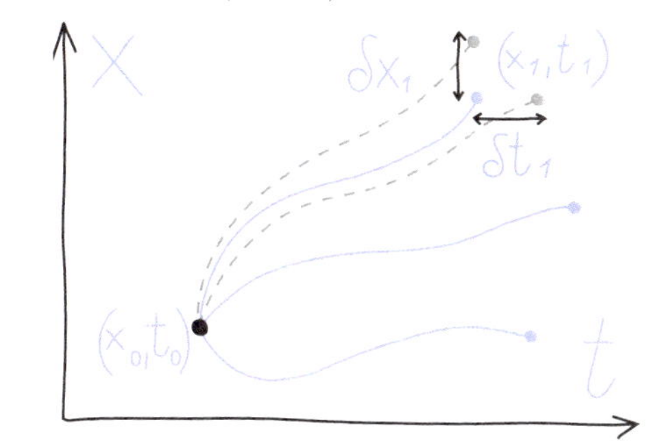

▶ **Equations for the action**

(a) Space variation $(\vec{x}_1, t_1) \to (\vec{x}_1 + \delta\vec{x}_1, t_1) \Rightarrow S \to (S + \delta S)$

$$\delta S = \int_{t_0}^{t_1} \Big(\frac{\partial \mathcal{L}}{\partial x_i}\delta x_i + \underbrace{\frac{\partial \mathcal{L}}{\partial \dot{x}_i}\delta \dot{x}_i}_{\frac{d}{dt}(\frac{\partial \mathcal{L}}{\partial \dot{x}_i}\delta x_i) - (\frac{d}{dt}\frac{\partial \mathcal{L}}{\partial \dot{x}_i})\delta x_i}\Big)dt = \underbrace{\left[\frac{\partial \mathcal{L}}{\partial \dot{x}_i}\delta x_i\right]_{t_0}^{t_1}}_{\frac{\partial \mathcal{L}}{\partial \dot{x}_i}\delta x_{1i}} + \int_{t_0}^{t_1} \underbrace{\left[\frac{\partial \mathcal{L}}{\partial x_i}\delta x_i - (\frac{d}{dt}\frac{\partial \mathcal{L}}{\partial \dot{x}_i})\delta x_i\right]}_{0} dt$$

$$\Rightarrow \frac{\partial S(\vec{x}_1, t_1)}{\partial x_{1i}} = \underbrace{\frac{\partial \mathcal{L}(\vec{x}, \dot{\vec{x}})}{\partial \dot{x}_i}}_{p_i}\Bigg|_{\substack{\vec{x}=\vec{x}_c(t_1)\\ \dot{\vec{x}}=\dot{\vec{x}}_c(t_1)}} \Rightarrow \boxed{\vec{\nabla}_{\vec{x}_1} S(\vec{x}_1, t_1) = \vec{p}_1}$$

(b) Time variation $(\vec{x}_1, t_1) \to (\vec{x}_1, t_1 + \delta t_1)$

$$\underbrace{\frac{dS}{dt_1}}_{\mathcal{L}[\vec{x}_c(t_1), \dot{\vec{x}}_c(t_1)]} = \frac{\partial S}{\partial t_1} + \underbrace{\frac{\partial S}{\partial x_{1i}}}_{p_{1i}} \underbrace{\dot{x}_{1i}}_{\dot{x}_{ci}(t_1)} \Rightarrow \frac{\partial S}{\partial t_1} = \underbrace{\left[\mathcal{L} - \vec{p}\cdot\dot{\vec{x}}\right]_{t=t_1}}_{-H(\vec{x}_1, \vec{p}_1, t_1)} \Rightarrow \boxed{\frac{\partial S(\vec{x}_1, t_1)}{\partial t_1} = -H(\vec{x}_1, \vec{p}_1, t_1)}$$

(c) Both equations together $\Rightarrow$ **Hamilton-Jacobi equation**

$$\frac{\partial}{\partial t_1} S(\vec{x}_1, t_1) + H[\vec{x}_1, \vec{\nabla}_{\vec{x}_1} S(\vec{x}_1, t_1), t_1] = 0 \quad \text{or shortly:} \quad \boxed{\frac{\partial}{\partial t} S + H(\vec{x}, \vec{\nabla} S, t) = 0}$$

This is a partial differential equation for $S(\vec{x}, t)$

Example: **particle in scalar potential**: $\boxed{\frac{\partial}{\partial t} S + \frac{1}{2M}(\vec{\nabla} S)^2 + V(\vec{x}) = 0}$

▶ **Time-independent Hamiltonian** $\Rightarrow$ energy conserved: $H = E = \text{const}$

$$\frac{\partial}{\partial t} S = -E \quad \Rightarrow \quad \boxed{S(\vec{x}, t) = W(\vec{x}) - Et} \quad \Rightarrow \quad H(\vec{x}, \vec{\nabla} W) = E$$

The generating function $W(\vec{x})$ can be determined as follows:

$$\vec{\nabla} W = \vec{p} \quad \Rightarrow \boxed{W(\vec{x}) = \int_{\vec{x}_0}^{\vec{x}} \vec{p} \cdot d\vec{x}'}$$

contour integral along a classical trajectory from arbitrary initial point $\vec{x}_0$

▶ **Interpretation**

$S(\vec{x}, t) \equiv$ solution Hamilton-Jacobi equation $\Rightarrow (\vec{p} = \vec{\nabla} S) \perp$ surfaces $S(\vec{x}, t) = \text{const}$
$\Rightarrow$ classical trajectories are like the **rays** associated with a "wave" $S(\vec{x}, t)$

◀ **Historical remark**
1827-37: W.R. Hamilton, C.G. Jacobi introduce ray formulation of class. mechanics

■ WKB equations, classical limit, pilot wave

We now jump into the derivation of the WKB equations. Soon, a link to the Hamilton-Jacobi theory will become apparent. We will then come to a branching point: one of the paths leads to an alternative formulation of QM in terms of some non-classical trajectories, the other (postponed to the next paragraph) to the quasiclassical approximation of standard QM.

▶ **Derivation** of WKB equations for single particle in a potential

Rewrite the Shrödinger equation $\boxed{\left[-\frac{\hbar^2}{2M}(\vec{\nabla})^2 + V(\vec{x})\right]\psi(\vec{x},t) = i\hbar\frac{\partial}{\partial t}\psi(\vec{x},t)}$

with substitution $\boxed{\psi(\vec{x},t) = \sqrt{\rho(\vec{x},t)}e^{\frac{i}{\hbar}S(\vec{x},t)}}$

$$-\frac{\hbar^2}{2M}\left[\Delta\sqrt{\rho} + \frac{2i}{\hbar}(\vec{\nabla}\sqrt{\rho})\cdot(\vec{\nabla}S) + \frac{i}{\hbar}\sqrt{\rho}\Delta S - \frac{1}{\hbar^2}\sqrt{\rho}(\vec{\nabla}S)^2\right]e^{\frac{i}{\hbar}S} + V\sqrt{\rho}e^{\frac{i}{\hbar}S} = i\hbar\left[\frac{\partial\sqrt{\rho}}{\partial t} + \frac{i}{\hbar}\sqrt{\rho}\frac{\partial S}{\partial t}\right]e^{\frac{i}{\hbar}S}$$

$$\text{Separate}\begin{cases}\text{Re part}: & -\frac{\hbar^2}{2M}\Delta\sqrt{\rho} + \frac{1}{2M}\sqrt{\rho}(\vec{\nabla}S)^2 + V\sqrt{\rho} = -\sqrt{\rho}\frac{\partial S}{\partial t}\\ \text{Im part}: & -\frac{\hbar}{M}(\vec{\nabla}\sqrt{\rho})\cdot(\vec{\nabla}S) - \frac{\hbar}{2M}\sqrt{\rho}\Delta S = \hbar\frac{\partial\sqrt{\rho}}{\partial t}\end{cases}$$

$$2\sqrt{\rho}\times\text{Im part} \Rightarrow \quad \frac{\partial\rho}{\partial t} + \frac{1}{M}\underbrace{\left[\rho\Delta S + (\vec{\nabla}\rho)\cdot(\vec{\nabla}S)\right]}_{\vec{\nabla}\cdot(\rho\vec{\nabla}S)} = 0 \quad \Rightarrow \quad \boxed{\frac{\partial\rho}{\partial t} + \vec{\nabla}\cdot\underbrace{\left(\rho\frac{\vec{\nabla}S}{M}\right)}_{\vec{j}=\rho\vec{v}} = 0}$$

continuity equation

Re part $\Rightarrow$

$$\boxed{\underbrace{-\frac{\hbar^2}{2M}\frac{1}{\sqrt{\rho}}\Delta\sqrt{\rho}}_{*} + \frac{1}{2M}(\vec{\nabla}S)^2 + V + \frac{\partial S}{\partial t} = 0}$$

Hamilton-Jacobi equation
+quantum correction $* \propto \hbar^2$

▶ **Classical limit**

Limit $\hbar \to 0 \Rightarrow$ the quantum correction term $\boxed{* \to 0}$ $\Rightarrow$ we obtain a coupled pair of classical equations:
(a) Hamilton-Jacobi equation for $S(\vec{x},t) \Rightarrow$ velocity field $\vec{v}(\vec{x},t) \equiv \frac{1}{M}\vec{\nabla}S(\vec{x},t)$
(b) continuity equation for $\rho(\vec{x},t)$, given $\vec{v}(\vec{x},t)$ determined in step (a)

These equations describe an **ensemble of classical particles** with initial space density $\rho(\vec{x},0)$ evolving in agreement with classical equations of motion.

▶ **"Pilot-wave" picture of QM**

In a general case ($\hbar \neq 0$) the quantum correction term $* \not\equiv 0$ may be considered as an addition to potential $V(\vec{x})$ ⇒ WKB equations interpreted in terms of classical-like motions of an ensemble of particles in total potential

$$V_{\rm tot}(\vec{x},t) = V(\vec{x}) \underbrace{- \frac{\hbar^2}{2M}\frac{1}{\sqrt{\rho}}\Delta\sqrt{\rho}}_{V_{\rm Q}(\vec{x},t)}$$

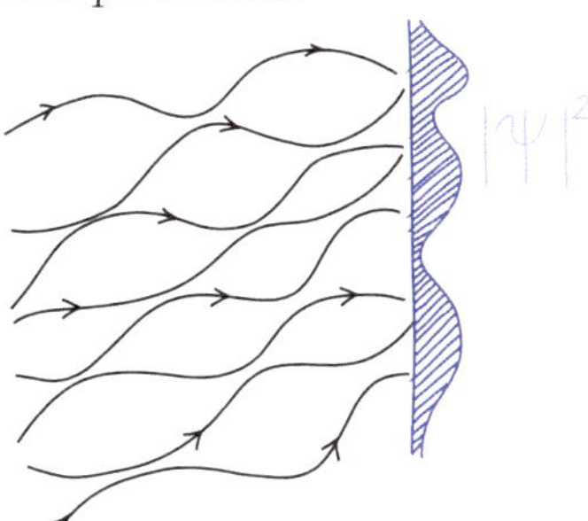

"Quantum potential" $V_{\rm Q}(\vec{x},t)$ depends on $|\psi(\vec{x},t)|^2$
⇒ force acting at places with *no classical field*
⇒ $\psi(\vec{x},t)$ plays the role of a "pilot wave" which "navigates" individual particle trajectories
⇒ interference patterns appear without abandoning the concept of trajectories

However, $V_{\rm Q}$ is a strange field (*not* an interaction with other particles of the ensemble ⇐ acts even for 1 particle) which turns out to have explicitly **non-local** character (⇒ non-local hidden-variable theory equivalent to QM).

Example: Gaussian wavepacket of free particle: $\rho(x,t) = \frac{1}{\sqrt{2\pi\sigma_x(t)^2}} e^{-\frac{[x-x_0(t)]^2}{2\sigma_x(t)^2}}$

$$\Rightarrow V_{\rm Q}(x,t) = \frac{\hbar^2}{4M\sigma_x(t)^2}\left\{1 - \frac{[x-x_0(t)]^2}{2\sigma_x(t)^2}\right\}$$

Force increasing with the wavepacket localization
⇒ consistent with the spreading process

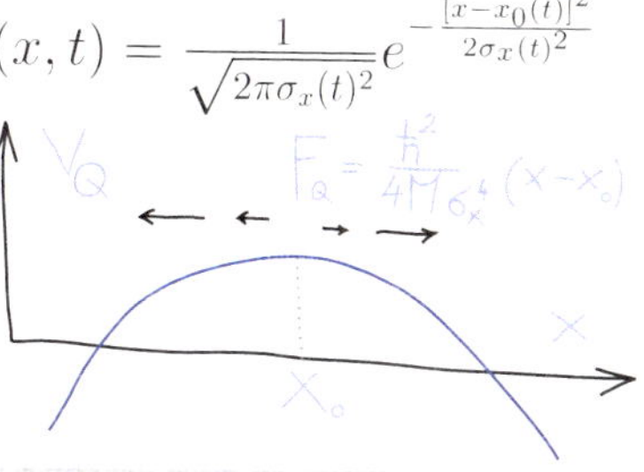

◀ **Historical remark**

1837, 1915, 1923: General foundations of the method (theory of dif. equations) elaborated by J. Liouville, G. Green, lord Rayleigh, H. Jeffreys *et al.*
1926: G. Wentzel, H.A. Kramers, L. Brillouin develop WKB for Schrödinger eq.
1927: Louis de Broglie formulates the basis of the pilot wave theory
1952: David Bohm uses the idea to formulate a hidden-variable alternative to QM

■ Quasiclassical approximation

Now we follow the path leading to quasiclassical QM. This will give us practical approximate expressions of quantum energies and wavefunctions, as well as some more insight into the quantum-classical correspondence.

▶ Conditions for use

The quasiclassical approximation follows from the **neglect of the quantum term** in the WKB equation:

$$\boxed{* = -\frac{\hbar^2}{2M}\frac{1}{\sqrt{\rho}}\Delta\sqrt{\rho}} = \frac{\hbar^2}{2M}\left[(\vec{\nabla}\ln\sqrt{\rho})^2 - \vec{\nabla}^2\ln\sqrt{\rho}\right] \sim \mathcal{O}(\hbar^2) \overset{?}{\ll} \text{terms} \sim \begin{cases}\mathcal{O}(\hbar^0)\\ \mathcal{O}(\hbar^1)\end{cases}$$

To derive the conditions, under which this step is justified, we proceed by **analogy**, comparing:

$$\left.\begin{array}{l}\text{(a) } \hbar^0(\vec{\nabla}S)^2\\ \text{(b) } \hbar^1\Delta S\end{array}\right\} \text{terms from} \begin{Bmatrix}\text{Re}\\ \text{Im}\end{Bmatrix} \text{parts of 1D WKB eqns.}$$

$$\boxed{S(x,t) = \pm \int_{x_0}^{x} \sqrt{2M[E - V(x')]}dx' - Et}$$

$$\Rightarrow \frac{\hbar \left|\frac{\partial^2 S}{\partial x^2}\right|}{\left(\frac{\partial S}{\partial x}\right)^2} = \frac{M\frac{dV}{dx}\frac{\hbar}{\sqrt{2M[E-V(x)]}}}{2M[E-V(x)]} = \frac{\frac{dV}{dx}\lambda_{\rm B}}{4\pi[E-V(x)]} = \frac{1}{4\pi}\frac{\Delta V|_{\Delta x\approx\lambda_{\rm B}}}{T_{\rm kin}} \overset{?}{\ll} 1$$

This condition is usually satisfied for sufficiently **high energy** E **except**:

(a) "wild" potentials $V(x)$ (with $\left|\frac{dV}{dx}\right|$ very large)

(b) regions near **return points** (where $\frac{dV}{dx}\neq 0$ and $T_{\rm kin}\to 0$)

(c) regions near generic **stationary points** (where $\frac{dV}{dx}, T_{\rm kin}\to 0, \frac{dV/dx}{T_{\rm kin}}\to\infty$)

► **Stationary states in 1D**

$$\underbrace{\frac{\partial\rho}{\partial t}}_{0} + \frac{1}{M}\frac{\partial}{\partial x}\left(\rho\frac{\partial S}{\partial x}\right) = 0 \quad \Rightarrow \quad \rho(x)\frac{\partial S(x)}{\partial x} = \text{const}$$

$$\Rightarrow \boxed{\rho(x) \propto \frac{1}{\sqrt{2M[E-V(x)]}} \propto \frac{1}{|v_{\rm clas}(x)|}} \quad \text{in the classical domain } V(x) < E$$

Quasiclassical wavefunction on both sides of class. turning point x_0 with $V(x_0)=E$:

$$\psi_{\rm I}(x,t) = \frac{\pm\mathcal{N}}{\left(2M[E-V(x)]\right)^{1/4}} e^{\pm\frac{i}{\hbar}\int_{x_0}^{x}\sqrt{2M[E-V(x')]}dx' - \frac{i}{\hbar}Et} \quad \text{for } V(x) < E \quad \text{(region I)}$$

$$\psi_{\rm II}(x,t) = \frac{\pm\mathcal{N}}{\left(2M[V(x)-E]\right)^{1/4}} e^{\pm\frac{1}{\hbar}\int_{x}^{x_0}\sqrt{2M[V(x')-E]}dx' - \frac{i}{\hbar}Et} \quad \text{for } V(x) > E \quad \text{(region II)}$$

How to connect these solutions at x_0 where $\psi\to\infty$? Bypassing x_0 from II to I in the complex plane $x\in\mathbb{C}$ along a half-circle with radius ε

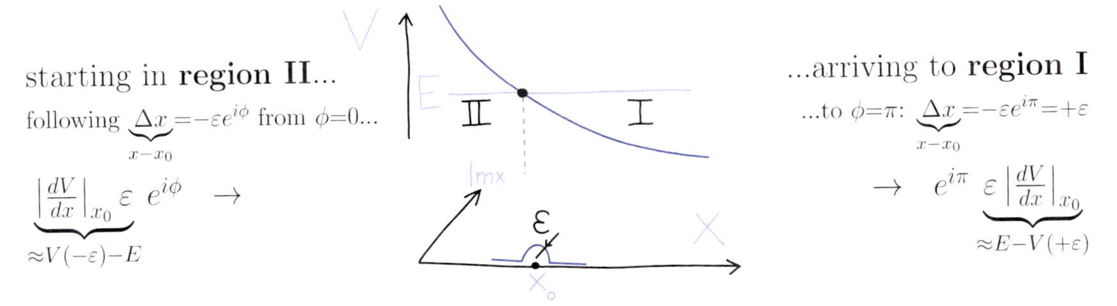

starting in **region II**...

following $\underbrace{\Delta x}_{x-x_0} = -\varepsilon e^{i\phi}$ from $\phi=0$...

$$\underbrace{\left|\frac{dV}{dx}\right|_{x_0}\varepsilon}_{\approx V(-\varepsilon)-E} e^{i\phi} \quad \rightarrow$$

...arriving to **region I**

...to $\phi=\pi$: $\underbrace{\Delta x}_{x-x_0} = -\varepsilon e^{i\pi} = +\varepsilon$

$$\rightarrow \quad e^{i\pi} \underbrace{\varepsilon\left|\frac{dV}{dx}\right|_{x_0}}_{\approx E-V(+\varepsilon)}$$

$\Rightarrow$ the wavefunction prefactor, exponent and whole ψ change as follows:

$$\left.\begin{aligned} \frac{\mathcal{N}}{\left(2M[V(x)-E)]\right)^{1/4}}\Bigg|_{x_0-\varepsilon} &\approx \frac{\mathcal{N}}{\left(2M\left|\frac{dV}{dx}\right|_{x_0}\varepsilon e^{i0}\right)^{1/4}} \\ \int_{x_0-\varepsilon}^{x_0}\sqrt{2M[V(x')-E]}dx' &\approx 0 \\ \psi_{\rm II}(x,t)\Big|_{x_0-\varepsilon} & \end{aligned}\right\} \text{II} \mathrel{\widehat{=}} \text{I} \left\{\begin{aligned} & \frac{\mathcal{N}}{\left(2M\left|\frac{dV}{dx}\right|_{x_0}\varepsilon e^{i\pi}\right)^{1/4}} \approx \frac{e^{-i\pi/4}\,\mathcal{N}}{\left(2M[E-V(x)]\right)^{1/4}}\Bigg|_{x_0+\varepsilon} \\ & 0 \approx \int_{x_0}^{x_0+\varepsilon}\sqrt{2M[E-V(x')]}dx' \\ & e^{-i\pi/4}\,\psi_{\rm I}(x,t)\Big|_{x_0+\varepsilon} \end{aligned}\right.$$

$\Rightarrow \psi_{\mathrm{I}}(x,t)$ given above receives an additional $\boxed{\text{phase factor } e^{-i\pi/4}}$

▶ Bound states in a 1D potential well

2 classical return points inside the well:

II	x_{01}	I	x_{02}	II′
forbidden	↑	allowed	↑	forbidden

Wavefunction in the allowed region can be connected to the left or right forbidden region II or II′:

$$\psi_{\mathrm{I}}(x) = \begin{cases} \pm\mathcal{N}\ (2M[E-V(x)])^{-1/4}\ e^{i\left[+\frac{1}{\hbar}\int_{x_{01}}^{x}\sqrt{2M[E-V(x')]}dx'-\frac{\pi}{4}\right]} & \text{using left return point } x_{01} \\ \pm\mathcal{N}\ (2M[E-V(x)])^{-1/4}\ e^{i\left[-\frac{1}{\hbar}\int_{x}^{x_{02}}\sqrt{2M[E-V(x')]}dx'+\frac{\pi}{4}\right]} & \text{using right return point } x_{02} \end{cases}$$

Consistency condition:

$$\left[+\frac{1}{\hbar}\int_{x_{01}}^{x}\sqrt{2M[E-V(x')]}dx'-\frac{\pi}{4}\right]-\left[-\frac{1}{\hbar}\int_{x}^{x_{02}}\sqrt{2M[E-V(x')]}dx'+\frac{\pi}{4}\right]=\pm n\pi_{\ n=0,1,2,3,\dots}$$

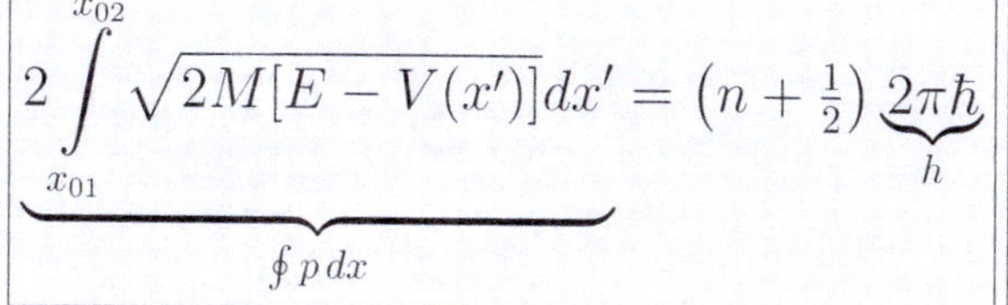

$$\underbrace{2\int_{x_{01}}^{x_{02}}\sqrt{2M[E-V(x')]}dx'}_{\oint p\,dx} = (n+\tfrac{1}{2})\underbrace{2\pi\hbar}_{h}$$

Bohr – Sommerfeld energy quantization (derived in old QM without the $\frac{1}{2}$ term)

▶ Examples (in which the WKB energies reproduce the exact QM results)

(a) 1D **harmonic oscillator**

$\frac{1}{2M}p^2+\frac{M\omega^2}{2}x^2=E \quad\Rightarrow\quad$ ellipse $\left(\frac{x}{a}\right)^2+\left(\frac{p}{b}\right)^2=1$ with area $S=\pi a\,b\equiv\oint p\,dx$

$=\pi\sqrt{\frac{2E}{M\omega^2}}\sqrt{2ME}=\left(n+\frac{1}{2}\right)h$

$\Rightarrow\quad E=\left(n+\frac{1}{2}\right)\hbar\omega$

(b) 1D **infinite well**

No access to region II

$\Rightarrow$ consistency condition reads as: $+\frac{1}{\hbar}\int_{x_{01}}^{x}\sqrt{2ME}dx'+\frac{1}{\hbar}\int_{x}^{x_{02}}\sqrt{2ME}dx'=k\pi$

with $k=1,2,3,\dots\ \Rightarrow\quad \oint p\,dx=2\sqrt{2ME}\ L=kh \quad\Rightarrow\quad E=\frac{(\pi\hbar)^2}{2ML^2}k^2$

▶ Transmission through a 1D barrier

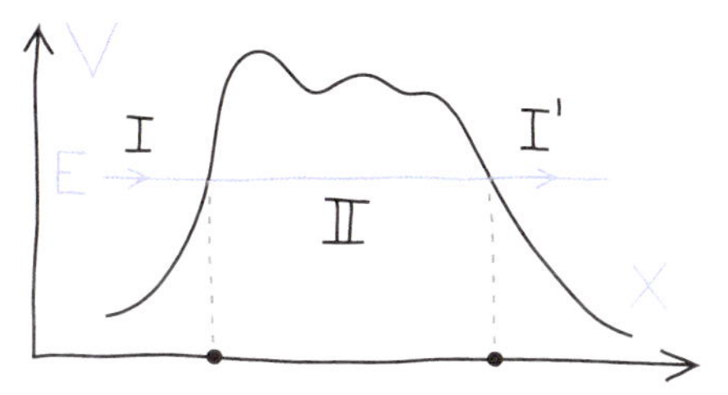

2 return points on both barrier sides:

I	x_{01}	II	x_{02}	I′
allowed	↑	forbidden	↑	allowed

Procedure: Assume single exponential in region I′ propagating to the right (flow $j_{\mathrm{I'}\rightarrow}$). Calculate solutions in regions II and I, determine the incoming flow $j_{\mathrm{I}\rightarrow}$.

WKB approximation of the transmission coefficient:

$$T_{\rm WKB} = \frac{|j_{\rm I'\to}|}{|j_{\rm I\to}|} \approx e^{-\frac{2}{\hbar}\int_{x_{01}}^{x_{02}} \sqrt{2M[V(x)-E]}dx}$$

◀ **Historical remark**
1913: Niels Bohr proposed a model of hydrogen based on semiclassical quantization
1919: Arnold Sommerfeld elaborates the semiclassical quantization ⇒ old QM
1924: George Gamow uses the WKB transmission coeff. to explain nuclear α-decay

3.3 Feynman integral

The method of path integration, called after its inventor Richard Feynman, represents an original reformulation of quantum theory which is tightly connected with the classical theory. The method elucidates the link between quantum and classical, but it also serves as a powerful computational tool for some more advanced problems of quantum theory. Here we just take a taste of this brilliant approach.

■ Formulation of quantum mechanics in terms of trajectories

When the classical trajectories of particles were replaced by quantum wavefunctions, we might believe that trajectories became irretrievably outmoded. Now they return with all their glory.

▶ **Infinitesimal single-particle propagator**

Free-particle propagator: $G_0[(\vec{x}+\Delta\vec{x})(t+\Delta t)|\vec{x}t] = \left(\frac{M}{2i\pi\hbar\Delta t}\right)^{\frac{3}{2}} e^{\frac{i}{\hbar}\overbrace{\frac{M}{2}\frac{(\Delta\vec{x})^2}{\Delta t}}^{\mathcal{L}_0\left(\vec{x},\frac{\Delta\vec{x}}{\Delta t}\right)\Delta t}}$

Infinitesimal $\Delta t\to 0$ propagator of **particle in potential** $V(\vec{x})$:

$\mathcal{L}\left(\vec{x},\frac{\Delta\vec{x}}{\Delta t}\right) = \frac{M}{2}\left(\frac{\Delta\vec{x}}{\Delta t}\right)^2 - V(\vec{x})$

$$G[(\vec{x}+\Delta\vec{x})(t+\Delta t)|\vec{x}t] = \left(\frac{M}{2i\pi\hbar\Delta t}\right)^{\frac{3}{2}} e^{\frac{i}{\hbar}\overbrace{\mathcal{L}\left(\vec{x},\frac{\Delta\vec{x}}{\Delta t}\right)\Delta t}^{dS}}$$

$$= G_0[(\vec{x}+\Delta\vec{x})(t+\Delta t)|\vec{x}t]\, e^{-\frac{i}{\hbar}V(\vec{x})\Delta t} \approx \left(\frac{M}{2i\pi\hbar\Delta t}\right)^{\frac{3}{2}} e^{\frac{i}{\hbar}\frac{M}{2}\frac{(\Delta\vec{x})^2}{\Delta t}}\left[1-\tfrac{i}{\hbar}V(\vec{x})\Delta t\right]$$

It must be so since the $\Delta t\to 0$ limit of evolution operator factorizes:

$$\hat{U}(\Delta t) = e^{-\frac{i}{\hbar}\left[-\frac{\hbar^2}{2M}\vec{\nabla}^2+V(\vec{x})\right]\Delta t} \approx \underbrace{e^{-\frac{i}{\hbar}\left[-\frac{\hbar^2}{2M}\vec{\nabla}^2\right]\Delta t}}_{\hat{U}_0(\Delta t)} e^{-\frac{i}{\hbar}V(\vec{x})\Delta t} \quad \text{(from } \left[\hat{T}\Delta t,\hat{V}\Delta t\right]\sim\mathcal{O}(\Delta t^2)\to 0)$$

▶ **Finite single-particle propagator**

Heisenberg representation: $G[\vec{x}t|\vec{x}_0t_0] \equiv \langle\vec{x}t|\vec{x}_0t_0\rangle$ $\quad|\vec{x}t\rangle \equiv$ eigenvector of $\hat{\vec{X}}_{\rm H}(t)$

$$G[\vec{x}t|\vec{x}_0t_0] = \int \overbrace{\langle\vec{x}t|\vec{x}_1t_1\rangle}^{G[\vec{x}t|\vec{x}_1t_1]}\overbrace{\langle\vec{x}_1t_1|\vec{x}_0t_0\rangle}^{G[\vec{x}_1t_1|\vec{x}_0t_0]}\, d\vec{x}_1 =$$

$$\int\!\dots\!\int \underbrace{G[\vec{x}t|\vec{x}_nt_n]}_{\left(\frac{M}{2i\pi\hbar\Delta t_n}\right)^{\frac{3}{2}}e^{\frac{i}{\hbar}\mathcal{L}\left(\vec{x}_n,\frac{\Delta\vec{x}_n}{\Delta t_n}\right)\Delta t_n}} \dots \underbrace{G[\vec{x}_{k+1}t_{k+1}|\vec{x}_kt_k]}_{\left(\frac{M}{2i\pi\hbar\Delta t_k}\right)^{\frac{3}{2}}e^{\frac{i}{\hbar}\mathcal{L}\left(\vec{x}_k,\frac{\Delta\vec{x}_k}{\Delta t_k}\right)\Delta t_k}} \dots \underbrace{G[\vec{x}_1t_1|\vec{x}_0t_0]}_{\left(\frac{M}{2i\pi\hbar\Delta t_0}\right)^{\frac{3}{2}}e^{\frac{i}{\hbar}\mathcal{L}\left(\vec{x}_0,\frac{\Delta\vec{x}_0}{\Delta t_0}\right)\Delta t_0}}\, d\vec{x}_n\dots d\vec{x}_k\dots d\vec{x}_1$$

Assume $\Delta t_k \equiv \Delta t = \frac{t-t_0}{n+1}$

$$\Rightarrow G[\vec{x}t|\vec{x}_0 t_0] = \overbrace{\int \ldots \int d\vec{x}_n \ldots d\vec{x}_1 \left[\frac{M}{2i\pi\hbar(\Delta t)}\right]^{\frac{3}{2}(n+1)}}^{\xrightarrow{n\to\infty} \int \mathcal{D}[\vec{x}(t)]} e^{\frac{i}{\hbar} \overbrace{\sum_{k=0}^{n} \mathcal{L}\left(\vec{x}_k, \frac{\vec{x}_{k+1}-\vec{x}_k}{\Delta t}\right)\Delta t}^{\xrightarrow{n\to\infty} \int \mathcal{L}\left(\vec{x},\dot{\vec{x}}\right) dt \;\equiv\; S[\vec{x}(t)]}}$$

$(\vec{x}',t') \equiv (\vec{x}_{n+1}, t_{n+1})$

Path integral:

$$\boxed{G[\vec{x}t|\vec{x}_0 t_0] = \int \mathcal{D}[\vec{x}(t)]\; e^{\frac{i}{\hbar}S[\vec{x}(t)]}}$$

$\equiv$ functional integral over the space of **all possible trajectories** $\vec{x}(t)$ satisfying $\vec{x}(t_0) = \vec{x}_0 \longrightarrow \vec{x} = \vec{x}(t)$

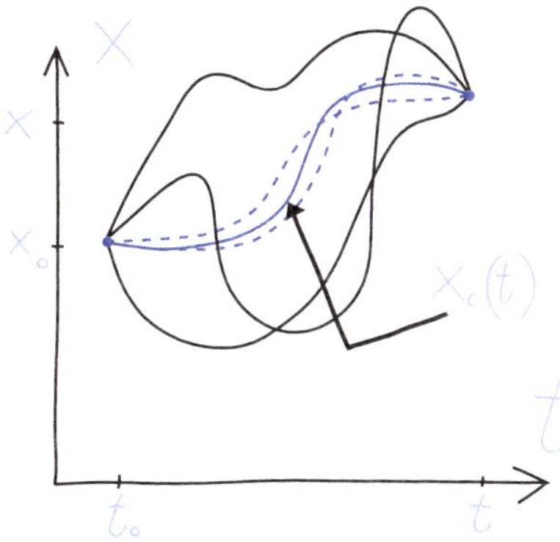

▶ **Classical correspondence**

The contribution to the functional integral is most significant for trajectories in a vicinity of the **classical trajectory** $\vec{x}_c(t)$, for which $\boxed{\delta S = 0}$ (these trajectories contribute "in phase" while the others tend to cancel each other)

Example: **free particle**

$$G_0[\vec{x}t|\vec{x}_0 t_0] = \left[\frac{M}{2\pi i\hbar(t-t_0)}\right]^{\frac{3}{2}} e^{\frac{i}{\hbar}\overbrace{\frac{M}{2}\frac{(\vec{x}-\vec{x}_0)^2}{t-t_0}}^{S_0[\vec{x}_c(t)]}} \Rightarrow \text{non-classical trajectories canceled out!}$$

For $\hbar \to 0$, this is true for any potential: $G[\vec{x}t|\vec{x}_0 t_0] \propto e^{\frac{i}{\hbar}S[\vec{x}_c(t)]}$

◀ **Historical remark**

1948: Richard Feynman derives the path-integral formulation of QM

■ Application I: Double-slit interference

To illustrate the method of path integration, we return to the double-slit experiment (see Introduction). A bonus of the present treatment will be an elegant explanation of the so-called Aharonov-Bohm effect—the fact that magnetic field confined in a compact domain between the slits causes a shift of the interference pattern irrespective of the particle absence in the field domain. Let us stress that the calculations in this paragraph are rather schematic.

▶ **Path-integral formulation** of the double-slit interference

$$\text{Emitter } \vec{x}_0 \equiv (\underbrace{-s}_{\to -\infty}, 0) \longrightarrow \text{Slits} \left\{ \begin{array}{l} \vec{x}_A \equiv (0, +\frac{d}{2}) \\ \vec{x}_B \equiv (0, -\frac{d}{2}) \end{array} \right\} \longrightarrow \text{Screen } \vec{x} \equiv (l, y)$$

Assume the initial state ($t \to -\infty$) a Gaussian wavepacket with average momentum $\vec{p} = (Mv, 0)$ and width $\sqrt{\sigma_p} \equiv \Delta p \ll p \quad \Rightarrow \quad$ on the slit plane we get a $\sim$planar wave with $\sim$sharp wavelength $\lambda_B = \frac{h}{Mv}$

Trajectories divided to disjunct subsets $\{\vec{x}_{\rm A}(t)\}$ & $\{\vec{x}_{\rm B}(t)\}$ passing slits A & B

$$G[\vec{x}t|\vec{x}_0t_0] = \int \mathcal{D}[\vec{x}_{\rm A}(t)]\; e^{\frac{i}{\hbar}S[\vec{x}_{\rm A}(t)]} + \int \mathcal{D}[\vec{x}_{\rm B}(t)]\; e^{\frac{i}{\hbar}S[\vec{x}_{\rm B}(t)]}$$

Only the classical trajectories contribute to the (almost) free propagation:

$$G[\vec{x}t|\vec{x}_0t_0] \propto \left[e^{\frac{i}{\hbar}S_{\rm A}} + e^{\frac{i}{\hbar}S_{\rm B}}\right] = e^{\frac{i}{\hbar}\frac{S_{\rm A}+S_{\rm B}}{2}}\left[e^{+\frac{i}{\hbar}\frac{S_{\rm A}-S_{\rm B}}{2}} + e^{-\frac{i}{\hbar}\frac{S_{\rm A}-S_{\rm B}}{2}}\right] \propto \cos\frac{S_{\rm A}-S_{\rm B}}{2\hbar}$$

$$\frac{S_{\rm A}-S_{\rm B}}{2\hbar} = \underbrace{\frac{M}{2\hbar}\frac{v_{\rm A}+v_{\rm B}}{2}}_{\frac{\pi}{\lambda_{\rm B}}}\;\underbrace{\Delta{\rm path}}_{\approx\frac{y}{l}d} \;\Rightarrow\; \boxed{\rho_{\rm scr}(y)\propto\cos^2\left(\frac{\pi}{\lambda_{\rm B}}\frac{d}{l}y\right)} \;\Rightarrow\; \Delta y = \frac{l}{d}\lambda_{\rm B}$$

interval between two minima/maxima

Classical limit $\Rightarrow$ $\Delta y\to 0$ (local averaging needed)

▶ Aharonov-Bohm effect

Consider an ideal electric coil placed in between both slits A & B. The coil is oriented perpendicularly to the plane defined by emitter & both slits, with the section area S. Magnetic flux $\Phi = B_\perp S$ is confined inside the coil. The area S can be made arbitrarily small and the coil can be shielded against the passage of particles—in this case the particles have no chance to experience the field $B_\perp$. In spite of this, the field has an influence on the interference pattern:

In general, vector potential $\vec{A}(\vec{x})\neq 0$ even where the field induction $\vec{B}(\vec{x})=0$.

For a cylindrical coil of radius R: $\vec{A}(\vec{x}) = \begin{cases} \frac{1}{2}Br\vec{e}_\varphi & r<R \text{ (region of } B\neq 0) \\ \frac{1}{2}BR^2\frac{1}{r}\vec{e}_\varphi & r\geq R \text{ (region of } B=0)\end{cases}$

Lagrangian of a charged particle: $\boxed{\mathcal{L}(\vec{x},\dot{\vec{x}}) \longrightarrow \mathcal{L}(\vec{x},\dot{\vec{x}}) + q\dot{\vec{x}}\cdot\vec{A}(\vec{x})}$

$$G[\vec{x}t|\vec{x}_0t_0] \propto \left[e^{\frac{i}{\hbar}\left(S_{\rm A}^{(0)}+q\int_{\rm A}\vec{v}_{\rm A}\cdot\vec{A}_{\rm A}dt\right)} + e^{\frac{i}{\hbar}\left(S_{\rm B}^{(0)}+q\int_{\rm B}\vec{v}_{\rm B}\cdot\vec{A}_{\rm B}dt\right)}\right] \propto \cos\left[\frac{1}{\hbar}\left(S_{\rm A}^{(0)}-S_{\rm B}^{(0)}+\frac{q\Phi}{2}\right)\right]$$

where $S_{\rm A}^{(0)}, S_{\rm B}^{(0)}$ are actions for zero field, and use was made of the relation:

$$\int_{\rm A}\vec{v}_{\rm A}\cdot\vec{A}_{\rm A}\,dt - \int_{\rm B}\vec{v}_{\rm B}\cdot\vec{A}_{\rm B}\,dt = \oint_{\rm AB}\vec{A}\cdot d\vec{x} = \Phi$$

$$\Rightarrow \boxed{\rho_{\rm scr}(y)\propto\cos^2\left(\frac{\pi}{\lambda_{\rm B}}\frac{d}{l}\,y + \frac{q\Phi}{2\hbar}\right)}$$

Interference pattern shifted although the particle may *never enter* the region $B\neq 0$

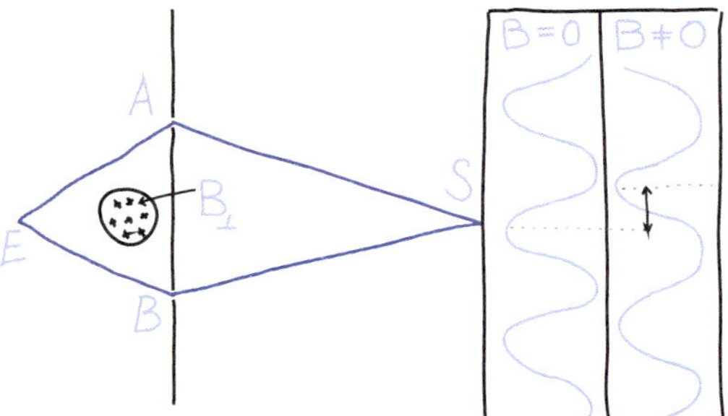

◀ Historical remark

1959: Yakir Aharonov & David Bohm discover the effect of elmg. potentials in QM

■ Application II: Quasiclassical approximation of quantum state density

This application concerns the evaluation of the density of discrete energy eigenstates for bound quantum systems. While the quantization of energy represents a genuinely quantum attribute of such systems, a density of levels in the quantized spectra turns out to be determined solely by *classical properties.*

▶ Density of energy eigenstates

$E_k \equiv$ discrete energy eigenvalues of a system

Density of states $\boxed{\varrho(E) = \sum_k \delta(E - E_k)}$

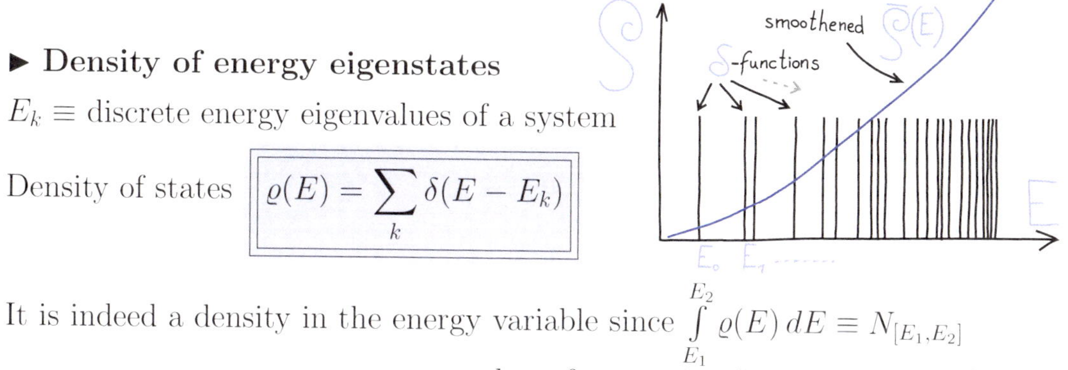

It is indeed a density in the energy variable since $\int_{E_1}^{E_2} \varrho(E)\, dE \equiv N_{[E_1,E_2]}$

$\equiv$ number of energy levels in the interval $[E_1, E_2]$

Exact state density contains complete information on the spectrum

$\Rightarrow$ also on the evolution operator of the system

▶ Relation of state density to propagator

Expression $\boxed{\delta(x) = \lim_{\varepsilon\to 0+} \frac{1}{\pi}\frac{\varepsilon}{\varepsilon^2+x^2}}$ $\Rightarrow$ $\varrho(E) = \lim_{\varepsilon\to 0+} \frac{1}{\pi}\sum_k \frac{\varepsilon}{\varepsilon^2+(E-E_k)^2} = \ldots$

$$= \lim_{\varepsilon\to 0+} \frac{1}{\pi}\Bigg[-\mathrm{Im}\underbrace{\sum_k \frac{1}{E-E_k+i\varepsilon}}_{\mathrm{Tr}\frac{1}{E-\hat{H}+i\varepsilon}}\Bigg] = -\frac{1}{\pi}\lim_{\varepsilon\to 0+} \mathrm{Im}\int \underbrace{\langle\vec{x}|\frac{1}{E-\hat{H}+i\varepsilon}|\vec{x}\rangle}_{-\frac{i}{\hbar}\int_0^{+\infty}\langle\vec{x}|\hat{G}(t)_\varepsilon|\vec{x}\rangle e^{+iEt/\hbar}dt}\, d\vec{x}$$

Green operator $\boxed{\hat{G}(t) = \lim_{\varepsilon\to 0+} \frac{i}{2\pi}\int \frac{1}{E-\hat{H}+i\varepsilon} e^{-iEt/\hbar}\, dE = \lim_{\varepsilon\to 0+} \hat{G}(t)_\varepsilon}$

$$\boxed{\varrho(E) = \frac{1}{\pi\hbar}\mathrm{Re}\left[\int\int_0^{+\infty} G(\vec{x}t\,|\,\vec{x}0)\; e^{+\frac{i}{\hbar}Et}\; dt\, d\vec{x}\right]}$$

Propagator

$\langle\vec{x}|\hat{G}(t)|\vec{x}_0\rangle \equiv G(\vec{x}t|\vec{x}_0 0)$

$\Rightarrow$ Feynman integral

▶ Expression through classical orbits

Level densities are usually determined in more or less smoothened forms. Such dependences contain components of different energy scales. In general, the exact level density can be decomposed into two basic parts which are evaluated separately: $\boxed{\varrho(E) = \varrho_0(E) + \varrho'(E)}$ where $\begin{cases} \varrho_0(E) \equiv & \textbf{smooth part} \\ \varrho'(E) \equiv & \textbf{oscillatory part} \end{cases}$

The smooth part (with a slow energy dependence) will be computed below. The oscillatory part (a fast energy dependence) is given in the form of so-called Gutzwiller formula (not derived here), which is a sum over **periodic orbits**:

$$\boxed{\varrho'(E) = \frac{1}{\pi\hbar}\sum_o\sum_{r=1}^{\infty}\frac{\tau_o}{|\boldsymbol{M}_o|}\cos\left[\frac{1}{\hbar}rS_o(E) + \phi_o\right]} \quad \text{where} \begin{cases} o \equiv \text{ identifier of periodic orbit} \\ r \equiv \text{ number of repetitions of } o \\ \tau_o \equiv \text{ basic time period of } o \\ |\boldsymbol{M}_o| \equiv \text{ a stability measure of } o \\ S_o(E) \equiv \text{ action along } o \\ \phi_o \equiv \text{ a phase connected with } o \end{cases}$$

For a cavity:

$$\frac{1}{\hbar}S_o(E)=\frac{1}{\hbar}\oint\vec{p}\cdot d\vec{x}=\frac{1}{\hbar}\sqrt{2ME}\,l_o=\frac{2}{\hbar}\underbrace{\tau_o(E)}_{2\pi/\Delta_o(E)}E \qquad \begin{cases} l_o\equiv \text{geometric length of orbit } o \\ \tau_o(E)\equiv \text{its time period at energy } E \\ \Delta_o(E)\equiv \text{variable wavelength of the energy oscillation: } \Delta_o(E)=\frac{\pi\hbar}{\tau_o(E)} \end{cases}$$

$\left.\begin{matrix}\text{Long } (l_o\gg L)\\ \text{Short } (l_o\approx L)\end{matrix}\right\}$ periodic orbits ($L\equiv$ cavity lin. size) cause $\left\{\begin{matrix}\text{short}\\ \text{long}\end{matrix}\right\}$ oscillations of $\varrho(E)$. Very long orbits, yielding $\Delta_o\lesssim(E_{i+1}-E_i)$, can be cut off. The summed $\varrho'(E)$ results from an interference of terms corresponding to several relevant (stable) orbits ($\Rightarrow$ beating patterns).

▶ Smooth part of the state density $\varrho_0(E)$

Derived from the contribution $G_{l=0}$ of **zero-length orbits** to the path-integral expression of $G(\vec{x}\,t\,|\,\vec{x}\,0)$, i.e. "orbits" corresponding to the **particle at rest** (for nonzero potential these are not classical orbits)

We will compute $\qquad G_{l=0}(\vec{x}\,t\,|\,\vec{x}\,0)=\lim\limits_{\Delta\vec{x}\to 0}\left(\frac{M}{2i\pi\hbar t}\right)^{\frac{3}{2}}e^{\frac{i}{\hbar}\left[\frac{M}{2}\frac{\Delta\vec{x}^2}{t}-V(\vec{x})\,t\right]}$

$$\varrho_0(E)=\frac{1}{\pi\hbar}\,\mathrm{Re}\left\{\int\int_0^{+\infty}G_{l=0}(\vec{x}\,t\,|\,\vec{x}\,0)\,e^{+\frac{i}{\hbar}Et}\,dt\,d\vec{x}\right\}=$$

$$=\frac{1}{\pi\hbar}\lim_{\Delta\vec{x}\to 0}\mathrm{Re}\left\{\left(\frac{M}{2i\pi\hbar}\right)^{\frac{3}{2}}\int\int_0^{+\infty}\underbrace{t^{-\frac{3}{2}}e^{\frac{i}{\hbar}\left[\frac{M}{2}\frac{\Delta\vec{x}^2}{t}\right]}}_{*}e^{\frac{i}{\hbar}[E-V(\vec{x})]t}\,dt\,d\vec{x}\right\}=\ldots$$

Trick: $\qquad *=\left(\frac{i}{2\pi\hbar M}\right)^{\frac{3}{2}}\int e^{\frac{i}{\hbar}\left[\vec{p}\cdot\Delta\vec{x}-\frac{\vec{p}^2}{2M}t\right]}d\vec{p} \qquad$ (Gaussian integral)

$$\ldots=\frac{1}{\pi\hbar}\frac{1}{(2\pi\hbar)^3}\lim_{\Delta\vec{x}\to 0}\mathrm{Re}\left\{\iint\int_0^{+\infty}e^{\frac{i}{\hbar}\left[\vec{p}\cdot\Delta\vec{x}-\frac{\vec{p}^2}{2M}t\right]}e^{\frac{i}{\hbar}[E-V(\vec{x})]t}\,dt\,d\vec{x}\,d\vec{p}\right\}$$

$$=\frac{1}{h^3}\iint\lim_{\Delta\vec{x}\to 0}\underbrace{\frac{1}{\pi\hbar}\mathrm{Re}\left(\int_0^{+\infty}e^{\frac{i}{\hbar}\left[E-\frac{\vec{p}^2}{2M}-V(\vec{x})\right]t}dt\right)}_{\delta\left(E-\frac{\vec{p}^2}{2M}-V(\vec{x})\right)}e^{\frac{i}{\hbar}\vec{p}\cdot\Delta\vec{x}}\,d\vec{x}\,d\vec{p}$$

$$\boxed{\varrho_0(E)=\frac{1}{h^3}\underbrace{\iint\delta\left[E-\frac{\vec{p}^2}{2M}-V(\vec{x})\right]d\vec{x}\,d\vec{p}}_{\frac{d}{dE}\Omega(E)}}$$

$$\Omega(E)\equiv\iint\Theta\left[E-\frac{\vec{p}^2}{2M}-V(\vec{x})\right]d\vec{x}\,d\vec{p}$$

$\equiv$ **phase-space volume** available for particle with energy $\leq E$

▶ Analogous result is valid for general systems with $2f$-dimensional phase space, e.g. for N-particle systems with $f=3N$

Number of states in interval $[E,E+dE]$ $\qquad \boxed{\varrho_0(E)dE=\frac{1}{h^f}\frac{d\Omega(E)}{dE}dE} \qquad$ Phase-space volume for energy interval $[E,E+dE]$ in units of **elementary quantum cell** $\hbar^f$

▶ "Cavities" of general dimension $f=\begin{cases}1\ \ldots\ \text{infinite square well}\\ 2\ \ldots\ \text{billiard}\\ 3\ \ldots\ \text{cavity}\end{cases}$

$$\frac{d}{dE}\Omega(E) = \iint \delta\left[E-\frac{p^2}{2M}\right] d\vec{x}\,d\vec{p} = \underbrace{V_f}_{\substack{\text{space}\\ \text{volume}}} \int\int \underbrace{\delta\left[E-\frac{p^2}{2M}\right]}_{\frac{M}{p_0}\delta(p-p_0)} p^{f-1}dp\,\underbrace{f(\theta)d\theta}_{\substack{\text{polar/spher.}\\ \text{angle(s)}}} = V_f\frac{M}{p_0}\underbrace{S_f p_0^{f-1}}_{\substack{\text{sphere}\\ \text{surface}}}$$

$$p_0 = \sqrt{2ME}$$

$$\varrho_0(E) \propto E^{\frac{f-2}{2}} = \begin{cases} E^{-1/2} & f=1 \\ E^{0} & f=2 \\ E^{+1/2} & f=3 \end{cases}$$

◀ **Historical remark**
1912: Hermann Weyl derives a formula for the density of resonances in a cavity
1927-30's: Development of semiclassical methods in the level-density evaluation
1970: Martin Gutzwiller derives the "periodic-orbit" formula

4. ANGULAR MOMENTUM

In Chapters 1 & 2, orbital and spin angular momenta of a single particle were discussed many times. We saw that angular momentum operators play an important general role in quantum theory, being generators of the 3D rotation group. However, the development of a complete formalism for angular momentum, including the theory of its coupling, has been postponed till now. In this chapter we are going to discover the importance as well as elegance of the "rotational segment" of QM.

4.1 General features of angular momentum

Employing just basic algebraic features of general angular-momentum operators (i.e., the well-known commutation relations between the components), one can derive a great majority of the relevant physical properties.

■ Eigenvalues of angular momentum projection & square

The debt of Chaps. 1 & 2 is now ready to be paid back: The familiar, frequently exploited properties of angular momentum eigensolutions will be finally derived!

▶ Angular-momentum ladder operators

General angular momentum operators $\hat{\vec{J}} \equiv (\hat{J}_1, \hat{J}_2, \hat{J}_3)$ satisfying:

$$\boxed{[\hat{J}_i, \hat{J}_j] = i\hbar\varepsilon_{ijk}\hat{J}_k} \quad\Rightarrow\quad [\hat{J}_i, \sum_{j=1}^{3}\hat{J}_j^2] = 0 \quad \text{where } \sum_{j=1}^{3}\hat{J}_j^2 \equiv \hat{J}^2$$

Simultaneous eigenvectors parametrized as
$$\boxed{\begin{aligned}\hat{J}^2|jm\rangle &= \hbar^2 j(j+1)|jm\rangle \\ \hat{J}_3|jm\rangle &= \hbar m|jm\rangle\end{aligned}}$$

Introduce operators $\boxed{\hat{J}_\pm = \hat{J}_1 \pm i\hat{J}_2}$

$$[\hat{J}^2, \hat{J}_\pm] = 0 \quad \Rightarrow \quad \hat{J}_\pm \text{ do not affect } j$$
$$[\hat{J}_3, \hat{J}_\pm] = \underbrace{[\hat{J}_3, \hat{J}_1]}_{i\hbar\varepsilon_{312}\hat{J}_2} \pm i \underbrace{[\hat{J}_3, \hat{J}_2]}_{i\hbar\varepsilon_{321}\hat{J}_1} = i\hbar(\hat{J}_2 \mp i\hat{J}_1) = \pm\hbar \underbrace{(\hat{J}_1 \pm i\hat{J}_2)}_{\hat{J}_\pm}$$

$\boxed{[\hat{J}_3, \hat{J}_\pm] = \pm\hbar\hat{J}_\pm}$ $\equiv$ general relation $[\hat{O}, \hat{T}_{\Delta o}] = \Delta o \hat{T}_{\Delta o}$ for ladder operators

$$\hat{J}_3\hat{J}_\pm|jm\rangle = \hat{J}_\pm \underbrace{\hat{J}_3|jm\rangle}_{\hbar m|jm\rangle} \pm\hbar\hat{J}_\pm|jm\rangle = \hbar(m\pm1)\hat{J}_\pm|jm\rangle \quad \text{shift by } \boxed{\Delta m = \pm 1}$$

▶ Possible values of quantum numbers

$\hat{J}_1^2 + \hat{J}_2^2 = \hat{J}^2 - \hat{J}_3^2$ $\equiv$ positively definite operator

$$(\hat{J}_1^2 + \hat{J}_2^2)|jm\rangle = \hbar^2 \underbrace{[j(j+1) - m^2]}_{\geq 0}|jm\rangle \quad \Rightarrow \quad -\sqrt{j(j+1)} \leq m \leq +\sqrt{j(j+1)}$$

$\Rightarrow$ $\exists$ values $m_{\rm min}$ and $m_{\rm max}$ such that $\boxed{\hat{J}_-|jm_{\rm min}\rangle = 0 = \hat{J}_+|jm_{\rm max}\rangle}$

To determine $m_{\rm min}$ and $m_{\rm max}$, we proceed as follows:

$$\hat{J}_+\hat{J}_-|jm_{\rm min}\rangle = 0 = \hat{J}_-\hat{J}_+|jm_{\rm max}\rangle$$
$$(\underbrace{\hat{J}_1^2 + \hat{J}_2^2}_{\hat{J}^2 - \hat{J}_3^2} + i \underbrace{[\hat{J}_2\hat{J}_1 - \hat{J}_1\hat{J}_2]}_{i\hbar\varepsilon_{213}\hat{J}_3})|jm_{\rm min}\rangle = 0 = (\underbrace{\hat{J}_1^2 + \hat{J}_2^2}_{\hat{J}^2 - \hat{J}_3^2} + i \underbrace{[\hat{J}_1\hat{J}_2 - \hat{J}_2\hat{J}_1]}_{i\hbar\varepsilon_{123}\hat{J}_3})|jm_{\rm max}\rangle$$
$$j(j+1) \overbrace{-m_{\rm min}^2 + m_{\rm min}}^{+m_{\rm min}(-m_{\rm min}+1)} = 0 = j(j+1) \overbrace{-m_{\rm max}^2 - m_{\rm max}}^{-m_{\rm max}(m_{\rm max}+1)}$$
$$\boxed{m_{\rm min} = -j} \qquad \boxed{m_{\rm max} = +j}$$

the other solutions $\notin [-\sqrt{j(j+1)}, \sqrt{j(j+1)}]$

Therefore, the action of $\hat{J}_\pm$ on $|jm\rangle$ proceeds according to the scheme:

$$0 \underset{\hat{J}_-}{\overset{\times}{\leftarrow}} |j \underbrace{m_{\rm min}}_{-j}\rangle \underset{\hat{J}_-}{\overset{\hat{J}_+}{\rightleftharpoons}} |j \underbrace{(m_{\rm min}+1)}_{-j+1}\rangle \underset{\hat{J}_-}{\overset{\hat{J}_+}{\rightleftharpoons}} \cdots\cdots \underset{\hat{J}_-}{\overset{\hat{J}_+}{\rightleftharpoons}} |j \underbrace{(m_{\rm max}-1)}_{+j-1}\rangle \underset{\hat{J}_-}{\overset{\hat{J}_+}{\rightleftharpoons}} |j \underbrace{m_{\rm max}}_{+j}\rangle \underset{\times}{\overset{\hat{J}_+}{\rightarrow}} 0$$

This chain is closed *iff* $\boxed{j = 0, \frac{1}{2}, 1, \frac{3}{2}, 2, \frac{5}{2}, \ldots}$

▶ Eigenstate normalization condition

We determine the normalization coefficients $\mathcal{N}_{jm}^\pm$ for the vectors obtained by the action of the ladder operators:

$$\hat{J}_\pm|jm\rangle = \mathcal{N}_{jm}^\pm|j(m\pm1)\rangle$$
$$1 = \langle j(m\pm1)|j(m\pm1)\rangle = \frac{1}{|\mathcal{N}_{jm}^\pm|^2}\langle jm|\overbrace{\hat{J}_\mp\hat{J}_\pm}^{\hat{J}^2 - \hat{J}_3^2 \mp \hbar\hat{J}_3}|jm\rangle = \frac{\hbar^2[j(j+1) - m(m\pm1)]}{|\mathcal{N}_{jm}^\pm|^2}$$

$$\boxed{\hat{J}_\pm|jm\rangle = \hbar\sqrt{j(j+1) - m(m\pm1)}\,|j(m\pm1)\rangle} \qquad \text{ensures } \hat{J}_\pm|j(\pm j)\rangle = 0$$

Addition of two angular momenta

Consider an angular momentum vector which is a sum of two partial angular momenta (like the total angular momentum obtained from spin and orbital momenta of a particle). The system can be characterized by eigenvectors of the total angular momentum as well as by eigenvectors of both partial angular momenta. The relation between both bases is just a unitary transformation.

▶ Separable angular-momentum basis

$$\mathcal{H}=\mathcal{H}^{(1)}\otimes\mathcal{H}^{(2)} \quad \text{with} \quad \left.\begin{matrix}\mathcal{H}^{(1)}\\ \mathcal{H}^{(2)}\end{matrix}\right\}\equiv \text{Hilbert space of} \left\{\begin{matrix}\vec{J}^{(1)}\\ \vec{J}^{(2)}\end{matrix}\right.$$

$$\boxed{[\hat{J}_i^{(m)},\hat{J}_j^{(n)}]=i\hbar\varepsilon_{ijk}\delta_{mn}\hat{J}_k^{(m)}} \quad \text{with } m,n=1,2$$

$$\left\{\hat{J}^{(1)2},\hat{J}_3^{(1)},\hat{J}^{(2)2},\hat{J}_3^{(2)}\right\}\equiv \text{complete set I} \Rightarrow \boxed{\left\{|j_1m_1\rangle|j_2m_2\rangle\right\}\equiv \text{basis I}}$$

▶ Coupled angular-momentum basis

Total angular-momentum operators $\boxed{\hat{\vec{J}}=\hat{\vec{J}}^{(1)}+\hat{\vec{J}}^{(2)}}$

$$\hat{J}_i=\hat{J}_i^{(1)}\otimes\hat{I}^{(2)}+\hat{I}^{(1)}\otimes\hat{J}_i^{(2)}$$

$$[\hat{J}_i,\hat{J}_j]=\overbrace{[\hat{J}_i^{(1)},\hat{J}_j^{(1)}]}^{i\hbar\varepsilon_{ijk}\hat{J}_k^{(1)}}+\overbrace{[\hat{J}_i^{(2)},\hat{J}_j^{(2)}]}^{i\hbar\varepsilon_{ijk}\hat{J}_k^{(2)}}=i\hbar\varepsilon_{ijk}\overbrace{(\hat{J}_k^{(1)}+\hat{J}_k^{(2)})}^{\hat{J}_k} \quad \text{standard commut. rel.}$$

$$\Rightarrow \quad [\hat{J}^2,\hat{J}_3]=0=\left\{\begin{matrix}[\hat{J}^2,\hat{J}^{(1)2}]=[\hat{J}_3,\hat{J}^{(1)2}]\\ [\hat{J}^2,\hat{J}^{(2)2}]=[\hat{J}_3,\hat{J}^{(2)2}]\end{matrix}\right. \quad \text{but} \quad \left.\begin{matrix}[\hat{J}^2,\hat{J}_3^{(1)}]\\ [\hat{J}^2,\hat{J}_3^{(2)}]\end{matrix}\right\}\neq 0$$

$$\left\{\hat{J}^{(1)2},\hat{J}^{(2)2},\hat{J}^2,\hat{J}_3\right\}\equiv \text{complete set II} \Rightarrow \boxed{\left\{|j_1j_2jm\rangle\right\}\equiv \text{basis II}}$$

▶ Possible values of total angular momentum

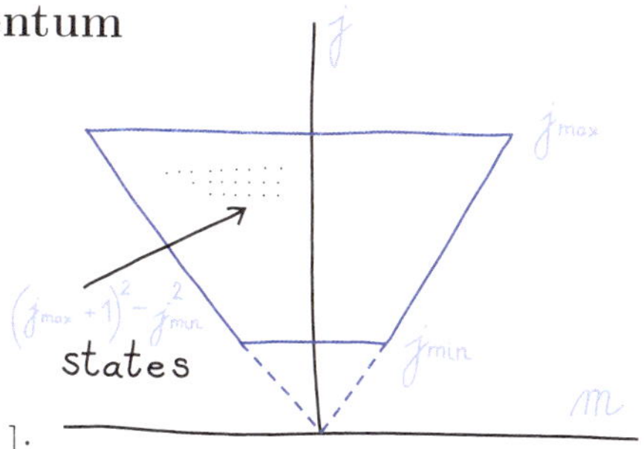

Allowed values of j obtained partly from dimension considerations

Basis I has dimension $\boxed{d=(2j_1+1)(2j_2+1)}$

$\Rightarrow$ the same dimension required for basis II

This helps to determine the bounds of square q. nums. of total ang. momentum $j\in[j_{\min},j_{\max}]$:

(a) $\hat{J}_3=\hat{J}_3^{(1)}+\hat{J}_3^{(2)} \quad \Rightarrow \quad m_{\max}=m_{\max 1}+m_{\max 2}=j_1+j_2 \quad \Rightarrow \boxed{j_{\max}=j_1+j_2}$

(b) The determination of minimal j from the dimension criterion:

Number of states for $j=\left\{\begin{matrix}0,1,\ldots\ldots\\ \frac{1}{2},\frac{3}{2},\ldots\ldots\end{matrix}\right. j_{\max}$ is $d_>=\lfloor(j_{\max}+1)^2\rfloor$

The surplus: $d_>-d=\lfloor(j_1+j_2+1)^2\rfloor-(2j_1+1)(2j_2+1)=\lfloor(j_1-j_2)^2\rfloor$

Number of states for $j=j_{\min}\ldots j_{\max}$ is $d=\lfloor(j_{\max}+1)^2\rfloor-\lfloor j_{\min}^2\rfloor \Rightarrow \boxed{j_{\min}=|j_1-j_2|}$

▶ Unitary transformation between bases I and II

$$|j_1 j_2 jm\rangle = \sum_{m_1=-j_1}^{+j_1} \sum_{m_2=-j_2}^{+j_2} C^{jm}_{j_1 m_1 j_2 m_2} |j_1 m_1\rangle |j_2 m_2\rangle$$

Clebsch–Gordan coefficients

$C^{jm}_{j_1 m_1 j_2 m_2} \equiv (j_1 m_1 j_2 m_2 | jm) \equiv \langle j_1 m_1 j_2 m_2 | j_1 j_2 jm\rangle$

$$\left.\begin{array}{l} m \neq m_1 + m_2 \quad \text{or} \\ j \notin [|j_1 - j_2|, j_1 + j_2] \end{array}\right\} \Rightarrow C^{jm}_{j_1 m_1 j_2 m_2} = 0$$

▶ Some properties of Clebsch-Gordan coefficients

(a) $C^{jm}_{j_1 m_1 j_2 m_2} \in \mathbb{R}$ (by convention)

(b) From reality we get: $\langle j_1 m_1 j_2 m_2 | j_1 j_2 jm\rangle = \langle j_1 j_2 jm | j_1 m_1 j_2 m_2\rangle$

$$\Rightarrow \quad |j_1 m_1\rangle |j_2 m_2\rangle = \sum_{j=|j_1-j_2|}^{j_1+j_2} \sum_{m=-j}^{+j} C^{jm}_{j_1 m_1 j_2 m_2} |j_1 j_2 jm\rangle$$

inverse relation

(c) Multiply $\langle j_1 j_2 j' m'| = \sum_{m_1', m_2'} C^{j'm'}_{j_1 m_1' j_2 m_2'} \langle j_1 m_1'| \langle j_2 m_2'|$ with $|j_1 j_2 jm\rangle = \sum_{m_1, m_2} C^{jm}_{j_1 m_1 j_2 m_2} |j_1 m_1\rangle |j_2 m_2\rangle$

$$\Rightarrow \quad \sum_{m_1, m_2} C^{jm}_{j_1 m_1 j_2 m_2} C^{j'm'}_{j_1 m_1 j_2 m_2} = \delta_{jj'} \delta_{mm'}$$

orthogonality relation I

(d) Multiply $\langle j_1 m_1'| \langle j_2 m_2'| = \sum_{j', m'} C^{j'm'}_{j_1 m_1' j_2 m_2'} \langle j_1 j_2 j' m'|$ with $|j_1 m_1\rangle |j_2 m_2\rangle = \sum_{j,m} C^{jm}_{j_1 m_1 j_2 m_2} |j_1 j_2 jm\rangle$

$$\Rightarrow \quad \sum_{j,m} C^{jm}_{j_1 m_1 j_2 m_2} C^{jm}_{j_1 m_1' j_2 m_2'} = \delta_{m_1 m_1'} \delta_{m_2 m_2'}$$

orthogonality relation II

The following relations we give here without the proofs:

(e) $C^{jm}_{j_1 m_1 j_2 m_2} = \underbrace{(-)^{j-j_1-j_2}}_{\pm} C^{jm}_{j_2 m_2 j_1 m_1}$ **exchange of indices I**

Special case: $C^{jm}_{j_1 m_1 j_1 m_1} = 0$ for $(j - 2j_1)$=odd

(f) $C^{jm}_{j_1 m_1 j_2 m_2} = \underbrace{(-)^{j_1-m_1}}_{\pm} \sqrt{\frac{2j+1}{2j_2+1}} C^{j_2(-m_2)}_{j_1 m_1 j(-m)}$ **exchange of indices II**

(g) $C^{jm}_{j_1 m_1 j_2 m_2} = \underbrace{(-)^{j-j_1-j_2}}_{\pm} C^{j(-m)}_{j_1(-m_1) j_2(-m_2)}$ **sign inversion**

Special case: $C^{j0}_{j_1 0 j_2 0} = 0$ for $(j - j_1 - j_2)$=odd

▶ 3j symbols

Definition: $\begin{pmatrix} j_1 & j_2 & j_3 \\ m_1 & m_2 & m_3 \end{pmatrix} \equiv \frac{(-)^{j_1-j_2-m_3}}{\sqrt{2j_3+1}} C^{j_3(-m_3)}_{j_1 m_1 j_2 m_2}$

These coefficients represent just a more symmetric form of CG coefficients:

$$\begin{pmatrix} j_1 & j_2 & j_3 \\ m_1 & m_2 & m_3 \end{pmatrix} = \varepsilon \begin{pmatrix} j_k & j_l & j_n \\ m_k & m_l & m_n \end{pmatrix} \text{ with } \varepsilon = \begin{cases} +1 & \text{for even permutation} \\ (-)^{j_1+j_2+j_3} & \text{for odd permutation} \end{cases}$$

$$\begin{pmatrix} j_1 & j_2 & j_3 \\ m_1 & m_2 & m_3 \end{pmatrix} = (-)^{j_1+j_2+j_3} \begin{pmatrix} j_1 & j_2 & j_3 \\ -m_1 & -m_2 & -m_3 \end{pmatrix}$$

▸ Construction of Clebsch-Gordan coefficients

The way how the CG coefficients can be calculated:

$$\hat{J}_\pm |j_1 j_2 j m\rangle = [\hat{J}_\pm^{(1)} \otimes \hat{I}^{(2)} + \hat{I}^{(1)} \otimes \hat{J}_\pm^{(2)}] \sum_{m_1,m_2} C^{jm}_{j_1 m_1 j_2 m_2} |j_1 m_1\rangle |j_2 m_2\rangle$$

$$\hbar\sqrt{j(j+1)-m(m\pm1)}|j_1 j_2 j (m\pm1)\rangle = \hbar \sum_{m_1,m_2} \sqrt{j_1(j_1+1)-m_1(m_1\pm1)} C^{jm}_{j_1 m_1 j_2 m_2} |j_1 (m_1\pm1)\rangle |j_2 m_2\rangle + \hbar \sum_{m_1,m_2} \sqrt{j_2(j_2+1)-m_2(m_2\pm1)} C^{jm}_{j_1 m_1 j_2 m_2} |j_1 m_1\rangle |j_2 (m_2\pm1)\rangle$$

Multiply by $\langle j_1 m_1'|\langle j_2 m_2'| \quad \Rightarrow$

$$\sqrt{j(j+1)-m(m\pm1)} C^{j(m\pm1)}_{j_1 m_1' j_2 m_2'} = \sum_{m_1,m_2} \sqrt{j_1(j_1+1)-m_1(m_1\pm1)} C^{jm}_{j_1 m_1 j_2 m_2} \delta_{m_1'(m_1\pm1)} \delta_{m_2' m_2} + \sum_{m_1,m_2} \sqrt{j_2(j_2+1)-m_2(m_2\pm1)} C^{jm}_{j_1 m_1 j_2 m_2} \delta_{m_1' m_1} \delta_{m_2'(m_2\pm1)}$$

After $\left.\begin{matrix} m_1' \\ m_2' \end{matrix}\right\} \mapsto \left\{\begin{matrix} m_1 \\ m_2 \end{matrix}\right.$ we get the following **recursive relation**

$$C^{j(m\pm1)}_{j_1 m_1 j_2 m_2} = \sqrt{\frac{j_1(j_1+1)-m_1(m_1\mp1)}{j(j+1)-m(m\pm1)}} C^{jm}_{j_1 (m_1\mp1) j_2 m_2} + \sqrt{\frac{j_2(j_2+1)-m_2(m_2\mp1)}{j(j+1)-m(m\pm1)}} C^{jm}_{j_1 m_1 j_2 (m_2\mp1)}$$

$$\boxed{C^{jm}_{j_1 m_1 j_2 m_2} = \sqrt{\frac{j_1(j_1+1)-m_1(m_1\mp1)}{j(j+1)-m(m\mp1)}} C^{j(m\mp1)}_{j_1 (m_1\mp1) j_2 m_2} + \sqrt{\frac{j_2(j_2+1)-m_2(m_2\mp1)}{j(j+1)-m(m\mp1)}} C^{j(m\mp1)}_{j_1 m_1 j_2 (m_2\mp1)}}$$

This relation enables one to construct the CG coefficients using the fact that

$$|j_1 j_2 j_{\max} \underbrace{(\pm j_{\max})}_{m}\rangle = |j_1 \underbrace{(\pm j_1)}_{m_1}\rangle |j_2 \underbrace{(\pm j_2)}_{m_2}\rangle \quad \Rightarrow \quad \boxed{C^{j_{\max}(\pm j_{\max})}_{j_1(\pm j_1) j_2(\pm j_2)} = 1}$$

▸ Example: coupling two spins $\frac{1}{2}$

$j_1 = j_2 = \frac{1}{2} \quad \Rightarrow \quad j_{\max} = 1, \; j_{\min} = 0$

$$|\tfrac{1}{2}\tfrac{1}{2}11\rangle = |\tfrac{1}{2}\tfrac{1}{2}\rangle_1 |\tfrac{1}{2}\tfrac{1}{2}\rangle_2 \quad \Rightarrow \quad \overbrace{\hat{J}_- |\tfrac{1}{2}\tfrac{1}{2}11\rangle}^{\sqrt{2}|\frac{1}{2}\frac{1}{2}10\rangle} = \overbrace{(\hat{J}_-^{(1)} |\tfrac{1}{2}\tfrac{1}{2}\rangle_1)}^{|\frac{1}{2}(-\frac{1}{2})\rangle_1} |\tfrac{1}{2}\tfrac{1}{2}\rangle_2 + |\tfrac{1}{2}\tfrac{1}{2}\rangle_1 \overbrace{(\hat{J}_-^{(2)} |\tfrac{1}{2}\tfrac{1}{2}\rangle_2)}^{|\frac{1}{2}(-\frac{1}{2})\rangle_2}$$

The state $|\frac{1}{2}\frac{1}{2}1(-1)\rangle$ known and $|\frac{1}{2}\frac{1}{2}00\rangle$ obtained from orthogonality to $|\frac{1}{2}\frac{1}{2}10\rangle$.

In summary:

$$\begin{aligned}
|\tfrac{1}{2}\tfrac{1}{2}11\rangle &= \underbrace{|\tfrac{1}{2}(+\tfrac{1}{2})\rangle_1 |\tfrac{1}{2}(+\tfrac{1}{2})\rangle_2}_{|\uparrow\rangle_1|\uparrow\rangle_2} \\
|\tfrac{1}{2}\tfrac{1}{2}10\rangle &= \tfrac{1}{\sqrt{2}} \underbrace{|\tfrac{1}{2}(-\tfrac{1}{2})\rangle_1 |\tfrac{1}{2}(+\tfrac{1}{2})\rangle_2}_{|\downarrow\rangle_1|\uparrow\rangle_2} + \tfrac{1}{\sqrt{2}} \underbrace{|\tfrac{1}{2}(+\tfrac{1}{2})\rangle_1 |\tfrac{1}{2}(-\tfrac{1}{2})\rangle_2}_{|\uparrow\rangle_1|\downarrow\rangle_2} \\
|\tfrac{1}{2}\tfrac{1}{2}1(-1)\rangle &= \underbrace{|\tfrac{1}{2}(-\tfrac{1}{2})\rangle_1 |\tfrac{1}{2}(-\tfrac{1}{2})\rangle_2}_{|\downarrow\rangle_1|\downarrow\rangle_2}
\end{aligned} \quad \Bigg\} \textbf{ triplet}$$

$$|\tfrac{1}{2}\tfrac{1}{2}00\rangle = \tfrac{1}{\sqrt{2}} \underbrace{|\tfrac{1}{2}(-\tfrac{1}{2})\rangle_1 |\tfrac{1}{2}(+\tfrac{1}{2})\rangle_2}_{|\downarrow\rangle_1|\uparrow\rangle_2} - \tfrac{1}{\sqrt{2}} \underbrace{|\tfrac{1}{2}(+\tfrac{1}{2})\rangle_1 |\tfrac{1}{2}(-\tfrac{1}{2})\rangle_2}_{|\uparrow\rangle_1|\downarrow\rangle_2} \quad \textbf{singlet}$$

■ Addition of three angular momenta

Coupling of $k>2$ angular momenta is trickier than the $k=2$ coupling. In general, the summed momentum operators $\hat{J}^2$ and $\hat{J}_3$ must be supplemented by $(2k-2)$ additional commuting operators to form a complete set. For $k=2$, as seen above, the two additional operators are just the $\hat{J}^{(1)2}$ and $\hat{J}^{(2)2}$ squares. However, for $k>2$ one has to find more than k additional operators; hence the squares of partial momenta do not suffice. The choice of the extra operators is not unique.

▶ Total and paired angular momenta

$$\mathcal{H}=\mathcal{H}^{(1)}\otimes\mathcal{H}^{(2)}\otimes\mathcal{H}^{(3)} \qquad \left.\begin{matrix}\mathcal{H}^{(1)}\\ \mathcal{H}^{(2)}\\ \mathcal{H}^{(3)}\end{matrix}\right\}\leftrightarrow\left\{\begin{matrix}\vec{J}^{(1)}\\ \vec{J}^{(2)}\\ \vec{J}^{(3)}\end{matrix}\right. \quad\Rightarrow\quad [\hat{J}^{(n)}_i,\hat{J}^{(l)}_i]=i\hbar\varepsilon_{ijk}\delta_{nl}\hat{J}^{(n)}_k$$

Total angular momentum: $\hat{\vec{J}}=\hat{\vec{J}}^{(1)}+\hat{\vec{J}}^{(2)}+\hat{\vec{J}}^{(3)}$

Paired angular momenta: $\hat{\vec{J}}^{(nl)}=\hat{\vec{J}}^{(n)}+\hat{\vec{J}}^{(l)} \quad\Rightarrow\quad \hat{\vec{J}}^{(12)},\hat{\vec{J}}^{(13)},\hat{\vec{J}}^{(23)}$

Commutation relations: $[\hat{J}_i,\hat{J}_j]=i\hbar\varepsilon_{ijk}\hat{J}_k \qquad \begin{cases}[\hat{J}^{(nl)}_i,\hat{J}^{(nl)}_j]=i\hbar\varepsilon_{ijk}\hat{J}^{(nl)}_k\\ [\hat{J}^{(nl)}_i,\hat{J}^{(n'l')}_j]\neq 0 \text{ for } nl\neq n'l'\end{cases}$

Compatibility:

$$[\hat{J}^2,\hat{J}_3]=0=\begin{cases}[\hat{J}^2,\hat{J}^{(1)2}]=[\hat{J}_3,\hat{J}^{(1)2}]=[\hat{J}^2,\hat{J}^{(23)2}]=[\hat{J}_3,\hat{J}^{(23)2}]\\ [\hat{J}^2,\hat{J}^{(2)2}]=[\hat{J}_3,\hat{J}^{(2)2}]=[\hat{J}^2,\hat{J}^{(13)2}]=[\hat{J}_3,\hat{J}^{(13)2}]\\ [\hat{J}^2,\hat{J}^{(3)2}]=[\hat{J}_3,\hat{J}^{(3)2}]=[\hat{J}^2,\hat{J}^{(12)2}]=[\hat{J}_3,\hat{J}^{(12)2}]\end{cases} \qquad \left.\begin{matrix}[\hat{J}^2,\hat{J}^{(1)}_3]\\ [\hat{J}^2,\hat{J}^{(2)}_3]\\ [\hat{J}^2,\hat{J}^{(3)}_3]\end{matrix}\right\}\neq 0\neq\left\{\begin{matrix}[\hat{J}^2,\hat{J}^{(23)}_3]\\ [\hat{J}^2,\hat{J}^{(13)}_3]\\ [\hat{J}^2,\hat{J}^{(12)}_3]\end{matrix}\right.$$

▶ Different coupling schemes

Several complete sets of commuting operators & associated bases:

$$\begin{aligned}
&\hat{J}^{(1)2},\hat{J}^{(1)}_3,\hat{J}^{(2)2},\hat{J}^{(2)}_3,\hat{J}^{(3)2},\hat{J}^{(3)}_3 &&\Rightarrow |j_1m_1\rangle|j_2m_2\rangle|j_3m_3\rangle &&\ldots\text{ basis I}\\
&\hat{J}^{(1)2},\hat{J}^{(2)2},\hat{J}^{(3)2},\hat{J}^{(23)2},\hat{J}^2,\hat{J}_3 &&\Rightarrow |j_1j_2j_3j_{23}jm\rangle &&\ldots\text{ basis II}\\
&\hat{J}^{(1)2},\hat{J}^{(2)2},\hat{J}^{(3)2},\hat{J}^{(13)2},\hat{J}^2,\hat{J}_3 &&\Rightarrow |j_1j_2j_3j_{13}jm\rangle &&\ldots\text{ basis III}\\
&\hat{J}^{(1)2},\hat{J}^{(2)2},\hat{J}^{(3)2},\hat{J}^{(12)2},\hat{J}^2,\hat{J}_3 &&\Rightarrow |j_1j_2j_3j_{12}jm\rangle &&\ldots\text{ basis IV}
\end{aligned}$$

Generation of the coupled bases (II,III,IV) from the uncoupled one (I):

$$\begin{aligned}
|j_1j_2j_3j_{23}jm\rangle &= \sum_{m_1,m_{23}} C^{jm}_{j_1m_1j_{23}m_{23}}|j_1m_1\rangle\sum_{m_2,m_3}C^{j_{23}m_{23}}_{j_2m_2j_3m_3}|j_2m_2\rangle|j_3m_3\rangle\\
&= \sum_{\substack{m_1,m_2,m_3\\ m_{23}}} C^{jm}_{j_1m_1j_{23}m_{23}}C^{j_{23}m_{23}}_{j_2m_2j_3m_3}|j_1m_1\rangle|j_2m_2\rangle|j_3m_3\rangle \qquad\ldots\text{II}
\end{aligned}$$

...similarly III & IV

Relation between coupled bases:

$$|j_1j_2j_3j_{23}jm\rangle=(-)^{j_1+j_2+j_3+j}\sum_{j_{12}}\sqrt{(2j_{23}+1)(2j_{12}+1)}\underbrace{\begin{Bmatrix}j_1 & j_2 & j_{12}\\ j_3 & j & j_{23}\end{Bmatrix}}_{\text{6j symbol}}|j_1j_2j_3j_{12}jm\rangle$$

◀ Historical remark

1866: A. Clebsch & P. Gordan introduce CG coefficients for spherical harmonics
1930: P. Dirac presents the algebraic treatment of angular-momentum operators
1940, 1942: E. Wigner & G. Racah analyze coupling of >2 angular momenta

4.2 Irreducible tensor operators

Transformations of the quantum Hilbert space induced by spatial rotations motivates the introduction of operators with a privileged form of transformation. These are spherical tensor operators of various ranks. The knowledge of the tensor calculus enables one to build up operators with required transformation properties (e.g., scalars) and substantially simplifies some calculations.

■ Irreducible representations of the rotation group

Any rotation, expressed by a 3D matrix $\mathbf{R}$, can be equivalently characterized either by axis $\vec{n}$ and angle ϕ, or by 3 Euler angles α, β, γ. Associated with $\mathbf{R}$ is a transformation operator $\hat{R}_{\mathbf{R}} \equiv \hat{R}_{\vec{n}\phi} \equiv \hat{R}(\alpha\beta\gamma)$ in $\mathcal{H}$. The action of this operator in angular-momentum eigenspaces spanned by vectors $|jm\rangle$ is described by a hierarchy of Wigner matrices, which for each fixed j form an irreducible representation of the rotation group. QM therefore provides a fundamental platform for the realization of this group.

▶ Factorization of rotation operators

Rotation around $\vec{n}$ by ϕ: operator $\boxed{\hat{R}_{\mathbf{R}} \equiv \hat{R}_{\vec{n}\phi} = e^{-\frac{i}{\hbar}(\hat{\vec{J}}\cdot\vec{n})\phi}} \neq \hat{R}_z \hat{R}_y \hat{R}_x$

Expression of a general rotation via **Euler angles**: 3 successive rotations

$$\left.\begin{array}{lll} (1) & \text{around } \vec{n}_z & \text{by } \alpha \\ (2) & \text{around } \vec{n}'_y \equiv \mathbf{R}_{\vec{n}_z\alpha}\vec{n}_y & \text{by } \beta \\ (3) & \text{around } \vec{n}'_z \equiv \mathbf{R}_{\vec{n}'_y\beta}\vec{n}_z & \text{by } \gamma \end{array}\right\} \Rightarrow \boxed{\underbrace{\hat{R}_{\vec{n}\phi}}_{\hat{R}(\alpha\beta\gamma)} = \underbrace{\hat{R}_{\vec{n}'_z\gamma}}_{\hat{R}_{z'}(\gamma)} \underbrace{\hat{R}_{\vec{n}'_y\beta}}_{\hat{R}_{y'}(\beta)} \underbrace{\hat{R}_{\vec{n}_z\alpha}}_{\hat{R}_z(\alpha)}}$$

Using identities

$$\left\{\begin{array}{l} \hat{R}_{z'}(\gamma) = \hat{R}_{y'}(\beta)\hat{R}_z(\gamma)\hat{R}^{-1}_{y'}(\beta) \\ \hat{R}_{y'}(\beta) = \hat{R}_z(\alpha)\hat{R}_y(\beta)\hat{R}^{-1}_z(\alpha) \end{array}\right\}$$

i.e., e.g., $\hat{R}_{y'}(\beta)\hat{R}_z(\alpha) = \hat{R}_z(\alpha)\hat{R}_y(\beta)$,

as shown on the right:

we obtain a **factorized formula in fixed coordinate system** xyz:

$$\boxed{\hat{R}(\alpha\beta\gamma) = \hat{R}_z(\alpha)\hat{R}_y(\beta)\hat{R}_z(\gamma)}$$

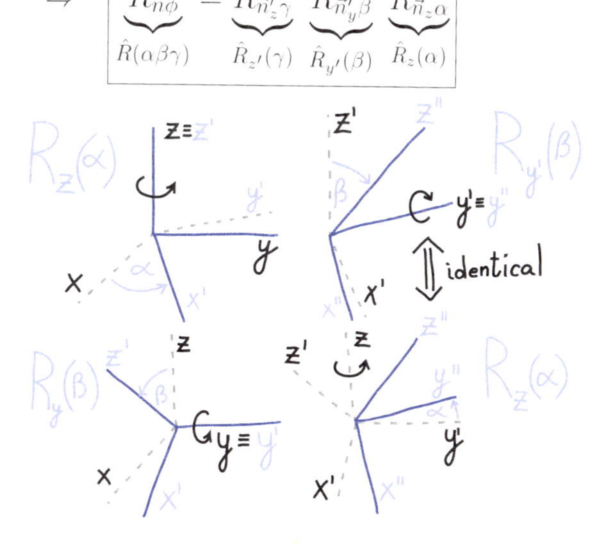

▶ Wigner functions

Action of rotation operators in the space spanned by angular-momentum eigenvectors $|jm\rangle \Rightarrow$ for each fixed j we get a representation of the rotation group

$$\boxed{\hat{R}(\alpha\beta\gamma)|jm\rangle = \sum_{j'm'} \underbrace{\langle j'm'|\hat{R}(\alpha\beta\gamma)|jm\rangle}_{\delta_{jj'}D^j_{m'm}(\alpha\beta\gamma)} |j'm'\rangle = \sum_{m'} \underbrace{D^j_{m'm}(\alpha\beta\gamma)}_{\text{Wigner functions}} |jm'\rangle}$$

$$\underbrace{D^j_{m'm}(\alpha\beta\gamma)}_{D^j_{m'm}(\vec{n}\phi)} \equiv \text{matrix of } \boxed{\text{dimension } 2j+1}$$

$$D^j_{m'm}(\vec{n}\phi) \equiv D^j_{m'm}(\mathbf{R}) = \langle jm'|\hat{R}_z(\alpha)\hat{R}_y(\beta)\hat{R}_z(\gamma)|jm\rangle = e^{-i(m'\alpha+m\gamma)}\overbrace{\langle jm'|\hat{R}_y(\beta)|jm\rangle}^{d^j_{m'm}(\beta)}$$

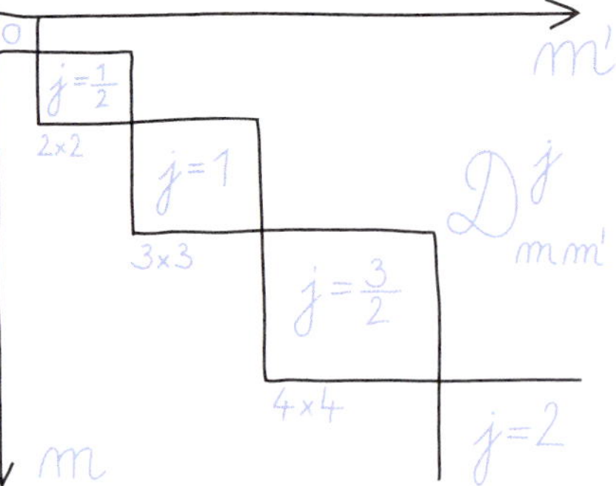

Matrices of Wigner functions form **irreps** of the rotational group for individual values

$$\boxed{j = 0, \tfrac{1}{2}, 1, \tfrac{3}{2}, 2, \tfrac{5}{2}, 3, \dots}$$

(a) identity $\phi=0$, (b) inverse $\phi=-\phi$,
(d) group multiplication ≡ matrix multipl.:

$$D^j_{m'm}(\mathbf{R}_2\mathbf{R}_1) = \sum_{m''} D^j_{m'm''}(\mathbf{R}_2) D^j_{m''m}(\mathbf{R}_1)$$

▶ **Clebsch-Gordan series for Wigner functions**

Action of rotation operators in the Hilbert space of coupled angular momenta:

$$\mathcal{H} = \underbrace{\mathcal{H}_1}_{\text{irrep } j_1} \otimes \underbrace{\mathcal{H}_2}_{\text{irrep } j_2} \quad \Rightarrow \quad \hat{R}(\alpha\beta\gamma) = \hat{R}_2(\alpha\beta\gamma) \otimes \hat{R}_1(\alpha\beta\gamma)$$

Separable basis $|j_1m_1\rangle|j_2m_2\rangle \equiv |j_1m_1j_2m_2\rangle$ and coupled basis $|j_1j_2jm\rangle$

$$\langle j_1m_1j_2m_2|\hat{R}|j_1m'_1j_2m'_2\rangle =$$

$$= \begin{cases} \underbrace{\langle j_1m_1|\hat{R}_1|j_1m'_1\rangle}_{D^{j_1}_{m_1m'_1}} \underbrace{\langle j_2m_2|\hat{R}_2|j_2m'_2\rangle}_{D^{j_2}_{m_2m'_2}} \\ \sum_{jm}\sum_{j'm'} \underbrace{\langle j_1m_1j_2m_2|j_1j_2jm\rangle}_{C^{jm}_{j_1m_1j_2m_2}} \underbrace{\langle j_1j_2jm|\hat{R}|j_1j_2j'm'\rangle}_{\delta_{jj'}D^j_{mm'}} \underbrace{\langle j_1j_2j'm'|j_1m'_1j_2m'_2\rangle}_{C^{j'm'}_{j_1m'_1j_2m'_2}} \end{cases}$$

$$\Rightarrow \boxed{D^{j_1}_{m_1m'_1}(\mathbf{R})\, D^{j_2}_{m_2m'_2}(\mathbf{R}) = \sum_{j=|j_1-j_2|}^{j_1+j_2} \sum_{\substack{m\\m'}=-j}^{+j} C^{jm}_{j_1m_1j_2m_2} C^{jm'}_{j_1m'_1j_2m'_2}\, D^j_{mm'}(\mathbf{R})}$$

This relation between Wigner functions determines the decomposition of the product rotation-group representation (coupling of irreps corresponding to j_1 & j_2) into a direct sum of irreps: $D^{j_1} \otimes D^{j_2} = D^{|j_1-j_2|} \oplus \cdots\cdots \oplus D^{(j_1+j_2)}$

■ Spherical tensors

We are ready now to understand and appreciate the introduction of spherical tensors, i.e., objects (in our case operators) which transform according to a single irreducible representation of the rotation group. Spherical tensors have some favorable properties that make them mathematically more convenient than the familiar Cartesian tensors.

▶ **Cartesian tensors** ⇔ Cartesian transformations under rotations

$n^{\rm th}$ rank tensor: $\underbrace{T_{ijk\ldots}}_{\substack{n \text{ indices} \\ i,j,k\cdots=1,2,3}} \mapsto$ $$T'_{ijk\ldots} = \sum_{i'j'k'\ldots} \underbrace{\mathbf{R}^{-1}_{ii'}\mathbf{R}^{-1}_{jj'}\mathbf{R}^{-1}_{kk'}\ldots}_{\text{Cartesian rot. matrices}} T_{i'j'k'\ldots}$$

Representation of the rotation group on Cartesian tensors is **reducible**

Example: 2nd rank tensor:

$$T_{ij} = \underbrace{\tfrac{1}{3}\mathrm{Tr}\,T\,\delta_{ij}}_{\text{scalar}} + \underbrace{\tfrac{1}{2}[T_{ij} - T_{ji}]}_{\text{antisymmetric tensor}} + \underbrace{\tfrac{1}{2}[T_{ij} + T_{ji}] - \tfrac{1}{3}\mathrm{Tr}T\,\delta_{ij}}_{\text{traceless symmetric tensor}} \Rightarrow \begin{cases} \text{each part} \\ \text{constitutes irrep} \end{cases}$$

► **Irreducible (spherical) tensors** ⇔ transformations by Wigner functions

$\lambda^{\rm th}$ rank spherical tensor

$$(T')^\lambda_\mu = \sum_{\mu'} D^\lambda_{\mu'\mu}(\mathbf{R})\,T^\lambda_{\mu'}$$

$\lambda^{\rm th}$ rank **spherical tensor operator**

$$\hat{R}_{\mathbf{R}}\,\hat{T}^\lambda_\mu\,\hat{R}^{-1}_{\mathbf{R}} = \sum_{\mu'} D^\lambda_{\mu'\mu}(\mathbf{R})\,\hat{T}^\lambda_{\mu'}$$

Infinitesimal rotation:

$$\overbrace{[\hat{I} - \tfrac{i}{\hbar}(\hat{\vec{J}}\cdot\vec{n})\delta\phi]}^{\delta\hat{R}}\,\hat{T}^\lambda_\mu\,\overbrace{[\hat{I} + \tfrac{i}{\hbar}(\hat{\vec{J}}\cdot\vec{n})\delta\phi]}^{\delta\hat{R}^{-1}} = \sum_{\mu'} \overbrace{\langle\lambda\mu'|[\hat{I} - \tfrac{i}{\hbar}(\hat{\vec{J}}\cdot\vec{n})\delta\phi]|\lambda\mu\rangle}^{D^\lambda_{\mu'\mu}(\delta\mathbf{R})}\,\hat{T}^\lambda_{\mu'}$$

$$\Rightarrow \quad \left[(\hat{\vec{J}}\cdot\vec{n}), \hat{T}^\lambda_\mu\right] = \sum_{\mu'}\langle\lambda\mu'|(\hat{\vec{J}}\cdot\vec{n})|\lambda\mu\rangle\hat{T}^\lambda_{\mu'}$$

⇒ An alternative (more useful) definition of the spherical tensor:

$$\left[\hat{J}_3, \hat{T}^\lambda_\mu\right] = \hbar\mu\hat{T}^\lambda_\mu \qquad \left[\hat{J}_\pm, \hat{T}^\lambda_\mu\right] = \hbar\sqrt{\lambda(\lambda+1) - \mu(\mu\pm1)}\,\hat{T}^\lambda_{\mu\pm1}$$

► **Example: Cartesian & spherical vector**

Cartesian vector operator $\hat{\vec{V}} \equiv (\hat{V}_1, \hat{V}_2, \hat{V}_3) \quad \Rightarrow \quad \hat{R}_{\mathbf{R}}\hat{V}_i\hat{R}^{-1}_{\mathbf{R}} = \sum_{j=1}^{3}\mathbf{R}^{-1}_{ij}\hat{V}_j$

Infinitesimal rotation around axis $\vec{n}$ leads to expression (sum. convention used):

$\hat{V}_i - \frac{i}{\hbar}\delta\phi[\hat{J}_k, \hat{V}_i]n_k = \hat{V}_i + \delta\phi\varepsilon_{ijk}\hat{V}_j n_k \quad \Rightarrow \quad \boxed{[\hat{J}_k, \hat{V}_i] = i\hbar\varepsilon_{kij}\hat{V}_j}$ alternative definition of Cartesian vector

Spherical components of the vector operator:

$$\begin{aligned} \hat{V}^1_{+1} &= -\tfrac{1}{\sqrt{2}}(\hat{V}_1 + i\hat{V}_2) \\ \hat{V}^1_0 &= \hat{V}_3 \\ \hat{V}^1_{-1} &= +\tfrac{1}{\sqrt{2}}(\hat{V}_1 - i\hat{V}_2) \end{aligned}$$

satisfy spherical tensor commut. relations

$$\left[\hat{J}_3, \hat{V}^1_0\right] = \left[\hat{J}_\pm, \hat{V}^1_{\pm1}\right] = 0$$

$$\left[\hat{J}_\pm, \hat{V}^1_{\mp1}\right] = \sqrt{2}\hbar\hat{V}^1_0, \quad \left[\hat{J}_\pm, \hat{V}^1_0\right] = \sqrt{2}\hbar\hat{V}^1_{\pm1}$$

Note: Relations given on p. 60 between (ψ_1, ψ_2, ψ_3) and $(\psi_{-1}, \psi_0, \psi_{+1})$ components of a vector wavefunction $\vec{\Psi}$ (spin-1 particle) differ in signs from the above formula because $\vec{\Psi}$ transforms as $\mathbf{R}\vec{\Psi}(\mathbf{R}^{-1}\vec{x})$ instead of $\mathbf{R}^{-1}\vec{\Psi}$.

► **Coupling of spherical tensors**

Let $\hat{A}^{\lambda_1}_{\mu_1}$ and $\hat{B}^{\lambda_2}_{\mu_2}$ be spherical tensors of ranks λ_1 and λ_2

$$\Rightarrow \quad \hat{T}^\lambda_\mu = \sum_{\mu_1,\mu_2} C^{\lambda\mu}_{\lambda_1\mu_1\lambda_2\mu_2}\hat{A}^{\lambda_1}_{\mu_1}\hat{B}^{\lambda_2}_{\mu_2} \equiv [\hat{A}^{\lambda_1}\times\hat{B}^{\lambda_2}]^\lambda_\mu$$ ≡ spherical tensor of rank λ

$$\hat{R}\,\hat{T}^{\lambda}_{\mu}\hat{R}^{-1}=\sum_{\mu_1,\mu_2}C^{\lambda\mu}_{\lambda_1\mu_1\lambda_2\mu_2}\underbrace{\hat{R}\,\hat{A}^{\lambda_1}_{\mu_1}\hat{R}^{-1}}_{\sum_{\mu'_1}D^{\lambda_1}_{\mu'_1\mu_1}\hat{A}^{\lambda_1}_{\mu'_1}}\underbrace{\hat{R}\,\hat{B}^{\lambda_2}_{\mu_2}\hat{R}^{-1}}_{\sum_{\mu'_2}D^{\lambda_2}_{\mu'_2\mu_2}\hat{B}^{\lambda_2}_{\mu'_2}}=\sum_{\mu_1,\mu_2\mu'_1,\mu'_2}\sum C^{\lambda\mu}_{\lambda_1\mu_1\lambda_2\mu_2}\underbrace{D^{\lambda_1}_{\mu'_1\mu_1}D^{\lambda_2}_{\mu'_2\mu_2}}_{\sum_{\lambda'}\sum_{\mu',\mu''}C^{\lambda'\mu'}_{\lambda_1\mu'_1\lambda_2\mu'_2}C^{\lambda'\mu''}_{\lambda_1\mu_1\lambda_2\mu_2}D^{\lambda'}_{\mu'\mu''}}\hat{A}^{\lambda_1}_{\mu'_1}\hat{B}^{\lambda_2}_{\mu'_2}$$

$$=\sum_{\lambda'}\sum_{\mu',\mu''}\sum_{\mu'_1,\mu'_2}\underbrace{\sum_{\mu_1,\mu_2}C^{\lambda\mu}_{\lambda_1\mu_1\lambda_2\mu_2}C^{\lambda'\mu''}_{\lambda_1\mu_1\lambda_2\mu_2}}_{\delta_{\lambda\lambda'}\delta_{\mu\mu''}}C^{\lambda'\mu'}_{\lambda_1\mu'_1\lambda_2\mu'_2}D^{\lambda'}_{\mu'\mu''}\hat{A}^{\lambda_1}_{\mu'_1}\hat{B}^{\lambda_2}_{\mu'_2}$$

$$=\sum_{\mu'}D^{\lambda}_{\mu'\mu}\overbrace{\sum_{\mu'_1,\mu'_2}C^{\lambda\mu'}_{\lambda_1\mu'_1\lambda_2\mu'_2}\hat{A}^{\lambda_1}_{\mu'_1}\hat{B}^{\lambda_2}_{\mu'_2}}^{\hat{T}^{\lambda}_{\mu'}}=\sum_{\mu'}D^{\lambda}_{\mu'\mu}\hat{T}^{\lambda}_{\mu'}$$

Conclusion: coupling of spherical tensors leads to other spherical tensors with ranks determined from the usual angular-momentum coupling relations.

Special case: **scalar coupling**

$$[\hat{A}^{\lambda}\times\hat{B}^{\lambda}]^0_0=\sum_{\mu}\underbrace{C^{00}_{\lambda\mu\lambda(-\mu)}}_{\frac{(-)^{\lambda-\mu}}{\sqrt{2\lambda+1}}}\hat{A}^{\lambda}_{\mu}\hat{B}^{\lambda}_{-\mu}=\boxed{\frac{(-)^{-\lambda}}{\sqrt{2\lambda+1}}\underbrace{\sum_{\mu}(-)^{\mu}\hat{A}^{\lambda}_{+\mu}\hat{B}^{\lambda}_{-\mu}}_{(\hat{A}^{\lambda}\cdot\hat{B}^{\lambda})}}$$

scalar product of tensor operators

■ Wigner-Eckart theorem

If spherical tensor operators are written in the angular-momentum eigenbasis, the corresponding matrix elements exhibit interesting properties: a large part of elements vanishes, the remaining ones satisfy certain relations. The rules behind this behavior come from the coupling of angular momenta. This is rather useful for instance if the amplitudes for a given multipolarity transition (represented by a tensorial transition operator of the respective rank) are computed.

▶ Properties of matrix elements of spherical tensors

$\{|ajm\rangle\}\equiv$ angular-momentum basis with a denoting remaining q. numbers

$\langle a'j'm'|\hat{T}^{\lambda}_{\mu}|ajm\rangle\equiv$ matrix elements of a general spherical tensor

Application of the definition properties of spherical tensors:

$$\text{(a)}\quad \langle a'j'm'|\underbrace{[\hat{J}_3,\hat{T}^{\lambda}_{\mu}]-\hbar\mu\hat{T}^{\lambda}_{\mu}}_{=0}|ajm\rangle=\hbar\underbrace{[(m'-m)-\mu)]}_{=0}\underbrace{\langle a'j'm'|\hat{T}^{\lambda}_{\mu}|ajm\rangle}_{\neq 0}\qquad\Rightarrow m+\mu=m'$$

$$\text{(b)}\quad \langle a'j'm'|\underbrace{[\hat{J}_{\pm},\hat{T}^{\lambda}_{\mu}]-\hbar\sqrt{\lambda(\lambda+1)-\mu(\mu\pm1)}\,\hat{T}^{\lambda}_{\mu\pm1}}_{=0}|ajm\rangle=0\qquad\Rightarrow$$

$$\sqrt{j'(j'+1)-m'(m'\mp1)}\langle a'j'(m'\mp1)|\hat{T}^{\lambda}_{\mu}|ajm\rangle-\sqrt{j(j+1)-m(m\pm1)}\langle a'j'm'|\hat{T}^{\lambda}_{\mu}|aj(m\pm1)\rangle=\sqrt{\lambda(\lambda+1)-\mu(\mu\pm1)}\langle a'j'm'|\hat{T}^{\lambda}_{\mu\pm1}|ajm\rangle$$

$$\langle a'j'(m'\mp1)|\hat{T}^{\lambda}_{\mu}|ajm\rangle=\sqrt{\frac{j(j+1)-m(m\pm1)}{j'(j'+1)-m'(m'\mp1)}}\langle a'j'm'|\hat{T}^{\lambda}_{\mu}|aj(m\pm1)\rangle+\sqrt{\frac{\lambda(\lambda+1)-\mu(\mu\pm1)}{j'(j'+1)-m'(m'\mp1)}}\langle a'j'm'|\hat{T}^{\lambda}_{\mu\pm1}|ajm\rangle$$

► The last relation can be compared with the **recursive relation for the Clebsch-Gordan coeffs.** (Sec. 4.1) with substitutions $\left.\begin{matrix} j_1,m_1 \\ j_2,m_2 \\ j,m \end{matrix}\right\} \mapsto \left\{\begin{matrix} j,m \\ \lambda,\mu \\ j',m' \end{matrix}\right.$

$$C^{j'(m'\mp 1)}_{jm\lambda\mu} = \sqrt{\frac{j(j+1)-m(m\pm 1)}{j'(j'+1)-m'(m'\mp 1)}}\, C^{j'm'}_{j(m\pm 1)\lambda\mu} + \sqrt{\frac{\lambda(\lambda+1)-\mu(\mu\pm 1)}{j'(j'+1)-m'(m'\mp 1)}}\, C^{j'm'}_{jm\lambda(\mu\pm 1)} \qquad \pm \mapsto \mp$$

With replacement $\boxed{\langle a'j'm'|\hat{T}^{\lambda}_{\mu}|ajm\rangle \leftrightarrow C^{j'm'}_{jm\lambda\mu}}$ both relations are *the same*

$\Rightarrow$ matrix elements $\langle a'j'm'|\hat{T}^{\lambda}_{\mu}|ajm\rangle$ for fixed j,λ,j' can be constructed from the same recursive relations as the corresponding Clebsch-Gordan coeffs. $C^{j'm'}_{jm\lambda\mu}$

$\Rightarrow \langle a'j'm'|\hat{T}^{\lambda}_{\mu}|ajm\rangle \propto C^{j'm'}_{jm\lambda\mu}$

► **Wigner-Eckart theorem**

$$\boxed{\langle a'j'm'|\hat{T}^{\lambda}_{\mu}|ajm\rangle = \underbrace{\langle a'j'||\hat{T}^{\lambda}||aj\rangle}_{\textbf{reduced matrix element}}\; C^{j'm'}_{jm\lambda\mu}}$$

This means the following:

(a) The dependence on projection q. numbers is just that of the CG coefficient

(b) The dependence on j, j', λ is involved in the so-called reduced matrix elements $\equiv \langle a'j'||\hat{T}^{\lambda}||aj\rangle$. Their values (independent of m, μ, m') cannot be determined just from the algebraic properties of angular-momentum operators but need to be evaluated for each particular case.

Selection rules for $\boxed{\langle a'j'm'|\hat{T}^{\lambda}_{\mu}|ajm\rangle \neq 0}$ are: $\boxed{\begin{matrix} |j-\lambda| \leq j' \leq (j+\lambda) \\ m+\mu = m' \end{matrix}}$

◄ **Historical remark**

1927: E. Wigner introduces D-matrices and applies the rotation group in QM

1930: C. Eckart publishes and applies his formulation of the W.-E. theorem

1942: G. Racah further extends the use of spherical tensors in spectroscopy

5. APPROXIMATION TECHNIQUES

As in any other branch of physics, realistic calculations can be seldom performed exactly. Various approximation techniques are of primary importance.

5.1 Variational method

In classical physics, variational principles represent an autonomous formulation of the fundamental laws of nature. The role of these principles in nonrelativistic quantum mechanics is not as important. Nevertheless, they constitute a very useful approximation method.

■ Dynamical variational principle

Let us start with a variational formulation of the dynamical Schrödinger equation. Trying to keep the formalism parallel—as much as possible—to that of classical mechanics, we employ the notion of independent bra and ket variations, which may seem a bit counterintuitive.

▶ We search a quantum analog of classical variational principle

$$\delta \int_{t_1}^{t_2} \mathcal{L}[\vec{x}(t), \dot{\vec{x}}(t)]\, dt = 0 \quad \text{with boundary conditions} \begin{cases} \delta\vec{x}(t_1) = 0 = \delta\vec{x}(t_2) \\ \delta\dot{\vec{x}}(t_1) \neq 0 \neq \delta\dot{\vec{x}}(t_2) \end{cases}$$

$$\underbrace{\delta \int_{t_1}^{t_2} \left\langle \psi(t) \middle| i\hbar\tfrac{d}{dt} - \hat{H} \middle| \psi(t) \right\rangle dt}_{\int_{t_1}^{t_2}\left[\langle\delta\psi'(t)|i\hbar\frac{d}{dt}-\hat{H}|\psi(t)\rangle+\langle\psi(t)|i\hbar\frac{d}{dt}-\hat{H}|\delta\psi(t)\rangle\right] dt} = 0 \quad \text{with} \begin{cases} \textbf{ket variation } |\delta\psi(t)\rangle \\ \boxed{|\delta\psi(t_1)\rangle = 0 = |\delta\psi(t_2)\rangle} \\ \textbf{bra variation } \langle\delta\psi'(t)| \\ \boxed{\langle\delta\psi'(t_1)| \neq 0 \neq \langle\delta\psi'(t_2)|} \end{cases}$$

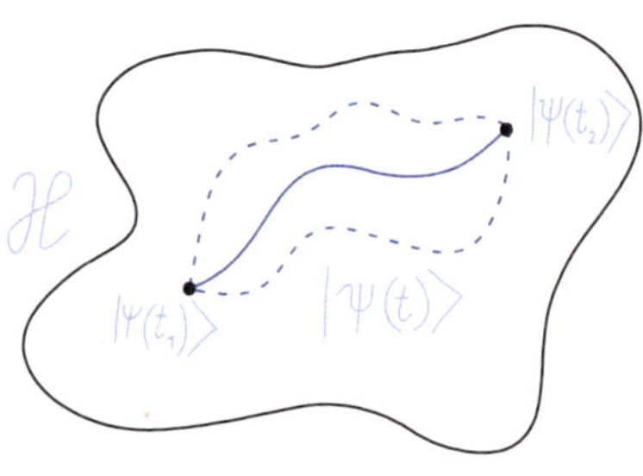

We consider the variations of kets and bras independently, distinguishing 4 different entities:
$\langle\delta\psi(t)| \leftrightarrow |\delta\psi(t)\rangle$ & $\langle\delta\psi'(t)| \leftrightarrow |\delta\psi'(t)\rangle$
The only correlation between $|\delta\psi(t)\rangle$ and $\langle\delta\psi'(t)|$ is through the conserved normalization $\langle\psi|\psi\rangle = 1$

$$\Rightarrow \delta\langle\psi|\psi\rangle = \boxed{\langle\delta\psi'(t)|\psi(t)\rangle + \langle\psi(t)|\delta\psi(t)\rangle = 0}$$

▶ Proof of the variational principle (we show that it implies Schrödinger eq.):

$$\int_{t_1}^{t_2}\left[\langle\delta\psi'(t)|i\hbar\tfrac{d}{dt} - \hat{H}|\psi(t)\rangle + \underbrace{\langle\psi(t)|i\hbar\tfrac{d}{dt} - \hat{H}|\delta\psi(t)\rangle}_{\langle\delta\psi(t)|i\hbar\frac{d}{dt}-\hat{H}|\psi(t)\rangle^*+i\hbar\frac{d}{dt}\langle\psi|\delta\psi\rangle}\right] dt$$

$$= \int_{t_1}^{t_2}\left[\langle\delta\psi'(t)|i\hbar\tfrac{d}{dt} - \hat{H}|\psi(t)\rangle + \langle\delta\psi(t)|i\hbar\tfrac{d}{dt} - \hat{H}|\psi(t)\rangle^*\right] dt + i\hbar\overbrace{\left[\langle\psi|\delta\psi\rangle\right]_{t_1}^{t_2}}^{0}$$

$$\stackrel{!}{=} 0 \quad \forall\, \langle\delta\psi'(t)|\ \&\ \langle\delta\psi(t)| \text{ (with the above constraints)} \boxed{\Rightarrow \left(i\hbar\tfrac{d}{dt} - \hat{H}\right)|\psi(t)\rangle = 0}$$

▶ **Note**: If $\langle\delta\psi'(t)|=\langle\delta\psi(t)|$ (kets & bras varied in the same way), we would only get $\mathrm{Re}\langle\delta\psi(t)|i\hbar\frac{d}{dt} - \hat{H}|\psi(t)\rangle = 0$, which would not imply Schrödinger eq. An alternative treatment of the variational principle (without independent bra & ket variations) is possible if the variation is performed only in kets (or bras): $\delta\langle\psi|i\hbar\frac{d}{dt} - \hat{H}|\psi\rangle \equiv \langle\psi|i\hbar\frac{d}{dt} - \hat{H}|\delta\psi\rangle$

■ Stationary variational procedure

The dynamical variational principle for nonrelativistic QM, derived in the previous paragraph, is not very impressive. Indeed, the Schrödinger equation can be recognized in it already before its formal derivation. On the other hand, the variational techniques are rather useful for stationary problems—in approximating the lowest eigenstates of complicated Hamiltonians.

▶ Transition to stationary problems

Assume $|\psi(t)\rangle = e^{-\frac{i}{\hbar}Et}|\psi\rangle \qquad \Rightarrow \qquad \begin{cases} |\delta\psi(t)\rangle = e^{-\frac{i}{\hbar}Et}|\delta\psi\rangle \\ \langle\delta\psi'(t)| = e^{+\frac{i}{\hbar}Et}\langle\delta\psi'| \end{cases}$

$$\int_{t_1}^{t_2}\left[\langle\delta\psi'(t)|i\hbar\tfrac{d}{dt}-\hat{H}|\psi(t)\rangle + \langle\psi(t)|i\hbar\tfrac{d}{dt}-\hat{H}|\delta\psi(t)\rangle\right]dt$$

$$=\int_{t_1}^{t_2}\Big[\underbrace{\langle\delta\psi'|E-\hat{H}|\psi\rangle + \langle\psi|E-\hat{H}|\delta\psi\rangle}_{\delta\langle\psi|E-\hat{H}|\psi\rangle}\Big]dt = \underbrace{(t_2-t_1)}_{\neq 0}\ \underbrace{\delta\langle\psi|E-\hat{H}|\psi\rangle}_{\overset{!}{=}0} = 0$$

▶ Variational principle for stationary problems

$$\boxed{\delta\langle\psi|\hat{H}-E|\psi\rangle = 0} \qquad \Leftrightarrow \qquad \boxed{\delta\langle\psi|\hat{H}|\psi\rangle = 0 \quad \& \quad \langle\psi|\psi\rangle = 1}$$

with a Lagrange multiplier $\delta\left[\langle\psi|\hat{H}|\psi\rangle - E\langle\psi|\psi\rangle\right] = 0$ — with explicit normalization constraint

If the above variational conditions are applied in the whole space $\mathcal{H}$, they yield the **ground state**. To obtain the first excited state, the conditions must be applied only within the orthogonal complement in $\mathcal{H}$ of the ground-state energy subspace. Increasing restrictions are needed to get higher excited states.

▶ More practical formulation

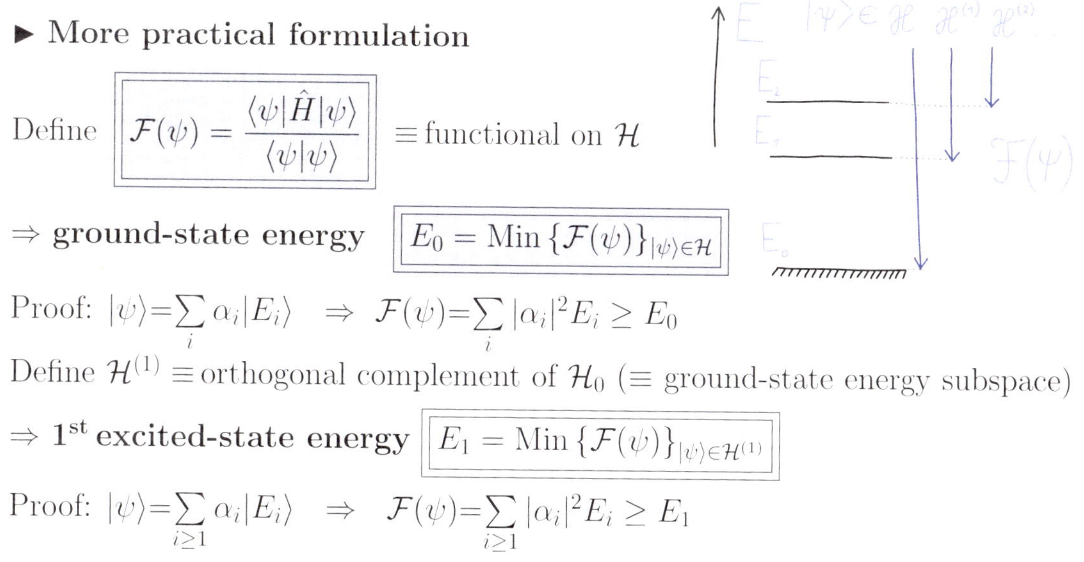

Define $\boxed{\mathcal{F}(\psi) = \dfrac{\langle\psi|\hat{H}|\psi\rangle}{\langle\psi|\psi\rangle}}$ $\equiv$ functional on $\mathcal{H}$

$\Rightarrow$ **ground-state energy** $\boxed{E_0 = \text{Min}\,\{\mathcal{F}(\psi)\}_{|\psi\rangle\in\mathcal{H}}}$

Proof: $|\psi\rangle = \sum_i \alpha_i|E_i\rangle \quad \Rightarrow \quad \mathcal{F}(\psi) = \sum_i |\alpha_i|^2 E_i \geq E_0$

Define $\mathcal{H}^{(1)} \equiv$ orthogonal complement of $\mathcal{H}_0$ ($\equiv$ ground-state energy subspace)

$\Rightarrow$ **1st excited-state energy** $\boxed{E_1 = \text{Min}\,\{\mathcal{F}(\psi)\}_{|\psi\rangle\in\mathcal{H}^{(1)}}}$

Proof: $|\psi\rangle = \sum_{i\geq 1} \alpha_i|E_i\rangle \quad \Rightarrow \quad \mathcal{F}(\psi) = \sum_{i\geq 1} |\alpha_i|^2 E_i \geq E_1$

Et cetera for **higher states**...

▶ Ritz variational method

Choose a suitable (for the given $\hat{H}$) subset of test vectors $|\psi(\boldsymbol{a})\rangle$ controlled by continuous real parameters $\boldsymbol{a} \equiv \{a_1, a_2, \ldots a_n\}$ forming a domain $\mathcal{D}_{\boldsymbol{a}} \subset \mathbb{R}^n$.

Functional $\mathcal{F}(\psi) \longmapsto$ Function $\boxed{\mathcal{F}(\boldsymbol{a}) = \dfrac{\langle\psi(\boldsymbol{a})|\hat{H}|\psi(\boldsymbol{a})\rangle}{\langle\psi(\boldsymbol{a})|\psi(\mathbf{a})\rangle}}$ on $\mathcal{D}_{\boldsymbol{a}}$

The search for an approximate ground state, and eventually also for approximate excited states, is performed within this set of vectors:

Ground state

$\mathrm{Min}\{\mathcal{F}(\boldsymbol{a})\}_{\boldsymbol{a}\in\mathcal{D}_{\boldsymbol{a}}} \equiv \tilde{E}_0 \gtrsim E_0$ ... estimate of the g.s. energy, and the corresp.
$|\psi(\boldsymbol{a}_0)\rangle \equiv |\tilde{\psi}_0\rangle \approx |\psi_0\rangle$ ... estimate of the g.s. eigenvector

Excited states

If the set of test vectors is sufficiently rich, select a subdomain $\mathcal{D}_{\boldsymbol{a}}^{(1)} \subset \mathcal{D}_{\boldsymbol{a}}$ such that $\langle\psi(\boldsymbol{a})|\psi(\boldsymbol{a}_0)\rangle = 0\ \forall \boldsymbol{a} \in \mathcal{D}_{\boldsymbol{a}}^{(1)}$

$\mathrm{Min}\{\mathcal{F}(\boldsymbol{a})\}_{\boldsymbol{a}\in\mathcal{D}_{\boldsymbol{a}}^{(1)}} \equiv \tilde{E}_1 \gtrsim E_1$ ... estimate of the 1st exc. energy, and the corresp.
$|\psi(\boldsymbol{a}_1)\rangle \equiv |\tilde{\psi}_1\rangle \approx |\psi_1\rangle$ ... estimate of the respective eigenvector

Et cetera for higher states...

◀ **Historical remark**
1909: W. Ritz publishes a method for solving variational problems
1926: E. Schrödinger uses variational arguments in derivation of stationary Sch. eq.
1930's: P. Dirac, J. Frenkel *et al.* formulate dynamical variational principle of QM

5.2 Stationary perturbation method

The stationary perturbation method is very useful if the actual Hamiltonian $\hat{H}$ is just a small modification of a simpler Hamiltonian $\hat{H}_0$, whose eigensolutions are known. The difference between both Hamiltonians represents a perturbation which is quantified by a dimensionless parameter λ. If expressing the eigensolutions of $\hat{H}$ as power series in λ, one may believe that high-power terms will naturally die out.

■ General setup & equations

The perturbation method is entirely based on a few general equations that can be easily derived.

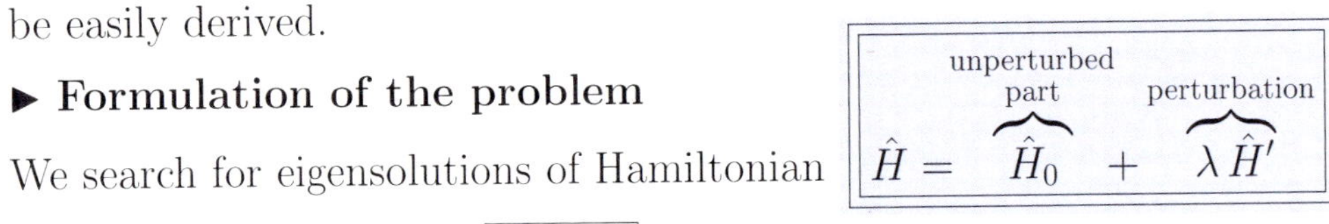

▶ **Formulation of the problem**

We search for eigensolutions of Hamiltonian $\hat{H} = \underbrace{\hat{H}_0}_{\text{unperturbed part}} + \underbrace{\lambda \hat{H}'}_{\text{perturbation}}$

Dimensionless parameter $\boxed{\lambda \ll 1}$ ⇒ we deal with a **"small perturbation"**

We assume that the scaling of $\hat{H}'$ by parameter λ was performed in such a way that its matrix elements in the unperturbed eigenbasis are of the same average size as matrix elements of $\hat{H}_0$: $\left\langle \langle\psi_{0i}|\hat{H}'|\psi_{0j}\rangle \right\rangle_{i,j} \sim \left\langle E_{0i} \right\rangle_i$

In general: $\boxed{[\hat{H}_0, \hat{H}'] \neq 0}$ ⇒ incompatible terms ⇒ nontrivial effect of $\hat{H}'$

For each level i we know the unperturbed energy E_{0i} and eigenvector $|\psi_{0i}\rangle$

Task: to express the effect of perturbation in the form of **power-law series**

$$\begin{aligned} E_i(\lambda) &= \lambda^0 E_{0i} + \underbrace{\lambda^1 E_{1i} + \lambda^2 E_{2i} + \lambda^3 E_{3i} + \cdots}_{E'_i(\lambda)} \\ |\psi_i(\lambda)\rangle &= \lambda^0|\psi_{0i}\rangle + \underbrace{\lambda^1|\psi_{1i}\rangle + \lambda^2|\psi_{2i}\rangle + \lambda^3|\psi_{3i}\rangle + \cdots}_{|\psi'_i(\lambda)\rangle} \end{aligned}$$

unnormalized vector

► **Orthogonality condition** for the eigenvector correction $\langle\psi_{0i}|\psi'_i(\lambda)\rangle = 0$

This is a natural requirement since the changes in the unperturbed vector component $|\psi_{0i}\rangle$ can be taken into account by normalization

► Term $\propto \lambda^n \equiv$ **n^th^order correction**
The sum of terms up to n^{th}order

$$E_i^{(n)}(\lambda) = E_{0i} + \overbrace{\sum_{n'=1}^{n} \lambda^{n'} E_{n'i}}^{E_i^{(n)\prime}(\lambda)} \qquad |\psi_i^{(n)}(\lambda)\rangle = |\psi_{0i}\rangle + \overbrace{\sum_{n'=1}^{n} \lambda^{n'} |\psi_{n'i}\rangle}^{|\psi_i^{(n)\prime}(\lambda)\rangle}$$

Since $\lambda \ll 1$, we may hope in sufficiently fast convergence to exact solution :

$$\begin{cases} E_i(\lambda) \approx E_i^{(n_{\text{up}})}(\lambda) \\ |\psi_i(\lambda)\rangle \approx |\psi_i^{(n_{\text{up}})}(\lambda)\rangle \end{cases}$$

► **Normalization** in n^{th}order: $|\psi_i^{(n)}(\lambda)\rangle = \overbrace{\frac{1}{\sqrt{1+\langle\psi_i^{(n)\prime}(\lambda)|\psi_i^{(n)\prime}(\lambda)\rangle}}}^{\mathcal{N}^{(n)}(\lambda)} \left[|\psi_{0i}\rangle + |\psi_i^{(n)\prime}(\lambda)\rangle\right]$

► The n^{th}order correction to eigenvectors expressed via **expansion in the unperturbed eigenbasis**: $|\psi_{ni}\rangle = \sum_{k\neq i} a_{nik}|\psi_{0k}\rangle$ $a_{nii} = 0$ following from the orthogonality condition

► **Equations for corrections of increasing order**

Schrödinger equation:
$$\left[\hat{H}_0 + \lambda\,\hat{H}'\right] \left(|\psi_{0i}\rangle + \lambda|\psi_{1i}\rangle + \lambda^2|\psi_{2i}\rangle + \lambda^3|\psi_{3i}\rangle + \cdots\right)$$
$$= \left[E_{0i} + \lambda E_{1i} + \lambda^2 E_{2i} + \lambda^3 E_{3i} + \cdots\right] \left(|\psi_{0i}\rangle + \lambda|\psi_{1i}\rangle + \lambda^2|\psi_{2i}\rangle + \lambda^3|\psi_{3i}\rangle + \cdots\right)$$

Comparison of different orders $\propto \lambda^n$:

$$\begin{aligned} \hat{H}_0|\psi_{0i}\rangle &= E_{0i}|\psi_{0i}\rangle & n=0 \\ \hat{H}_0|\psi_{1i}\rangle + \hat{H}'|\psi_{0i}\rangle &= E_{0i}|\psi_{1i}\rangle + E_{1i}|\psi_{0i}\rangle & n=1 \\ \hat{H}_0|\psi_{2i}\rangle + \hat{H}'|\psi_{1i}\rangle &= E_{0i}|\psi_{2i}\rangle + E_{1i}|\psi_{1i}\rangle + E_{2i}|\psi_{0i}\rangle & n=2 \\ \cdots \quad &\quad \cdots & \\ \hat{H}_0|\psi_{ni}\rangle + \hat{H}'|\psi_{(n-1)i}\rangle &= \sum_{n'=0}^{n} E_{n'i}|\psi_{(n-n')i}\rangle & \text{general } n \end{aligned}$$

▶ Two **possibilities**:

the unperturbed level $E_{0i} \equiv \begin{cases} \text{(a) } \textbf{non-degenerate} \\ \quad |\psi_{0i}\rangle \text{ unique} \\ \text{(b) } \textbf{degenerate} \\ \quad |\psi_{0i}\rangle \equiv \{|\psi_{0i;1}\rangle, |\psi_{0i;2}\rangle, \ldots |\psi_{0i;d_i}\rangle\} \end{cases}$

■ Nondegenerate case

The nondegenerate case is easier to start with.

▶ Zeroth-order solution

The n=0 eq. $\Rightarrow$ the 0thorder solutions $\equiv$ unperturbed energy & eigenvector
This yields unique solution only in the nondegenerate case!

▶ First-order correction to energy

Multiply the n=1 eq. by $\langle\psi_{0i}|$

$$\Rightarrow \underbrace{\langle\psi_{0i}|\hat{H}_0|\psi_{1i}\rangle}_{E_{0i}\langle\psi_{0i}|\psi_{1i}\rangle} + \langle\psi_{0i}|\hat{H}'|\psi_{0i}\rangle = E_{0i}\langle\psi_{0i}|\psi_{1i}\rangle + E_{1i}\overbrace{\langle\psi_{0i}|\psi_{0i}\rangle}^{1}$$

$$\Rightarrow \quad \boxed{E_{1i} = \langle\psi_{0i}|\hat{H}'|\psi_{0i}\rangle}$$

▶ First-order correction to eigenvector

The n=1 eq. $\Rightarrow$

$$\left[\hat{H}_0 - E_{0i}\right]\underbrace{\left(\sum_{k\neq i} a_{1ik}|\psi_{0k}\rangle\right)}_{|\psi_{1i}\rangle} = \left[\langle\psi_{0i}|\hat{H}'|\psi_{0i}\rangle - \hat{H}'\right]|\psi_{0i}\rangle$$

Multiply by $\langle\psi_{0j}|$ for $j \neq i$

$$\Rightarrow \quad \sum_{k\neq i}(E_{0j} - E_{0i})\, a_{1ik}\overbrace{\langle\psi_{0j}|\psi_{0k}\rangle}^{\delta_{jk}} = \langle\psi_{0i}|\hat{H}'|\psi_{0i}\rangle\overbrace{\langle\psi_{0j}|\psi_{0i}\rangle}^{0} - \langle\psi_{0j}|\hat{H}'|\psi_{0i}\rangle$$

$$\Rightarrow \quad a_{1ij} = -\frac{\langle\psi_{0j}|\hat{H}'|\psi_{0i}\rangle}{E_{0j} - E_{0i}} \quad \Rightarrow \quad \boxed{|\psi_{1i}\rangle = \sum_{j\neq i}\frac{\langle\psi_{0j}|\hat{H}'|\psi_{0i}\rangle}{E_{0i} - E_{0j}}|\psi_{0j}\rangle}$$

Multiplication by $\langle\psi_{0i}|$ yields just identity 0=0 $\Rightarrow a_{1ii}$ undetermined
$\Rightarrow$ consistent with the above setting $a_{1ii} = 0$

▶ Second-order correction to energy

Multiply the n=2 eq. by $\langle\psi_{0i}|$

$$\Rightarrow \quad E_{0i}\langle\psi_{0i}|\psi_{2i}\rangle + \langle\psi_{0i}|\hat{H}'|\psi_{1i}\rangle = E_{0i}\langle\psi_{0i}|\psi_{2i}\rangle + E_{1i}\overbrace{\langle\psi_{0i}|\psi_{1i}\rangle}^{0} + E_{2i}\overbrace{\langle\psi_{0i}|\psi_{0i}\rangle}^{1}$$

$$\Rightarrow \quad E_{2i} = \langle\psi_{0i}|\hat{H}'|\psi_{1i}\rangle \quad \Rightarrow \quad \boxed{E_{2i} = \sum_{j\neq i}\frac{|\langle\psi_{0j}|\hat{H}'|\psi_{0i}\rangle|^2}{E_{0i} - E_{0j}}}$$

▶ General-order correction to energy

Multiply the general-n eq. by $\langle\psi_{0i}|$

$$E_{0i}\underbrace{\langle\psi_{0i}|\psi_{ni}\rangle}_{0} + \langle\psi_{0i}|\hat{H}'|\psi_{(n-1)i}\rangle = \sum_{n'=0}^{n} E_{n'i}\underbrace{\langle\psi_{0i}|\psi_{(n-n')i}\rangle}_{\delta_{nn'}} \quad \Rightarrow \quad \boxed{E_{ni} = \langle\psi_{0i}|\hat{H}'|\psi_{(n-1)i}\rangle}$$

The $n^{\rm th}$order correction to energy determined from $(n-1)^{\rm th}$order correction to the eigenvector

▶ **General-order correction to eigenvector**

The general-n eq. $\Rightarrow$ $$\left[\hat{H}_0 - E_{0i}\right]\underbrace{\left(\sum_{k\neq i} a_{nik}|\psi_{0k}\rangle\right)}_{|\psi_{ni}\rangle} + \hat{H}'\underbrace{\left(\sum_{k\neq i} a_{(n-1)ik}|\psi_{0k}\rangle\right)}_{|\psi_{(n-1)i}\rangle} = \sum_{n'=1}^{n} E_{n'i}\underbrace{\left(\sum_{k\neq i} a_{(n-n')ik}|\psi_{0k}\rangle\right)}_{|\psi_{(n-n')i}\rangle}$$

Multiply by $\langle\psi_{0j}|$ for $j\neq i$

$$\Rightarrow \quad [E_{0j}-E_{0i}]\,a_{nij} + \sum_{k\neq i}\langle\psi_{0j}|\hat{H}'|\psi_{0k}\rangle a_{(n-1)ik} = \sum_{n'=1}^{n-1} E_{n'i}a_{(n-n')ij}$$

$$\Rightarrow \quad \boxed{a_{nij} = \frac{1}{E_{0j}-E_{0i}}\left[\sum_{n'=1}^{n-1} E_{n'i}a_{(n-n')ij} - \sum_{k\neq i}\langle\psi_{0j}|\hat{H}'|\psi_{0k}\rangle a_{(n-1)ik}\right]}$$

The $n^{\rm th}$order correction to the eigenvector determined from the corrections to energy & eigenvector of all lower orders $1,2,\ldots,(n-1)$

If $a_{n'ii}=0\ \forall\, n'\leq(n-1)$, the multiplication by $\langle\psi_{ni}|$ yields just identity 0=0 $\Rightarrow a_{nii}$ undetermined $\Rightarrow$ consistent with the above setting $a_{nii}=0$

■ Degenerate case

In the above-derived corrections for a nondegenerate level we noticed the denominators containing the differences of unperturbed energies. These imply that if levels of the unperturbed system come close together, the size of corrections quickly increases. In other words, a generic perturbation gets more efficient in denser parts of the spectrum. But what about if a particular level becomes exactly degenerate? In that case, the derivation presented above fails and must be redone from the scratch.

▶ E_{0i} has degeneracy subspace $\{|\psi_{0i:1}\rangle, |\psi_{0i:2}\rangle, \ldots |\psi_{0i:d_i}\rangle\}$ with $\langle\psi_{0i:l}|\psi_{0i:k}\rangle=\delta_{kl}$ dimension d_i $\Rightarrow$ The n=0 eq. does not determine which of the eigenvectors $|\psi_{0i:k}\rangle$ represents the zeroth-order solution. Assume $\boxed{|\psi_{0i}\rangle = \sum_{k=1}^{d_i}\alpha_k|\psi_{0i:k}\rangle}$

$\alpha_k \equiv$ unknown coeffs.

▶ **First-order solution**

The n=1 eq.: $$\hat{H}_0|\psi_{1i}\rangle + \sum_{k=1}^{d_i}\alpha_k\hat{H}'|\psi_{0i:k}\rangle = E_{0i}|\psi_{1i}\rangle + E_{1i}\sum_{k=1}^{d_i}\alpha_k|\psi_{0i:k}\rangle$$

Multiply by $\langle\psi_{0i:l}|$:

$$E_{0i}\langle\psi_{0i:l}|\psi_{1i}\rangle + \sum_{k=1}^{d_i}\alpha_k\langle\psi_{0i:l}|\hat{H}'|\psi_{0i:k}\rangle = E_{0i}\langle\psi_{0i:l}|\psi_{1i}\rangle + E_{1i}\sum_{k=1}^{d_i}\alpha_k\overbrace{\langle\psi_{0i:l}|\psi_{0i:k}\rangle}^{\delta_{kl}} \quad \Rightarrow$$

$$\sum_{k=1}^{d_i} \langle \psi_{0i;l} | \hat{H}' | \psi_{0i;k} \rangle \alpha_k = E_{1i} \alpha_l$$

$\Rightarrow$ matrix form of this equation:

$$\begin{pmatrix} \langle \psi_{0i;1} | \hat{H}' | \psi_{0i;1} \rangle & \langle \psi_{0i;1} | \hat{H}' | \psi_{0i;2} \rangle & \dots \\ \langle \psi_{0i;2} | \hat{H}' | \psi_{0i;1} \rangle & \langle \psi_{0i;2} | \hat{H}' | \psi_{0i;2} \rangle & \dots \\ \vdots & & \ddots \end{pmatrix} \begin{pmatrix} \alpha_1 \\ \alpha_2 \\ \vdots \end{pmatrix} = E_{1i} \begin{pmatrix} \alpha_1 \\ \alpha_2 \\ \vdots \end{pmatrix}$$

$\equiv$ **diagonalization of the perturbation matrix** in the degeneracy subspace

Note: the degeneracy subspace is *not* in general invariant under $\hat{H}'$ (since $[\hat{H}_0, \hat{H}'] \neq 0$), but the above formula implicitly **projects** the action of $\hat{H}'$ to the degeneracy subspace prior the diagonalization.

▶ Zeroth-order eigenstates & first-order energies

In general, d_i energy solutions of polynomial eq. $\quad \mathrm{Det} \begin{pmatrix} H'_{11}-E_{1i} & H'_{12} & \dots \\ H'_{21} & H'_{22}-E_{1i} & \dots \\ \vdots & & \ddots \end{pmatrix} = 0$

$\Rightarrow \boxed{E_{1i} \mapsto E_{1i;k}} \quad k = 1, 2, \dots d_i \quad$ **degeneracy lifting** in 1storder correction

$$\begin{pmatrix} H'_{11} & H'_{12} & \dots \\ H'_{21} & H'_{22} & \dots \\ \vdots & & \ddots \end{pmatrix} \begin{pmatrix} \alpha_{1;k} \\ \alpha_{2;k} \\ \vdots \end{pmatrix} = E_{1i;k} \begin{pmatrix} \alpha_{1;k} \\ \alpha_{2;k} \\ \vdots \end{pmatrix}$$

this eq. selects the eigenvector associated with the correction $E_{1i;k}$ $\quad \boxed{|\psi_{0i}\rangle \mapsto |\tilde{\psi}_{0i;k}\rangle}$

Eigenfunction in 0thorder: $\quad$ Energy up to 1storder:

$$\boxed{|\tilde{\psi}_{0i;k}\rangle = \sum_{l=1}^{d_i} \alpha_{l;k} |\psi_{0i;l}\rangle} \quad \Leftrightarrow \quad \boxed{E^{(1)}_{i;k}(\lambda) = E_{0i} + \lambda\, E_{1i;k}} \quad k = 1, 2, \dots d_i$$

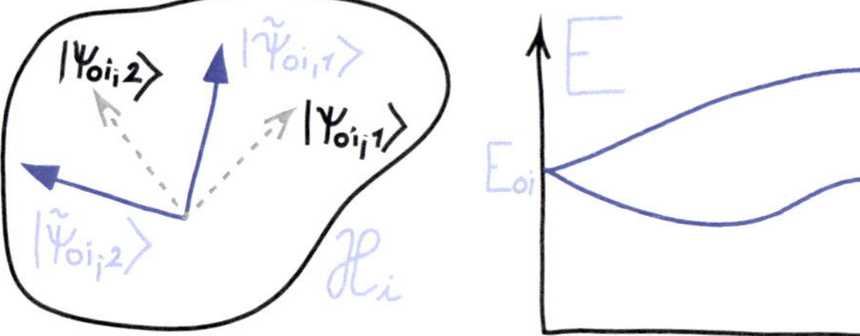

▶ Special case $d=2$

$$\begin{pmatrix} H'_{11} & H'_{12} \\ H'_{21} & H'_{22} \end{pmatrix} \begin{pmatrix} \alpha_1 \\ \alpha_2 \end{pmatrix} = E_{1i} \begin{pmatrix} \alpha_1 \\ \alpha_2 \end{pmatrix} \quad \Rightarrow \quad \mathrm{Det} \begin{pmatrix} H'_{11}-E_{1i} & H'_{12} \\ H'_{21} & H'_{22}-E_{1i} \end{pmatrix} = 0$$

$$\Rightarrow \text{2 solutions:} \quad E_{1i\pm} = \frac{H'_{11}+H'_{22}}{2} \pm \overbrace{\sqrt{\left(\frac{H'_{11}-H'_{22}}{2}\right)^2 + H'_{12} H'_{21}}}^{D}$$

$$\begin{pmatrix} H'_{11} & H'_{12} \\ H'_{21} & H'_{22} \end{pmatrix} \begin{pmatrix} \alpha_{1\pm} \\ \alpha_{2\pm} \end{pmatrix} = E_{1i\pm} \begin{pmatrix} \alpha_{1\pm} \\ \alpha_{2\pm} \end{pmatrix}$$

Lowest-order eigenfunctions & energies: $\begin{cases} |\tilde{\psi}_{0i\pm}\rangle = \alpha_{1\pm} |\psi_{0i;1}\rangle + \alpha_{2\pm} |\psi_{0i;2}\rangle \\ E^{(1)}_{i\pm}(\lambda) = E_{0i} + \lambda E_{1i\pm} \end{cases}$

▶ Higher-order corrections

Diagonalize the perturbation $\hat{H}'$ in the degeneracy subspace of every level

$\Rightarrow$ orthonormal basis $\left\{ \left\{ |\tilde{\psi}_{0i;k}\rangle \right\}_{k=1}^{d_i} \right\}_i \quad \Rightarrow \quad \langle \tilde{\psi}_{0i;l} | \hat{H}' | \tilde{\psi}_{0i;k} \rangle = 0$ for $l \neq k$

$\Rightarrow$ the procedure used in non-degenerate case can be reiterated without problems with zero energy denominators

$$E_{i;k}(\lambda) = E_{0i} + \lambda\langle\tilde{\psi}_{0i;k}|\hat{H}'|\tilde{\psi}_{0i;k}\rangle + \lambda^2 \sum_{j(\neq i)} \sum_{l=1}^{d_j} \frac{|\langle\tilde{\psi}_{0j;l}|\hat{H}'|\tilde{\psi}_{0i;k}\rangle|^2}{E_{0i}-E_{0j}} + \mathcal{O}(\lambda^3)$$

$$|\psi_{i;k}(\lambda)\rangle = |\tilde{\psi}_{0i;k}\rangle + \lambda \sum_{j(\neq i)} \sum_{l=1}^{d_j} \frac{\langle\tilde{\psi}_{0j;l}|\hat{H}'|\tilde{\psi}_{0i;k}\rangle}{E_{0i}-E_{0j}} |\tilde{\psi}_{0j;l}\rangle + \mathcal{O}(\lambda^2)$$

◄ **Historical remark**

1860's: Ch.-E. Delaunay introduces a perturb. analysis of Earth-Moon-Sun problem
1894: Lord Rayleigh studies harmonic vibrations in presence of small inhomogenities
1926: E. Schrödinger introduces the perturbation theory to QM

■ Application in atomic physics

The primary domain of application of the perturbation theory in the old-day quantum theory was atomic physics. Indeed, the plain hydrogen Hamiltonian needs to be corrected for some subtle internal effects as well as in presence of external electric or magnetic fields. The Hamiltonian of multielectron atoms (starting from helium) must contain (besides the electron-nucleus interactions) also all the electron-electron interactions. All these corrections are naturally treated in terms of the perturbation theory.

► Alternative eigensolutions of the hydrogen atom

Plain hydrogen Hamiltonian: $\boxed{\hat{H}_0 = -\frac{\hbar^2}{2M}\Delta - \frac{e^2}{4\pi\epsilon_0}\frac{1}{r}}$

$a_{\rm B} = \frac{4\pi\epsilon_0\hbar^2}{Me^2} \doteq 0.53\cdot 10^{-10}\,\text{m}$ Bohr radius $\qquad E_n = -\overbrace{\frac{e^2}{4\pi\epsilon_0}\frac{1}{a_{\rm B}}}^{\sim\frac{Mc^2}{137^2}}\frac{1}{2n^2}$ $\quad (n{=}1,2,3\ldots)$

$[\hat{L}_i, \hat{H}_0] = 0 = [\hat{S}_i, \hat{H}_0] \quad \Rightarrow \quad [\underbrace{\hat{L}_i + \hat{S}_i}_{\hat{J}_i}, \hat{H}_0] = 0$

for $i{=}1,2,3$

Uncoupled eigenstates: $\boxed{|\psi_{nlm_lm_s}\rangle} \equiv \overbrace{R_{nl}(r)Y_{lm_l}(\vartheta,\varphi)}^{\psi_{nlm_l}(\vec{r})}\ \overbrace{|\tfrac{1}{2}m_s\rangle}^{|\uparrow\rangle \text{ or } |\downarrow\rangle}$

Level sequence: $nl_{m_s} \equiv \underbrace{1s_{\uparrow\downarrow}}_{E_1}, \underbrace{2s_{\uparrow\downarrow}, 2p_{\uparrow\downarrow}}_{E_2}, \underbrace{3s_{\uparrow\downarrow}, 3p_{\uparrow\downarrow}, 3d_{\uparrow\downarrow}}_{E_3}, \underbrace{4s_{\uparrow\downarrow}, 4p_{\uparrow\downarrow}, 4d_{\uparrow\downarrow}, 4f_{\uparrow\downarrow}}_{E_4} \ldots$

Coupled eigenstates (total angular momentum):

$$\boxed{|\Psi_{nljm_j}\rangle} = \underbrace{C^{jm_j}_{l(m_j-\frac{1}{2})\frac{1}{2}(+\frac{1}{2})}}_{\pm\sqrt{\frac{l\pm m_j+\frac{1}{2}}{2l+1}}}\underbrace{\psi_{nl(m_j-\frac{1}{2})}(\vec{r})}_{R_{nl}Y_{l(m_j-\frac{1}{2})}}\underbrace{|\uparrow\rangle}_{\binom{1}{0}} + \underbrace{C^{jm_j}_{l(m_j+\frac{1}{2})\frac{1}{2}(-\frac{1}{2})}}_{\sqrt{\frac{l\mp m_j+\frac{1}{2}}{2l+1}}}\underbrace{\psi_{nl(m_j+\frac{1}{2})}(\vec{r})}_{R_{nl}Y_{l(m_j+\frac{1}{2})}}\underbrace{|\downarrow\rangle}_{\binom{0}{1}}$$

$\boxed{j = l \pm \frac{1}{2}}$

$$= R_{nl}(r)\underbrace{\frac{1}{\sqrt{2l+1}}\begin{pmatrix} \pm\sqrt{l\pm m_j+\frac{1}{2}}\ Y_{l(m_j-\frac{1}{2})}(\vartheta,\varphi) \\ \sqrt{l\mp m_j+\frac{1}{2}}\ Y_{l(m_j+\frac{1}{2})}(\vartheta,\varphi)\end{pmatrix}}_{\mathcal{Y}_{ljm_j}(\vartheta,\varphi)}$$ **spinor spherical functions**

Nomenclature: $nl_j \equiv \underbrace{1s_{\frac{1}{2}}}_{E_1}, \underbrace{2s_{\frac{1}{2}}, 2p_{\frac{1}{2}}, 2p_{\frac{3}{2}}}_{E_2}, \underbrace{3s_{\frac{1}{2}}, 3p_{\frac{1}{2}}, 3p_{\frac{3}{2}}, 3d_{\frac{3}{2}}, 3d_{\frac{5}{2}}}_{E_3}, \ldots$

▶ Stark effect

Hydrogen atom in an external **electric field** of intensity $\vec{\mathcal{E}}_\lambda \equiv \lambda \mathcal{E}_1 \vec{n}_z$ (we introduce a dimensionless factor λ to scale the intensity):

$\hat{H} = \hat{H}_0 + \lambda \hat{H}'$ $\boxed{\hat{H}' = e\mathcal{E}_1 z} \equiv \hat{T}^1_0$ component of spherical vector

Unperturbed hydrogen solutions expressed in the uncoupled basis $|\psi_{nlm_lm_s}\rangle$

Selection rules for matrix elements:

(a) $\langle\psi_{nlm_lm_s}|\hat{H}'|\psi_{nlm_lm_s}\rangle = 0 \quad \Leftarrow \quad$ parity conservation ($\int \overbrace{|\psi_{nlm_l}(\vec{r})|^2}^{\text{even}} z\, d\vec{r} = 0$)

(b) $\langle\psi_{n'l'm_l'm_s'}|\hat{H}'|\psi_{nlm_lm_s}\rangle = 0$ for $m_l' \neq m_l$ or $m_s' \neq m_s$ or $|l - l'| > 1$
$\Leftarrow$ Wigner-Eckart theorem

We disregard spin quantum number m_s as the interaction does not affect it

Ground-state: the 1st order term vanishes

Correction up to **2nd order**: $E_1^{(2)} = E_1 + (e\mathcal{E}_\lambda)^2 \sum_{n=2}^{\infty} \sum_{l=0}^{n-1} \sum_{m_l=-l}^{+l} \frac{|\langle\psi_{nlm_l}|z|\psi_{100}\rangle|^2}{E_1 - E_n} < E_1$

Reasoning: any state with a good parity has null electric dipole moment $\Rightarrow$ no linear effect of an electric field

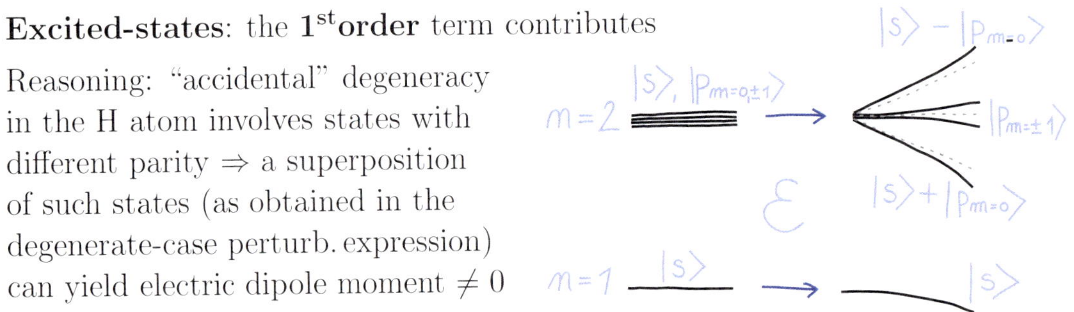

Excited-states: the **1st order** term contributes

Reasoning: "accidental" degeneracy in the H atom involves states with different parity $\Rightarrow$ a superposition of such states (as obtained in the degenerate-case perturb. expression) can yield electric dipole moment $\neq 0$

Example: $n = 2$ shell:

According to the selection rules, the only nonzero matrix element is the following one (its calculation is not presented here):

$$\langle\psi_{210}|\hat{H}'|\psi_{200}\rangle = \langle\psi_{200}|\hat{H}'|\psi_{210}\rangle = -3a_{\rm B}e\mathcal{E}_1$$

$$-3a_{\rm B}e\mathcal{E}_1 \begin{pmatrix} 0&0&1&0\\0&0&0&0\\1&0&0&0\\0&0&0&0 \end{pmatrix} \begin{pmatrix} \alpha_{200}\\ \alpha_{21(-1)}\\ \alpha_{210}\\ \alpha_{21(+1)} \end{pmatrix} = \Delta E \begin{pmatrix} \alpha_{200}\\ \alpha_{21(-1)}\\ \alpha_{210}\\ \alpha_{21(+1)} \end{pmatrix} \quad \Rightarrow$$

$\lvert\psi_{0;k}\rangle$		$E_2^{(1)}$
$\frac{1}{\sqrt{2}}(\lvert\psi_{200}\rangle + \lvert\psi_{210}\rangle)$	$\Rightarrow$	$E_2 - 3a_{\rm B}e\mathcal{E}_\lambda$
$\frac{1}{\sqrt{2}}(\lvert\psi_{200}\rangle - \lvert\psi_{210}\rangle)$	$\Rightarrow$	$E_2 + 3a_{\rm B}e\mathcal{E}_\lambda$
$\lvert\psi_{21(-1)}\rangle$	$\Rightarrow$	E_2
$\lvert\psi_{21(+1)}\rangle$	$\Rightarrow$	E_2

◀ Historical remark

1913: J. Stark & A. Lo Surdo discover the effect of electric field on atomic levels
1916: P. Epstein & K. Schwarzschild calculate the effect using the old QM
1926: E. Schrödinger provides a QM calculation based on the perturbation method

▶ Zeeman effect

Hydrogen atom in magnetic field of induction $\vec{B}_\lambda \equiv \lambda B_1 \vec{n}_z$, where we again introduce a dimensionless field scaling factor λ:

$$\hat{H} = \hat{H}_0 + \lambda \hat{H}' \quad \text{with} \quad \boxed{\hat{H}' = -\tfrac{e}{2M} B_1 (\hat{L}_z + 2\hat{S}_z)} = -\tfrac{e}{2M} B_1 (\hat{J}_z + \hat{S}_z)$$

The above formula takes into account both orbital and spin gyromagnetic ratios. Perturbation **diagonalized in uncoupled basis**:

$$\langle \psi_{nlm_lm_s} | \hat{H}' | \psi_{nlm_lm_s} \rangle = -\underbrace{\tfrac{e\hbar}{2M}}_{\mu_B} B_1 (m_l + 2m_s) \qquad \text{exact solution}$$

The **1st order** effect in the **coupled basis** (just an exercise):

$$\langle \Psi_{nljm_j} | \hat{H}' | \Psi_{nljm_j} \rangle = -\mu_B B_1 \left[m_j + \tfrac{1}{2} (C^{jm_j}_{l(m_j-\frac{1}{2})\frac{1}{2}(+\frac{1}{2})})^2 - \tfrac{1}{2} (C^{jm_j}_{l(m_j+\frac{1}{2})\frac{1}{2}(-\frac{1}{2})})^2 \right]$$

$$= \begin{cases} -\mu_B B_1 \frac{2l+2}{2l+1} m_j & \text{for } j = l + \frac{1}{2} \\ -\mu_B B_1 \frac{2l}{2l+1} m_j & \text{for } j = l - \frac{1}{2} \end{cases}$$

◀ Historical remark

1897: P. Zeeman discovers the splitting of atomic lines in magnetic field
1925: So-called anomalous Zeeman effect contributes to the discovery of spin

▶ Spin-orbital coupling

Correction caused by an interaction of the spin **magnetic moment** of electron with the magnetic field generated by its orbital motion. In the electron's rest frame, this can be seen as an interaction of its magnetic moment with the magnetic field produced by a "moving" nucleus (then a relativistic effect, so-called Thomas precession, must be taken into account). The resulting perturbation of the hydrogen Hamiltonian reads as follows:

$$\boxed{\hat{H}' = \tfrac{e^2}{4\pi\epsilon_0} \tfrac{1}{2M^2c^2} \tfrac{1}{r^3} (\hat{\vec{L}} \cdot \hat{\vec{S}}) = \tfrac{e^2}{4\pi\epsilon_0} \tfrac{1}{4M^2c^2} \tfrac{1}{r^3} \left(\hat{\vec{J}}^2 - \hat{\vec{L}}^2 - \hat{\vec{S}}^2 \right)}$$

1st order effect in the **coupled basis**:

$$\langle \Psi_{nljm_j} | \hat{H}' | \Psi_{nljm_j} \rangle = \tfrac{e^2}{4\pi\epsilon_0} \tfrac{\hbar^2}{4M^2c^2} \left[j(j+1) - l(l+1) - \tfrac{3}{4} \right] \overbrace{\langle \Psi_{nljm_j} | \tfrac{1}{r^3} | \Psi_{nljm_j} \rangle}^{\int_0^\infty |R_{nl}(r)|^2 \frac{1}{r^3} r^2 dr \equiv \langle r^{-3} \rangle_{nl}}$$

$$= \begin{cases} +\frac{e^2}{4\pi\epsilon_0} \frac{\hbar^2}{4M^2c^2} \langle r^{-3} \rangle_{nl}\, l & \text{for } j = l + \frac{1}{2} \\ -\frac{e^2}{4\pi\epsilon_0} \frac{\hbar^2}{4M^2c^2} \langle r^{-3} \rangle_{nl}\, (l+1) & \text{for } j = l - \frac{1}{2} \end{cases}$$

▶ Relativistic correction

Rel. **kinetic energy**: $T = \underbrace{\sqrt{(Mc^2)^2 + (pc)^2}}_{Mc^2\sqrt{1+(\frac{p}{Mc})^2}} - Mc^2 \approx \underbrace{\tfrac{p^2}{2M} - \tfrac{1}{2Mc^2} \left(\tfrac{p^2}{2M} \right)^2 + \ldots}_{\sqrt{1+\delta} = 1 + \frac{\delta}{2} - \frac{\delta^2}{8} + \ldots}$

The effect of this correction can be treated within the non-relativistic QM, adding to $\hat{H}_0$ a perturbation term: $\boxed{\hat{H}' = -\tfrac{1}{2Mc^2} \left(\tfrac{\hat{p}^2}{2M} \right)^2 = -\tfrac{1}{2Mc^2} (\hat{H}_0 - \hat{V})^2}$

$1^{\rm st}$**order** effect in the **coupled basis**:

$$\langle\Psi_{nljm_j}|\hat{H}'|\Psi_{nljm_j}\rangle = -\frac{1}{2Mc^2}\Bigg[E_n^2 - 2E_n\overbrace{\langle\Psi_{nljm_j}|\hat{V}|\Psi_{nljm_j}\rangle}^{\frac{e^2}{4\pi\epsilon_0}\langle r^{-1}\rangle_{nl}} + \overbrace{\langle\Psi_{nljm_j}|\hat{V}^2|\Psi_{nljm_j}\rangle}^{\left(\frac{e^2}{4\pi\epsilon_0}\right)^2\langle r^{-2}\rangle_{nl}}\Bigg]$$

▶ Comparison of atomic corrections

After evaluation of the radial integrals $\langle r^{-k}\rangle_{nl} \propto a_{\rm B}^{-k}$ it turns out that the following effects are of the same order of magnitude: (a) the spin-orbit coupling, (b) relativistic correction, (c) Zeeman splitting for $B \in (1,10)$ T

(a) + (b) = **fine structure** of atomic levels: $\boxed{\Delta E_{\rm FS} \approx -\frac{Mc^2}{137^4}\frac{1}{4n^4}\left(\frac{2n}{j+\frac{1}{2}} - \frac{3}{2}\right)}$

◀ Historical remark

1916: A. Sommerfeld introduces the fine-structure constant; he calculates the relativistic splitting of hydrogen levels within the old QM
1925-6: Electron spin & mag. moment taken into account; L. Thomas computes atomic LS-interaction including the relativistic effect of inter-frame transformation

▶ Helium atom

Besides the kinetic terms & Coulomb interaction of both electrons with the nucleus one has to consider also Coulomb interaction between the 2 electrons:

$$\hat{H} = \overbrace{-\frac{\hbar^2}{2M}(\Delta_1 + \Delta_2) - \frac{2e^2}{4\pi\epsilon_0}\left(\frac{1}{|\vec{x}_1|} + \frac{1}{|\vec{x}_2|}\right)}^{\hat{H}_0} + \overbrace{\frac{e^2}{4\pi\epsilon_0}\frac{1}{|\vec{x}_1 - \vec{x}_2|}}^{\hat{H}'} \quad \text{2-electron Hamiltonian}$$

The calculation can be performed in the coupled spin basis of both electrons:

$$|S, M_S\rangle = \begin{cases} |0,0\rangle & \textbf{singlet} \text{ (antisymmetric under exchange)} \\ |1,M_S\rangle & \textbf{triplet} \text{ (symmetric under exchange)} \end{cases}$$

As the total 2-electron wavefunction must be antisymmetric under the exchange (fermions), the orbital part associated with spin singlet/triplet is symmetric/antisymmetric:

$$\Psi_{0\pm}(\vec{x}_1, \vec{x}_2) = \frac{1}{\sqrt{2}}\Big[\overbrace{\psi_{n_1l_1m_1}(\vec{x}_1)}^{\psi_1(\vec{x}_1)}\overbrace{\psi_{n_2l_2m_2}(\vec{x}_2)}^{\psi_2(\vec{x}_2)} \pm \overbrace{\psi_{n_2l_2m_2}(\vec{x}_1)}^{\psi_2(\vec{x}_1)}\overbrace{\psi_{n_1l_1m_1}(\vec{x}_2)}^{\psi_1(\vec{x}_2)}\Big]$$

Define $$E_{12}^{\rm A} = \begin{cases} \frac{e^2}{4\pi\epsilon_0}\int \psi_1^*(\vec{x}_1)\psi_2^*(\vec{x}_2)\frac{1}{|\vec{x}_1 - \vec{x}_2|}\psi_1(\vec{x}_1)\psi_2(\vec{x}_2)\, d\vec{x}_1 d\vec{x}_2 \\ \frac{e^2}{4\pi\epsilon_0}\int \psi_2^*(\vec{x}_1)\psi_1^*(\vec{x}_2)\frac{1}{|\vec{x}_1 - \vec{x}_2|}\psi_2(\vec{x}_1)\psi_1(\vec{x}_2)\, d\vec{x}_1 d\vec{x}_2 \end{cases}$$

and $$E_{12}^{\rm B} = \begin{cases} \frac{e^2}{4\pi\epsilon_0}\int \psi_1^*(\vec{x}_1)\psi_2^*(\vec{x}_2)\frac{1}{|\vec{x}_1 - \vec{x}_2|}\psi_2(\vec{x}_1)\psi_1(\vec{x}_2)\, d\vec{x}_1 d\vec{x}_2 \\ \frac{e^2}{4\pi\epsilon_0}\int \psi_2^*(\vec{x}_1)\psi_1^*(\vec{x}_2)\frac{1}{|\vec{x}_1 - \vec{x}_2|}\psi_1(\vec{x}_1)\psi_2(\vec{x}_2)\, d\vec{x}_1 d\vec{x}_2 \end{cases}$$

$1^{\rm st}$**order energy correction**:

Singlet & triplet spin states are degenerate, but $\hat{H}'$ is diagonal in these states ⇒ the non-degenerate expression applicable:

$$\langle\Psi_{0\pm}|\hat{H}'|\Psi_{0\pm}\rangle = \tfrac{1}{2}(E_{12}^{\rm A} + E_{12}^{\rm A} \pm E_{12}^{\rm B} \pm E_{12}^{\rm B}) = E_{12}^{\rm A} \pm E_{12}^{\rm B}$$

For spin $\left\{\begin{array}{c}\text{singlet}\\ \text{triplet}\end{array}\right\}$ states the correction $E_i^{(1)\prime} = \left\{\begin{array}{l} E_{12}^{\mathrm{A}} + E_{12}^{\mathrm{B}} \\ E_{12}^{\mathrm{A}} - E_{12}^{\mathrm{B}} \end{array}\right.$

The splitting of singlet & triplet ("**parahelium**" & "**orthohelium**") states is a direct consequence of indistinguishability!

◀ **Historical remark**

1892: F. Paschen & C. Runge discover the splitting of He spectrum

1926: W. Heisenberg provides interpretation through (anti)symmetric wavefunctions

■ Application to level dynamics

So far it was assumed that the parameter λ, weighting the perturbation term in the Hamiltonian, has a fixed (small) value. However, one may think of Hamiltonians $\hat{H}(\lambda)$ for which λ is a *variable* control parameter. There is a huge class of such Hamiltonians, we just require their *linear* dependence on the parameter. The energy spectrum $E_i(\lambda)$ changes (nonlinearly) with running λ and one may use the perturbation theory to write down a set of differential equations governing these changes. In this way, the spectral variations are treated as if λ were time and level energies $E_i(\lambda)$ positions of moving 1D particles. This provides an interesting interpretation of the parameter-induced "level dynamics".

▶ Hamiltonian with a linear parametric dependence

$$\boxed{\hat{H}(\lambda) = \hat{H}_0 + \lambda\,\hat{H}'} \qquad \lambda \in (-\infty, +\infty)$$

Perturbative treatment at any λ: $\hat{H}(\lambda+\delta\lambda) = \overbrace{\hat{H}_0 + \lambda\,\hat{H}'}^{\hat{H}(\lambda)} + (\delta\lambda)\,\hat{H}'$

Level dynamics:

evolving energy levels $\boxed{E_i(\lambda) \quad \longleftrightarrow \quad x_i(t)}$ "particle trajectories" in 1D

▶ Local equations obtained from the perturbation theory

$$\tfrac{d}{d\lambda}E_i(\lambda) = \langle\psi_i(\lambda)|\hat{H}'|\psi_i(\lambda)\rangle \quad \Rightarrow \quad \boxed{\dot{E}_i = H'_{ii}} \qquad \text{velocity}$$

$$\tfrac{d^2}{d\lambda^2}E_i(\lambda) = 2\sum_{j(\neq i)} \tfrac{|\langle\psi_j(\lambda)|\hat{H}'|\psi_i(\lambda)\rangle|^2}{E_i(\lambda)-E_j(\lambda)} \Rightarrow \boxed{\ddot{E}_i = 2\sum_{j(\neq i)} \frac{|H'_{ji}|^2}{E_i-E_j}}$$

acceleration $\propto$ repulsive **Coulomb (2D) – like force between levels**

$$\tfrac{d}{d\lambda}\langle\psi_j(\lambda)|\hat{H}'|\psi_i(\lambda)\rangle = \langle\tfrac{d}{d\lambda}\psi_j(\lambda)|\hat{H}'|\psi_i(\lambda)\rangle + \langle\psi_j(\lambda)|\hat{H}'|\tfrac{d}{d\lambda}\psi_i(\lambda)\rangle =$$
$$\sum_{k(\neq j)} \tfrac{\langle\psi_j(\lambda)|\hat{H}'|\psi_k(\lambda)\rangle}{E_j(\lambda)-E_k(\lambda)} \langle\psi_k(\lambda)|\hat{H}'|\psi_i(\lambda)\rangle + \sum_{k(\neq i)} \langle\psi_j(\lambda)|\hat{H}'|\psi_k(\lambda)\rangle \tfrac{\langle\psi_k(\lambda)|\hat{H}'|\psi_i(\lambda)\rangle}{E_i(\lambda)-E_k(\lambda)}$$

$$\Rightarrow \boxed{\dot{H}'_{ji} = \sum_{k(\neq j)} \frac{H'_{jk}H'_{ki}}{E_j-E_k} + \sum_{k(\neq i)} \frac{H'_{jk}H'_{ki}}{E_i-E_k}}$$

$\Rightarrow$ evolution of product charge $|H'_{ji}|^2$ but no individual charges : $|H'_{ji}|^2 \neq Q_j Q_i$

Known $H'_{ij}(0)$ & $E_i(0)$ $(\forall\, i,j)$ $\Rightarrow$ we can calculate $H'_{ij}(\lambda)$ & $E_i(\lambda)$ for any λ

▶ **"Integrals of motion"** (in the sense λ≡time) For instance:

$P = \mathrm{Tr}\hat{H}' = \sum_i H'_{ii} = \sum_i \dot{E}_i = \mathrm{const}$

$W = \frac{1}{2}\mathrm{Tr}(\hat{H}')^2 = \frac{1}{2}\sum_{i,j} H'_{ij}H'_{ji} = \frac{1}{2}\sum_i \dot{E}_i^2 + \frac{1}{2}\sum_{i\neq j}|H'_{ij}|^2 = \mathrm{const}$

There exist many more, in fact, as many that the above system of differential equations is integrable!

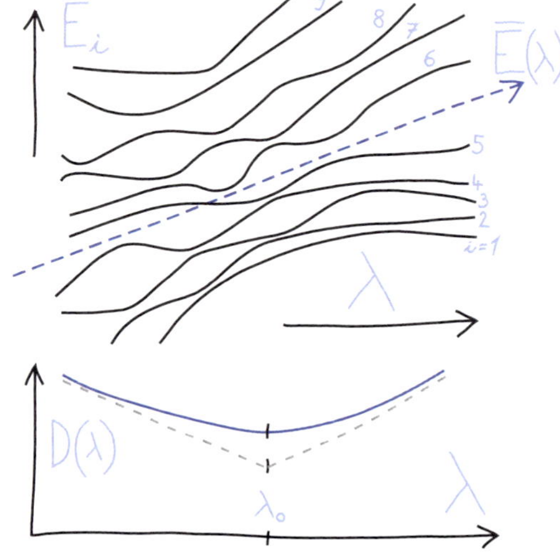

▶ **Global properties of spectrum**

for finite dimension n

"Center of mass":

$$\bar{E}(\lambda) \equiv \frac{1}{n}\sum_i E_i(\lambda) = \frac{1}{n}\mathrm{Tr}\hat{H}(\lambda) = \left[\frac{1}{n}\mathrm{Tr}\hat{H}_0\right] + \lambda\left[\frac{1}{n}\hat{H}'\right]$$

"Spread":

$$D(\lambda) \equiv \sqrt{\frac{1}{n}\sum_i [E_i(\lambda) - \bar{E}(\lambda)]^2} = \sqrt{\frac{1}{n}\sum_i E_i(\lambda)^2 - \bar{E}^2(\lambda)} =$$

$$\sqrt{\left[\frac{1}{n}\mathrm{Tr}\hat{H}_0^2 - \frac{1}{n^2}\mathrm{Tr}^2\hat{H}_0\right] + \lambda\left[\frac{2}{n}\mathrm{Tr}(\hat{H}_0\hat{H}') - \frac{2}{n^2}\mathrm{Tr}\hat{H}_0\mathrm{Tr}\hat{H}'\right] + \lambda^2\left[\frac{1}{n}\mathrm{Tr}(\hat{H}')^2 - \frac{1}{n^2}\mathrm{Tr}^2\hat{H}'\right]}$$

$D(\lambda)$ minimal at certain λ_0 (maximal compression of the spectrum)
For $\lambda \to \pm\infty$ the spectrum freely expands: $D(\lambda) \propto \lambda$

▶ **No-crossing rule**

The equation for $\ddot{E}_i$ corresponds to a repulsive "force" between levels, which is analogous to the Coulomb force in 2D ($F_{\mathrm{2D}} \propto r^{-1}$). This force prevents crossings of levels. For an actual crossing of two levels at a certain $\lambda_\times$ one needs to simultaneously satisfy 2 equations: $\boxed{E_i(\lambda_\times) = E_j(\lambda_\times)}$ (levels coalesce) & $\boxed{H'_{ji}(\lambda_\times) = 0}$ (force vanishes). This is not achievable with just a single control parameter λ to vary (except of some accidental, extremely rare cases).

Instead of real crossings there exist numerous so-called **avoided crossings** of energy levels. At such places, the corresponding eigenfunctions change very rapidly, as can be seen from the "survival probability" given by overlap formula:

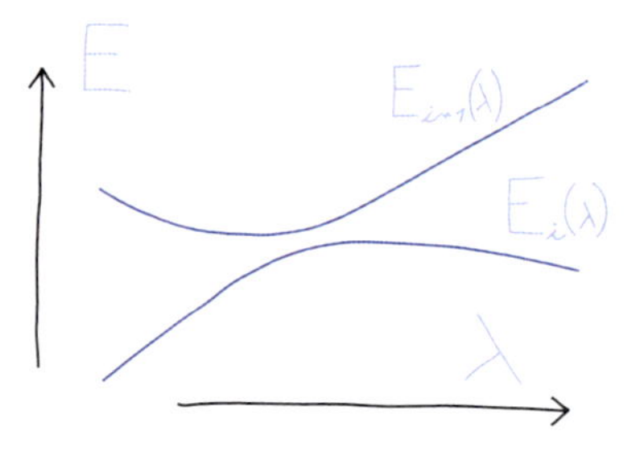

$$P_i(\lambda, \delta\lambda) \equiv |\langle\psi_i(\lambda+\delta\lambda)|\psi_i(\lambda)\rangle|^2 \approx 1 - (\delta\lambda)^2 \sum_{j(\neq i)} \frac{|\langle\psi_j(\lambda)|\hat{H}'|\psi_i(\lambda)\rangle|^2}{[E_i(\lambda) - E_j(\lambda)]^2}$$

Proof of the overlap formula:

$$|\langle\psi_i(\lambda+\delta\lambda)|\psi_i(\lambda)\rangle|^2=\langle\psi_i(\lambda+\delta\lambda)\overbrace{|\psi_i(\lambda)\rangle\langle\psi_i(\lambda)|}^{\hat{I}-\sum_{j(\neq i)}|\psi_j(\lambda)\rangle\langle\psi_j(\lambda)|}\psi_i(\lambda+\delta\lambda)\rangle$$

$$=1-\sum_{j(\neq i)}\overbrace{|\langle\psi_j(\lambda)|\psi_i(\lambda+\delta\lambda)\rangle|^2}^{(\delta\lambda)^2\frac{|\langle\psi_j(\lambda)|\hat{H}'|\psi_i(\lambda)\rangle|^2}{[E_i(\lambda)-E_j(\lambda)]^2}}$$

On the other hand, real crossings are abundant if the perturbation matrix element H'_{ji} vanishes identically (e.g., for some symmetry reasons, like in case of levels with different total angular momenta in rotationally invariant field).

◀ **Historical remark**

1929: J. von Neumann & E. Wigner formulate the no-crossing rule
1932: L. Landau & C. Zener calculate transition rate for a 2-level avoided crossing
1980's: P. Pechukas & T. Yukawa elaborate the Coulomb analogy for level dynamics

■ Driven systems and adiabatic approximation

We will now briefly deal with problems combining parameter-dependent and time-dependent descriptions. The aim will be to analyze the dynamics of systems whose Hamiltonian parameters are *driven* (i.e., varied with a given, externally controlled time dependence). This can be done with the aid of methods which are closely related to the stationary perturbation theory.

▶ Time-dependent Hamiltonian

Consider Hamiltonian $\hat{H}(\vec{G})$ depending on a set of parameters $\vec{G}\equiv(G_1,G_2,\dots)$. Choose a curve $\vec{G}(g)$ in the parameter space described by a single parameter $g\in\mathbb{R}$, whose value varies with time according to $g(t)=\dot{g}t$ (with $\dot{g}\equiv$constant). This turns the original parameter-dependent Hamiltonian into a nonstationary Hamiltonian:

$$\boxed{\hat{H}(\vec{G})\longrightarrow\hat{H}(g)=\hat{H}(\dot{g}t)}$$

Task: to describe evolution induced by $\hat{H}(\dot{g}t)$

▶ Equations for instantaneous eigenvectors

Stationary Schr. eq.: $\hat{H}(g)|\psi_i(g)\rangle=E_i(g)|\psi_i(g)\rangle$

$$\Rightarrow\quad\frac{d\hat{H}(g)}{dg}|\psi_i(g)\rangle+\hat{H}(g)|\frac{d\psi_i}{dg}(g)\rangle=\frac{dE_i(g)}{dg}|\psi_i(g)\rangle+E_i(g)|\frac{d\psi_i}{dg}(g)\rangle$$

Multiply by $\langle\psi_j(g)|$ for $j\neq i$:

$$\langle\psi_j|\frac{d\hat{H}}{dg}|\psi_i\rangle+E_j\langle\psi_j|\frac{d\psi_i}{dg}\rangle=\frac{dE_i}{dg}\overbrace{\langle\psi_j|\psi_i\rangle}^{0}+E_i\langle\psi_j|\frac{d\psi_i}{dg}\rangle$$

$$\Rightarrow\quad\boxed{\langle\psi_j(g)|\frac{d\psi_i}{dg}(g)\rangle=\frac{\langle\psi_j(g)|\frac{d\hat{H}}{dg}(g)|\psi_i(g)\rangle}{E_i(g)-E_j(g)}}\quad\text{for } j\neq i$$

For $j=i$ we use: $\frac{d}{dg}\langle\psi_i|\psi_i\rangle=\langle\frac{d\psi_i}{dg}|\psi_i\rangle+\langle\psi_i|\frac{d\psi_i}{dg}\rangle=2\,\mathrm{Re}\langle\psi_i|\frac{d\psi_i}{dg}\rangle=0$

$$\Rightarrow\quad\boxed{\langle\psi_i(g)|\frac{d\psi_i}{dg}(g)\rangle=i\phi_i(g)}\quad\text{with } \phi_i(g)\in\mathbb{R}$$

With substitutions $g \to \dot{g}t$ and $\frac{d}{dg} \to \frac{1}{\dot{g}}\frac{d}{dt}$ the above formulas become t-dependent

▶ Time evolution by the driven Hamiltonian

Expansion in the instantaneous eigenbasis:

$$|\Psi(t)\rangle = \sum_j \alpha_j(t)|\psi_j(\dot{g}t)\rangle$$

Nonstationary Schr. eq.: $i\hbar\frac{d}{dt}|\Psi(t)\rangle = \hat{H}(\dot{g}t)|\Psi(t)\rangle$

$$\Rightarrow i\hbar\sum_j\left(\dot{\alpha}_j(t)|\psi_j(\dot{g}t)\rangle + \alpha_j(t)\tfrac{d}{dt}|\psi_j(\dot{g}t)\rangle\right) = \sum_j \alpha_j(t)E_j(\dot{g}t)|\psi_j(\dot{g}t)\rangle$$

Multiply by $\langle\psi_i(\dot{g}t)|$: $\quad i\hbar\dot{\alpha}_i(t) + i\hbar\sum_j \alpha_j(t)\underbrace{\langle\psi_i(\dot{g}t)|\tfrac{d}{dt}\psi_j(\dot{g}t)\rangle} = \alpha_i(t)E_i(\dot{g}t)$

Here we use the previously derived result and obtain a system of coupled differential equations for coeffs. $\alpha_i(t)$:

$\dot{g}\frac{\langle\psi_j(g)\|\frac{d\hat{H}}{dg}(g)\|\psi_i(g)\rangle}{E_i(g)-E_j(g)}$	$i\dot{g}\phi_i(g)$
for $i\neq j$	for $i=j$

$$\frac{d\alpha_i}{dt}(t) = \left[-\tfrac{i}{\hbar}E_i(\dot{g}t) + i\dot{g}\phi_i(\dot{g}t)\right]\alpha_i(t) + \dot{g}\sum_{j(\neq i)}\frac{\langle\psi_j(\dot{g}t)|\frac{d\hat{H}}{dg}(\dot{g}t)|\psi_i(\dot{g}t)\rangle}{E_i(\dot{g}t)-E_j(\dot{g}t)}\alpha_j(t) \qquad \frac{d}{dt} = \dot{g}\frac{d}{dg}, \quad \dot{g}t = g$$

▶ Adiabatic approximation

The initial state is one of $g{=}0$ eigenstates: $|\Psi(t{=}0)\rangle \equiv |\psi_i(g{=}0)\rangle$ $\Rightarrow \alpha_j(0){=}\delta_{ij}$

and $\dot{g}$ is very small $\to 0$ (adiabatic limit)

The offdiag. terms of the above system of equations yield contributions $\mathcal{O}(\dot{g}^2)$ in the solution $\Rightarrow$ can be neglected.

So the adiabatic solution reads as:

$$\alpha_i(t) = \underbrace{e^{-\frac{i}{\hbar}\int_0^t E_i(\dot{g}t')dt'}}_{\text{dynamical}}\ \underbrace{e^{i\int_0^g \phi_i(g')dg'}}_{\text{geometrical}}$$

phase factor

▶ Conclusions

(1) In the limit $\dot{g}{\to}0$ the system remains in the instantaneous eigenstate $|\psi_i(g)\rangle$

This is known as the **adiabatic theorem**

Remark: For $\dot{g}$ small but $\neq 0$ this remains a good approx. *iff* the levels do not come too close to each other (see neglected term in the above eq.)

(2) The adiabatic evolution leads to the occurrence of **two phase factors**:

(a) dynamical phase derived from the standard evolution of energy eigenstates (taking into account variations of energy with the parameter),

(b) the geometrical phase (also called **Berry phase**) depends only on the geometrical path in the space of parameters $\vec{G}$. It can yield a nonzero value even if the path returns back to the initial point.

◀ Historical remark

1928: M. Born & V. Fock formulate the adiabatic theorem

1956, 1984: S. Pancharatnam & M. Berry discover the geometrical phase factor

5.3 Nonstationary perturbation method

Having digested a bit of nonstationary QM from the end of the previous section, we now devote ourselves fully to deriving a suitable perturbative approximation of quantum dynamical problems. Our task will be to calculate the rates of transitions between various eigenstates of the principal Hamiltonian $\hat{H}_0$ induced by a small supplement $\hat{H}'(t)$ (nonstationary, in general). Note that the nonstationary perturbation technique, which we are going to outline in the following, represents the prevailing treatment of nonstationary problems in quantum theory.

■ General formalism

The nonstationary perturbation method can describe a vast variety of quantum processes, which are running with a characteristic time scale placed somewhere in between **two limiting time scales**: The long time scale, $T_>$, is derived from the *total energy width* of the initial eigenstate of the unperturbed Hamiltonian $\hat{H}_0$. The short time scale, $T_<$, is more difficult to specify. For problems with a *discrete spectrum* of initial states (and we implicitly deal with this type of problems in the following), $T_<$ corresponds to the *average energy spacing* between the eigenstates of $\hat{H}_0$ around the initial state.

Here we focus mainly on the general formulation of the method, postponing the treatment of concrete applications to more specialized courses of QM. Indeed, realistic calculations are often hindered by a difficult structure of the corresponding Hilbert spaces. This is so particularly in decay and scattering processes, in which one typically deals with composite objects and intricate mixtures of discrete & continuous energy spectra.

▸ Setup

Total Hamiltonian assumed in the form $\boxed{\hat{H}(t) = \hat{H}_0 + \lambda \hat{H}'(t)}$ where:

$$\left.\begin{array}{lcl} \hat{H}_0 & \equiv & \text{free stationary Hamiltonian} \\ \hat{H}'(t) & \equiv & \text{generally time dependent perturbation} \\ \lambda & \equiv & \text{dimensionless parameter} \end{array}\right\} \begin{array}{l} \text{matrix elements of } \hat{H}_0 \\ \text{and } \hat{H}'(t) \text{ are of about} \\ \text{the same size, } \lambda \ll 1 \end{array}$$

Task: to evaluate probabilities of transitions between eigenstates of $\hat{H}_0$ as a function of time in the form of a power-law series in λ

▸ Typology of applications

Example I (stimulated transition): $\boxed{A \leftrightarrow A^*}$. Hamiltonian $\hat{H}_0$ describes a bound system with discrete spectrum $\{E_{00}, E_{01}, \ldots\}$ and $\lambda \hat{H}'(t)$ is a nonstationary external field inducing transitions between unperturbed eigenstates.

Example II (spontaneous decay): $\boxed{A^* \to A + \gamma}$. Hamiltonian $\hat{H}_0 = \hat{H}_a + \hat{H}_\gamma$ describes a bound system (atom, nucleus) with discrete spectrum $\{E_{00}, E_{01}, \ldots\}$

and free elmag. field (photons) with continuous spectrum $E_\gamma \in [0, +\infty)$. The perturbation $\lambda\hat{H}'$ represents the atom-photon interaction. Initial state $|\psi_{0i}\rangle \equiv |E_{0i}\rangle_{\rm a} \otimes |0\rangle_\gamma$. Final state $|\psi_{0j}\rangle \equiv |E_{0j}\rangle_{\rm a} \otimes |\vec{k}\nu\rangle_\gamma$ with $j < i$ and $|\vec{k}\nu\rangle_\gamma \equiv$ single-photon plane wave with given polarization.

Example III (scattering): Process $\boxed{a + A \to B + b}$. Free Hamiltonian $\hat{H}_0$ describes a system of non-interacting particles a, A, B, b with indefinite particle numbers and continuous spectrum (bases created from $|\vec{p}_i, \psi_i\rangle_\bullet \equiv$ plane waves times relevant internal freedom degrees of individual particles). The term $\lambda\hat{H}'$ stands for interactions of all particles involved. Initial state $|\psi_{0i}\rangle \equiv |\vec{p}_i, \sigma_i\rangle_a \otimes |-\vec{p}_i, \alpha_i\rangle_A \otimes |0\rangle_B \otimes |0\rangle_b$. Final state $|\psi_{0j}\rangle \equiv |0\rangle_a \otimes |0\rangle_A \otimes |\vec{p}_j, \beta_j\rangle_B \otimes |-\vec{p}_j, \sigma_j\rangle_b$.

▶ Dyson series in the interaction (Dirac) picture

It is favorable now to move to the Dirac picture of time evolution, identifying the free Hamiltonian with $\hat{H}_0$. This immediately yields the desired power-law series in the perturbation.

$$\text{Operators:}\quad \hat{A}_{\rm D}(t) = \hat{U}_0^\dagger(t)\,\hat{A}_{\rm S}\,\overbrace{\hat{U}_0(t)}^{e^{-i\frac{\hat{H}_0 t}{\hbar}}} \quad\Rightarrow\quad \begin{cases} \hat{H}_{0\rm D} = \hat{H}_{0\rm S} \equiv \hat{H}_0 \\ \hat{H}'_{\rm D}(t) = \hat{U}_0^\dagger(t)\hat{H}'(t)\hat{U}_0(t) \end{cases}$$

$$\text{Vectors:}\quad |\psi(t)\rangle_{\rm D} = \hat{U}_0^\dagger(t)|\psi(t)\rangle_{\rm S} \quad\Rightarrow\quad i\hbar\tfrac{d}{dt}|\psi(t)\rangle_{\rm D} = \hat{H}'_{\rm D}(t)|\psi(t)\rangle_{\rm D}$$

Dyson series for evolution operator:

$$\boxed{\begin{aligned}\hat{U}_{\rm D}(t,t_0) = {} & \hat{I} + \left(-\tfrac{i}{\hbar}\lambda\right)^1 \int_{t_0}^{t} \hat{H}'_{\rm D}(t_1)\,dt_1 + \left(-\tfrac{i}{\hbar}\lambda\right)^2 \int_{t_0}^{t}\int_{t_0}^{t_2} \hat{H}'_{\rm D}(t_2)\hat{H}'_{\rm D}(t_1)\,dt_1 dt_2 + \ldots \\ & + \left(-\tfrac{i}{\hbar}\lambda\right)^n \int_{t_0}^{t}\int_{t_0}^{t_n}\cdots\int_{t_0}^{t_2} \hat{H}'_{\rm D}(t_n)\hat{H}'_{\rm D}(t_{n-1})\cdots\hat{H}'_{\rm D}(t_1)\,dt_1\ldots dt_{n-1}dt_n + \ldots\end{aligned}}$$

▶ Estimate of the upper time scale

Fast convergence of the above series is expected if the time difference $(t-t_0)$ is *much smaller* than a characteristic time scale of the exact state evolution:

$$\boxed{(t-t_0) \ll T_>}$$

The scale $T_>$ given by total energy width of the initial state in the eigenbasis of the full Hamiltonian: $\boxed{T_> \sim \hbar\,\langle\langle E^2\rangle\rangle_{\psi(0)}^{-1/2}}$

Example: for a decay process, $T_> \sim T_{\rm life} \equiv$ aver. lifetime of the decaying state

Note: For nonstationary perturbation, one should evaluate $T_>$ from a maximal energy width acquired during the evolution: $T_> \sim {\rm Min}\left\{\hbar\,\langle\langle E^2\rangle\rangle_{\psi(t')}^{-1/2}\right\}_{t'\in[t_0,t]}$

▶ Dyson series for transition amplitudes

We rewrite the above Dyson series for the evolution operator to evaluate transition amplitudes (initial time t_0 $\to$ final time t) between individual eigenstates of the unperturbed Hamiltonian $\hat{H}_0$:

wavefunction	$\|\psi_{0i}\rangle \longrightarrow \|\psi_{0j}\rangle$
unperturbed energy	$E_{0i} \longrightarrow E_{0j}$
transition frequency	$\omega_{ji} = \frac{E_{0j}-E_{0i}}{\hbar}$

$$S_{ji}(t,t_0) \equiv \langle\psi_{0j}|\hat{U}_{\rm D}(t,t_0)|\psi_{0i}\rangle = \delta_{ij} + \left(-\tfrac{i}{\hbar}\lambda\right)^1 \int_{t_0}^{t} \underbrace{\langle\psi_{0j}|\hat{H}'_{\rm D}(t_1)|\psi_{0i}\rangle}_{H'_{ji}(t_1)e^{i\omega_{ji}t_1}} dt_1 +$$

$$+ \left(-\tfrac{i}{\hbar}\lambda\right)^2 \int_{t_0}^{t}\int_{t_0}^{t_2} \sum_k \overbrace{\langle\psi_{0j}|\hat{H}'_{\rm D}(t_2)|\psi_{0k}\rangle}^{H'_{ji}(t_1)e^{i\omega_{jk}t_2}} \overbrace{\langle\psi_{0k}|\hat{H}'_{\rm D}(t_1)|\psi_{0i}\rangle}^{H'_{ji}(t_1)e^{i\omega_{ki}t_1}} dt_1 dt_2 + \ldots$$

$$+ \left(-\tfrac{i}{\hbar}\lambda\right)^n \int_{t_0}^{t}\int_{t_0}^{t_n}\cdots\int_{t_0}^{t_2} \sum_{k_{n-1}}\sum_{k_{n-2}}\cdots\sum_{k_1} \overbrace{\langle\psi_{0j}|\hat{H}'_{\rm D}(t_n)|\psi_{0k_{n-1}}\rangle}^{H'_{jk_{n-1}}(t_n)e^{i\omega_{jk_{n-1}}t_n}} \overbrace{\langle\psi_{0k_{n-1}}|\hat{H}'_{\rm D}(t_{n-1})|\psi_{0k_{n-2}}\rangle}^{H'_{k_{n-1}k_{n-2}}(t_{n-1})e^{i\omega_{k_{n-1}k_{n-2}}t_{n-1}}} \cdots$$

$$\cdots \underbrace{\langle\psi_{0k_1}|\hat{H}'_{\rm D}(t_1)|\psi_{0i}\rangle}_{H'_{k_1 i}(t_1)e^{i\omega_{k_1 i}t_1}} dt_1\cdots dt_{n-1}dt_n + \ldots\ldots$$

► Perturbation series

$$\begin{array}{ll}
S_{ji}(t,t_0) = & \delta_{ij} \qquad\qquad [n\!=\!0] \\
[n\!=\!1] & + \left(-\tfrac{i}{\hbar}\lambda\right)^1 \int_{t_0}^{t} H'_{ji}(t_1)e^{i\omega_{ji}t_1}dt_1 \\
[n\!=\!2] & + \left(-\tfrac{i}{\hbar}\lambda\right)^2 \int_{t_0}^{t}\int_{t_0}^{t_2}\sum_k H'_{jk}(t_2)e^{i\omega_{jk}t_2}H'_{ki}(t_1)e^{i\omega_{ki}t_1}dt_1dt_2 \\
 & + \ldots \\
[\text{general}\, n] & + \left(-\tfrac{i}{\hbar}\lambda\right)^n \int_{t_0}^{t}\int_{t_0}^{t_n}\cdots\int_{t_0}^{t_2}\sum_{k_{n-1}}\sum_{k_{n-2}}\cdots\sum_{k_1} H'_{jk_{n-1}}(t_n)e^{i\omega_{jk_{n-1}}t_n} \\
 & \cdots H'_{k_{n-1}k_{n-2}}(t_{n-1})e^{i\omega_{k_{n-1}k_{n-2}}t_{n-1}} \cdots H'_{k_1 i}(t_1)e^{i\omega_{k_1 i}t_1}dt_1 \cdot\cdot dt_{n-1}dt_n \\
 & + \ldots\ldots
\end{array}$$

► Estimate of the lower time scale

The transition amplitudes $S_{ji}(t,t_0)$ depend in general on both initial & final times t_0 & t. To make the perturbative expressions simpler and more universal, one usually assumes that the time difference $(t-t_0)$ is *much larger* than a characteristic time scale of the system's *internal* or *single-particle* dynamics:

$$\boxed{(t-t_0) \gg T_<}$$

For systems with *discrete spectra*, the short time scale is determined by the average density of unperturbed energy eigenstates, $\rho_0 \equiv \langle |E_{0(i+1)}-E_{0i}|^{-1}\rangle_i$, around the initial state: $\boxed{T_< \sim \hbar\,\varrho_0(E_{0i})}$

Examples: For decay processes of composite objects, $T_<$ represents a characteristic period of motions of internal particles. For scattering of particles with a short-range interaction, $T_<$ associated with the time spent by the colliding particles within the interaction distance.

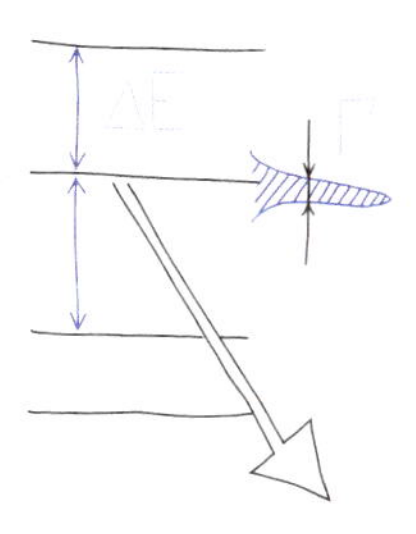

Consequence: In case of discrete spectrum, time window $(T_<,T_>)$ for application of t-dependent perturb. technique exists *iff* the total energy width of the initial state is (much) less than the spacing of unperturbed levels: $\boxed{\Gamma \ll \Delta E}$

▶ **S-matrix**

The dependence of $S_{ji}(t,t_0)$ on times t_0 & t can be removed by considering an *asymptotic time limit* with respect to the short time scale $T_<$. The resulting so-called scattering matrix (S-matrix) includes asymptotic-time amplitudes of the $i \to j$ transitions:

$$S_{ji}(t,t_0) \mapsto \boxed{S_{ji} \equiv \begin{cases} \lim\limits_{t\to\infty} S_{ji}(+t,0) & (\text{with } t_0=0) \quad (\text{a}) \\ \lim\limits_{t\to\infty} S_{ji}(+t,-t) & (\text{with } t_0=-\infty) \quad (\text{b}) \end{cases}}$$

Case (a) applied if the interaction is "homogeneous" in time (decay processes)
Case (b) applied if interaction $\hat{H}'_{\mathrm{D}}(t)$ can be "centered" at $t=0$ (scattering)

■ Step perturbation

Consider first the case in which the perturbation is switched on abruptly, in a step-like fashion, at time $t=0$. This is, in fact, the same as if we describe the $t>0$ effects of a *stationary perturbation* $\hat{H}'$ on a system, which was prepared at $t=0$ in the initial eigenstate $|\psi_{0i}\rangle$ of $\hat{H}_0$.

▶ **Perturbation Hamiltonian**

$$\boxed{\hat{H}'(t) = \begin{cases} 0 & \text{for } t<0 \\ \hat{H}' & \text{for } t\geq 0 \end{cases}}$$

Initial state $|\psi_{0i}\rangle$ prepared at any $t_0 \leq 0$.
All cases between $t_0{=}0$ and $t_0 \to -\infty$ are equivalent.
The case $t_0{=}0$ describes the effect of a constant perturbation $\hat{H}'(t){=}\hat{H}'$.
We consider transitions $\boxed{|\psi_{0i}\rangle \to |\psi_{0j}\rangle \text{ for } j \neq i}$

▶ **Transition amplitude & probability up to 1st order contribution**

$$S^{(1)}_{ji}(t) = -\tfrac{i}{\hbar}\lambda H'_{ji} \int\limits_0^t e^{i\omega_{ji}t_1}dt_1 = \lambda H'_{ji}\,\tfrac{1}{\hbar}\,\tfrac{1-e^{i\omega_{ji}t}}{\omega_{ji}}$$

$$P^{(1)}_{ji}(t) = |S^{(1)}_{ji}(t)|^2 = \tfrac{1}{\hbar^2}|\lambda H'_{ji}|^2 \underbrace{\tfrac{(1-\cos\omega_{ji}t)^2+\sin^2\omega_{ji}t}{\omega_{ji}^2}}_{\frac{4\sin^2\left(\frac{\omega_{ji}}{2}t\right)}{\omega_{ji}^2}}$$

$$= \boxed{\tfrac{1}{\hbar^2}|\lambda H'_{ji}|^2\,\frac{\sin^2\left(\frac{\omega_{ji}}{2}t\right)}{\left(\frac{\omega_{ji}}{2}t\right)^2}\,t^2}$$

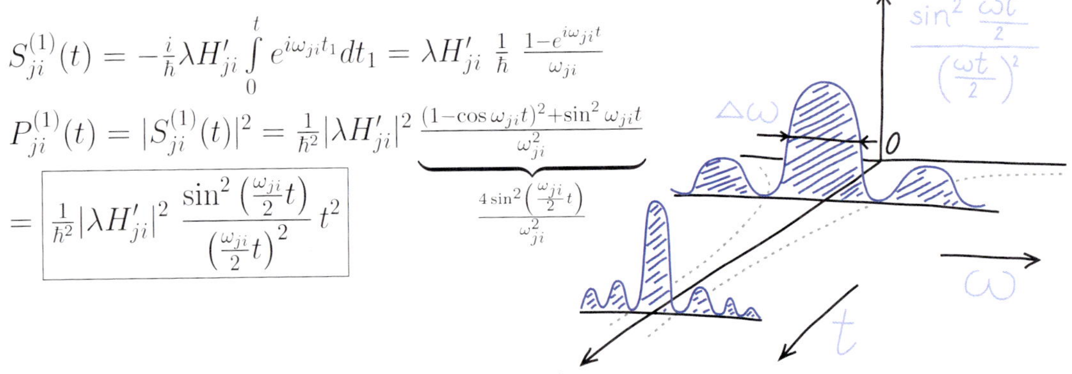

▶ The right way of treating this expression:

(a) Consider long time $t \gg \frac{\hbar}{\Delta E}$ (see above) $\Rightarrow \boxed{t\to\infty} \Rightarrow \boxed{\frac{\sin^2\left(\frac{\omega_{ji}}{2}t\right)}{\left(\frac{\omega_{ji}}{2}t\right)^2}\,t \approx 2\pi\,\delta(\omega_{ji})}$

$$\text{using: } \lim_{\alpha\to\infty}\frac{\sin^2(\alpha x)}{\alpha x^2} = \pi\delta(x) \quad \Leftarrow \quad \frac{\sin^2(\alpha x)}{\alpha x^2} = \begin{cases} \alpha & \text{for } x=0 \\ 0 & \text{for } x=\frac{\pi}{\alpha},\cdots \end{cases} \quad \int\limits_{-\infty}^{+\infty}\frac{\sin^2(\alpha x)}{\alpha x^2}dx = \pi$$

$$\Rightarrow P^{(1)}_{ji}(t) \approx \tfrac{2\pi}{\hbar}\,|\lambda H'_{ji}|^2\,\delta(E_{0j}-E_{0i})\,t$$

(b) Instead of transition probability calculate the **transition rate**

$$\boxed{W_{ji}(t) \equiv \tfrac{d}{dt}P_{ji}(t)} \qquad \Rightarrow \qquad W^{(1)}_{ji} = \tfrac{2\pi}{\hbar}|\lambda H'_{ji}|^2\delta(E_{0j}-E_{0i})$$

(c) **Sum over all final states** at energy $E_f{=}E_{0i}$ making use of averaging with respect to an ϵ-smoothened density of final states $\rho_f(E_f)_\epsilon \equiv \sum_j \underbrace{\delta_\epsilon(E_f-E_{0j})}_{\overset{\text{e.g.}}{=}\frac{1}{\pi}\frac{\epsilon}{\epsilon^2+(E_f-E_{0j})^2}}$

Remark: The density of *final states* at energy E_{0i} differs in general from the density of *initial states* at the same energy. This is so because initial and final states can often be treated as vectors in different Hilbert spaces. Consider, e.g., elmg. decay $A^* \to A+\gamma$ of an excited system A^* (see Example II above): General states are described within the product space $\mathcal{H} \equiv \mathcal{H}_{\rm a}\otimes\mathcal{H}_\gamma$ of the decying system times the elmg. field. While, the initial state $|\psi_{0i}\rangle \equiv |E_{0i}\rangle_{\rm a}\otimes|0\rangle_\gamma$ can be directly reduced only to the space $\mathcal{H}_{\rm a}$, the final state $|\psi_{0j}\rangle \equiv |E_{0j}\rangle_{\rm a}\otimes|\vec{k}\nu\rangle_\gamma$ belongs to $\mathcal{H}_{\rm a}\otimes\mathcal{H}^{(1)}_\gamma$, where $\mathcal{H}^{(1)}_\gamma$ stands for a single-photon subspace of $\mathcal{H}_\gamma$. Therefore, the density of *final* states at energy $E_f \equiv E_{0j}+E_\gamma = E_{0i}$ must be calculated in the larger space.

The summation over final states leads to the following general expression:

$$\boxed{W_{ji}(t) \mapsto W_{fi}(t)} = \tfrac{2\pi}{\hbar}\underbrace{\sum_j |\lambda H'_{ji}|^2 \underbrace{\delta(E_{0j}-E_{0i})}_{\mapsto\,\delta_\epsilon(E_{0j}-E_{0i})}}_{\approx\langle|\lambda H'_{ji}|^2\rangle_f\sum_j\delta_\epsilon(E_{0j}-E_{0i})} = \tfrac{2\pi}{\hbar}\underbrace{\langle|\lambda H'_{ji}|^2\rangle_f}_{\substack{\text{average with}\\ \text{respect to }\rho_f(E_f)_\epsilon}}\underbrace{\rho_f(E_f{=}E_{0i})_\epsilon}_{\substack{\text{to be replaced}\\ \text{by }\rho_f(E_f)}}$$

► **Fermi golden rule**

The above derivation is summarized in a very useful and famous formula, whose validity turns out to be much wider than in the presently studied case:

$$\boxed{W^{(1)}_{fi} = \frac{2\pi}{\hbar}\left\langle|\lambda H'_{ji}|^2\right\rangle_f \rho_f(E_{0i})} \quad \text{where} \quad \begin{cases} \langle|\lambda H'_{ji}|^2\rangle_f \equiv \text{squared matrix element} \\ \quad \text{averaged over available final states} \\ \rho_f(E_{0i}) \equiv \text{density of final states} \\ \quad \text{at final energy } E_f = E_{0i} \end{cases}$$

► **2$^{\rm nd}$order correction**

$$S_{2ji}(t) = \left(-\tfrac{i}{\hbar}\lambda\right)^2 \sum_k H'_{jk}H'_{ki} \underbrace{\int_0^t\!\!\int_0^{t_2} e^{i\omega_{jk}t_2}e^{i\omega_{ki}t_1}dt_1dt_2}_{-\left[\frac{e^{i\omega_{ji}t}-1}{\omega_{ki}\omega_{ji}}-\frac{e^{i\omega_{jk}t}-1}{\omega_{jk}\omega_{ki}}\right]} = \lambda^2\sum_k H'_{jk}H'_{ki}\left[\frac{e^{\frac{i}{\hbar}(E_{0j}-E_{0i})t}-1}{(E_{0k}-E_{0i})(E_{0j}-E_{0i})} - \frac{e^{\frac{i}{\hbar}(E_{0j}-E_{0k})t}-1}{(E_{0j}-E_{0k})(E_{0k}-E_{0i})}\right]$$

$$\omega_{ki}+\omega_{jk}=\omega_{ji} \quad \Rightarrow$$

Assuming $E_{0j} \neq E_{0k} \neq E_{0i}$ (so $H'_{jk}H'_{ki} \approx 0$ for equal energies) we may neglect the 2$^{\rm nd}$ time-dependent term (otherwise special treatment needed). The 1$^{\rm st}$ term yields the same dependence on $(E_{0j}-E_{0i})$ as the 1$^{\rm st}$-order correction $\Rightarrow$

$$\boxed{W^{(2)}_{fi} = \frac{2\pi}{\hbar}\left\langle\left|\lambda H'_{ji} + \lambda^2\sum_k \frac{H'_{jk}H'_{ki}}{E_{0i}-E_{0k}}\right|^2\right\rangle_f \rho_f(E_{0i})} \quad \text{“direct”+“virtual” transitions}$$

◀ **Historical remark**
1927-30: Paul Dirac derives the 1st and 2nd order perturbative expressions. He applies the theory to calculation of electromagnetic transition rates in atoms and nuclei
1950: Enrico Fermi coins the name "golden rule" for the general 1st order expression

■ Exponential perturbation

Another perturbation type, for which the Dyson series can be calculated explicitly, is the one with an exponential dependence on time. In this case, we move the initial time to $-\infty$ and look at the effect of a slowly rising term $\hat{H}'(t)$.

▶ Perturbation Hamiltonian

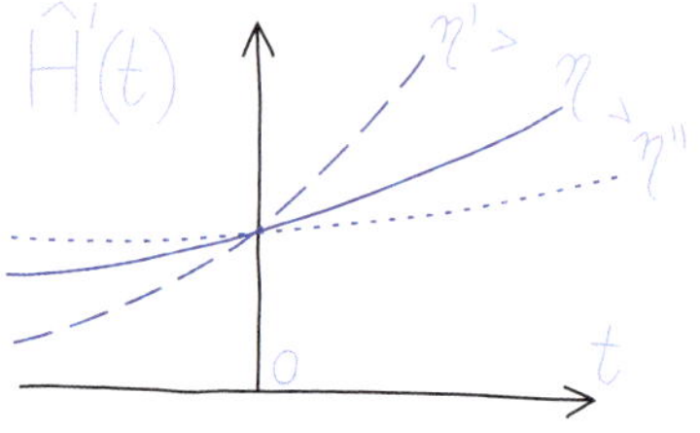

$$\boxed{\hat{H}'(t) = e^{\eta t}\hat{H}'} \qquad \eta \geq 0$$

Initial state $|\psi_{0i}\rangle$ prepared at $t_0 \to -\infty$
Final state $|\psi_{0j}\rangle$ with $j \neq i$ measured at any t

▶ Transition rate up to 1st order contribution

Transition amplitude: $S^{(1)}_{ji}(t) = -\frac{i}{\hbar}\lambda H'_{ji} \int\limits_{-\infty}^{t} e^{(\eta+i\omega_{ji})t_1} dt_1 = -\frac{i}{\hbar}\lambda H'_{ji} \frac{e^{(\eta+i\omega_{ji})t}}{\eta+i\omega_{ji}}$

Transition probability: $P^{(1)}_{ji}(t) = |S^{(1)}_{ji}(t)|^2 = \frac{1}{\hbar^2}|\lambda H'_{ji}|^2 \frac{e^{2\eta t}}{\eta^2+\omega_{ji}^2}$

Transition rate:

$$\frac{d}{dt}P^{(1)}_{ji}(t) = \boxed{W^{(1)}_{ji}(t) = \frac{2\pi}{\hbar^2}|\lambda H'_{ji}|^2 \underbrace{\frac{1}{\pi}\frac{\eta}{\eta^2+\omega_{ji}^2}}_{\hbar\Omega^{\rm BW}_{\Gamma}(E_{0j}-E_{0i})} e^{2\eta t}}$$

$\Omega^{\rm BW}_{\Gamma} \equiv$ **Breit–Wigner** energy distribution (Sec. 1.5) with the width $\boxed{\Gamma = 2\hbar\eta}$

▶ Adiabatic limit $(\eta, \Gamma \to 0)$

$$\lim_{\Gamma\to 0}\Omega^{\rm BW}_{\Gamma}(E_{0j}-E_{0i}) = \delta(E_{0j}-E_{0i}) \quad \Rightarrow \quad \lim_{\eta\to 0} W^{(1)}_{ji} = \frac{2\pi}{\hbar}|\lambda H'_{ji}|^2\delta(E_{0j}-E_{0i})$$

This is consistent with the previous result on constant $\hat{H}' \Rightarrow$ golden rule

■ Periodic perturbation

Expressions similar to those derived above come out also for $T = \frac{2\pi}{\omega}$ periodic perturbations. In this case, however, the perturbation induces transitions up and down to final energies $E_i + \hbar\omega$ or $E_i - \hbar\omega$.

▶ Perturbation Hamiltonian

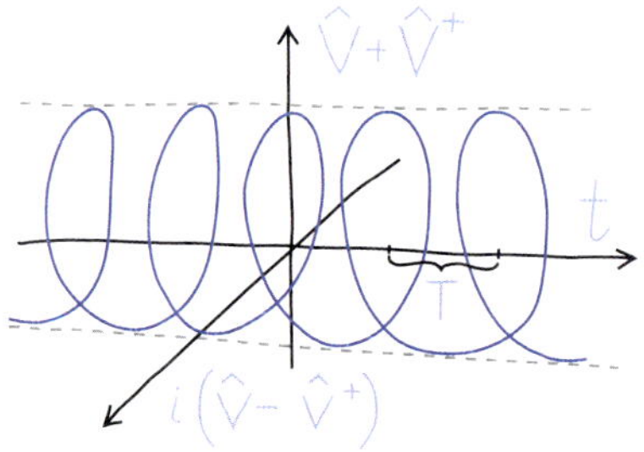

$$\boxed{\hat{H}'(t) = \hat{V}e^{+i\omega t} + \hat{V}^\dagger e^{-i\omega t}} = \begin{cases} (\hat{V}+\hat{V}^\dagger)\cos(\omega t) \\ +i(\hat{V}-\hat{V}^\dagger)\sin(\omega t) \end{cases}$$

Initial state $|\psi_{0i}\rangle$ at $t_0 = 0$

▶ Up/down transition rates to 1st order

Transition amplitude to $|\psi_{0j}\rangle (j \neq i)$:

$$S^{(1)}_{ji}(t) = -\frac{i\lambda}{\hbar}\left[V_{ji}\int\limits_0^t e^{i(\omega_{ji}+\omega)t_1}dt_1 + V^*_{ij}\int\limits_0^t e^{i(\omega_{ji}-\omega)t_1}dt_1\right]$$
$$= \frac{\lambda}{\hbar}\left[V_{ji}\frac{1-e^{i(\omega_{ji}+\omega)t}}{\omega_{ji}+\omega} + V^*_{ij}\frac{1-e^{i(\omega_{ji}-\omega)t}}{\omega_{ji}-\omega}\right]$$

Transition probability:

$$P^{(1)}_{ji}(t)=\frac{\lambda^2}{\hbar^2}\Bigg[|V_{ji}|^2\frac{\sin^2\left(\frac{\omega_{ji}+\omega}{2}t\right)}{\left(\frac{\omega_{ji}+\omega}{2}t\right)^2}t^2+|V_{ij}|^2\frac{\sin^2\left(\frac{\omega_{ji}-\omega}{2}t\right)}{\left(\frac{\omega_{ji}-\omega}{2}t\right)^2}t^2+2\mathrm{Re}\Big(V_{ji}V_{ij}\overbrace{\frac{1-e^{i(\omega_{ji}+\omega)t}}{\omega_{ji}+\omega}\frac{1-e^{-i(\omega_{ji}-\omega)t}}{\omega_{ji}-\omega}}^{-2e^{i\omega t}\frac{\cos\omega_{ji}t-\cos\omega t}{\omega_{ji}^2-\omega^2}}\Big)\Bigg]$$

The first 2 terms yield: $2\pi t\,\delta(\omega_{ji}+\omega)$ & $2\pi t\,\delta(\omega_{ji}-\omega)$

The last term for $\omega_{ji}=\pm\omega\underbrace{+\epsilon}_{\to 0}$ is negligible relative to the previous terms:

$$\propto -\frac{\cos\omega_{ji}t-\cos\omega t}{\omega_{ji}^2-\omega^2}=-\frac{\cos(\pm\omega+\epsilon)t-\cos\omega t}{(\pm\omega+\epsilon)^2-\omega^2}\xrightarrow{\epsilon\to 0}t\,\frac{\sin\omega t}{2\omega}\sim 0$$

► **Transition rates**

$$W^{(1)}_{fi}=\begin{cases}\frac{2\pi}{\hbar}\langle|\lambda V_{ji}|^2\rangle_f\,\rho_f(E_{0i}-\hbar\omega) & \textbf{stimulated emission}\\ \frac{2\pi}{\hbar}\langle|\lambda V_{ij}|^2\rangle_f\,\rho_f(E_{0i}+\hbar\omega) & \textbf{absorption}\end{cases}$$

◄ **Historical remark**

1916: A. Einstein theoretically discovers stimulated emission and discusses the detailed balance between absorption and emission processes

1950's: Application of these ideas in the construction of laser

■ **Application to stimulated electromagnetic transitions**

Results of the periodic-field perturbation theory can be directly applied to atoms or nuclei interacting with external electromagnetic waves of appropriate wavelengths. We outline these issues, leaving the description of the *spontaneous* elmg. emissions to Chapter 6 (after the quantization of elmg. field).

► **Hamiltonian of particles in external field**

Ensemble of N charged particles in an external classical elmg. field

$$\hat{H}=\sum_{k=1}^{N}\frac{1}{2M_k}\left[\hat{\vec{p}}_k-q_k\vec{A}(\hat{\vec{x}}_k,t)\right]^2+\sum_{k=1}^{N}V(\hat{\vec{x}}_k,t)$$

Neglecting $q_k^2\vec{A}(\hat{\vec{x}}_k,t)^2$ and assuming Coulomb gauge condition $\vec{\nabla}\cdot\vec{A}(\hat{\vec{x}},t)=0\Rightarrow$

$$\hat{H}\approx\underbrace{\sum_{k=1}^{N}\frac{1}{2M_k}\hat{\vec{p}}_k^{\,2}+V(\hat{\vec{x}}_k,t)}_{\hat{H}_0}\underbrace{-\sum_{k=1}^{N}\frac{q_k}{M_k}\left[\vec{A}(\hat{\vec{x}}_k,t)\cdot\hat{\vec{p}}_k\right]}_{-\int\vec{A}(\vec{x},t)\cdot\vec{j}(\vec{x})\,d\vec{x}\,\equiv\,\hat{H}'(t)}$$

► **Planar elmg. wave**

$$\vec{A}(\vec{x},t)=A_0\,\vec{\varepsilon}\cos\big(\overbrace{\tfrac{\omega}{c}\vec{n}}^{\vec{k}}\cdot\vec{x}-\omega t\big)\qquad\text{with }\boxed{\vec{\varepsilon}\cdot\vec{n}=0}$$

$$\Rightarrow\begin{pmatrix}\vec{E}(\vec{x},t)\\ \vec{B}(\vec{x},t)\end{pmatrix}=-A_0\begin{pmatrix}\omega\,\vec{\varepsilon}\,\sin(\vec{k}\cdot\vec{x}-\omega t)\\ \frac{\omega}{c}[\vec{n}\times\vec{\varepsilon}]\,\sin(\vec{k}\cdot\vec{x}-\omega t)\end{pmatrix}\qquad\text{el. \& mg. field intensities}$$

Averaged energy density: $\langle w\rangle = \frac{1}{2}[\epsilon_0\langle\vec{E}^2(\vec{x},t)\rangle + \mu_0^{-1}\langle\vec{B}^2(\vec{x},t)\rangle] = \frac{1}{2}\epsilon_0 A_0^2\omega^2$

Averaged energy flow: $\langle P\rangle = \langle w\rangle c = \frac{1}{2}\epsilon_0 A_0^2\omega^2 c$

▶ **Periodic perturbation Hamiltonian**

$$\boxed{\hat{H}'(t) = -\frac{A_0}{2}\sum_{k=1}^{N}\frac{q_k}{M_k}\Big[\underbrace{e^{-i\frac{\omega}{c}\vec{n}\cdot\hat{\vec{x}}_k}(\vec{\varepsilon}\cdot\hat{\vec{p}}_k)}_{\propto\hat{V}\ \leftrightarrow\ \text{stimul. emission}}\, e^{+i\omega t} + \underbrace{e^{+i\frac{\omega}{c}\vec{n}\cdot\hat{\vec{x}}_k}(\vec{\varepsilon}\cdot\hat{\vec{p}}_k)}_{\propto\hat{V}^\dagger\ \leftrightarrow\ \text{absorption}}\, e^{-i\omega t}\Big]}$$

Emission: $\hbar\omega = E_{0i} - E_{0j}$ **Absorption**: $\hbar\omega = E_{0j} - E_{0i}$

▶ **Absorption cross section**

In the following, we focus on the absorption processes (the procedure for stimulated emission is analogous). We define the absorption cross section, which can be seen as an *area* on the plane perpendicular to the incident wave propagation direction. The elmg. energy passing through this area is being continuously transferred to the system: $\boxed{\sigma_{ji}^{\rm abs} = \frac{\text{energy absorbed in unit time}}{\text{incoming energy flow}} = \frac{\hbar\omega\, W_{ji}}{\frac{1}{2}\epsilon_0 A_0^2\omega^2 c}}$

Perturbation theory prediction (1st order):

$$\sigma_{ji}^{\rm abs} \approx \frac{\pi}{\epsilon_0\omega c}\left|\left\langle\psi_{0j}\left|\sum_{k=1}^{N}\frac{q_k}{M_k}e^{+i\frac{\omega}{c}\vec{n}\cdot\hat{\vec{x}}_k}(\vec{\varepsilon}\cdot\hat{\vec{p}}_k)\right|\psi_{0i}\right\rangle\right|^2 \delta(E_{0i}+\hbar\omega-E_{0j})$$

▶ **Dipole approximation**

Assume that the atom/nucleus size $\boxed{R \ll \lambda}$ radiation wavelength

$$\Rightarrow e^{+i\frac{\omega}{c}\vec{n}\cdot\hat{\vec{x}}_k} = 1+\sum_{n=1}^{\infty}\frac{1}{n!}\left(i\frac{\omega}{c}\vec{n}\cdot\hat{\vec{x}}_k\right)^n \approx 1$$

$$\left\langle\psi_{0j}\left|\sum_{k=1}^{N}\frac{q_k}{M_k}\,e^{+i\frac{\omega}{c}\vec{n}\cdot\hat{\vec{x}}_k}(\vec{\varepsilon}\cdot\hat{\vec{p}}_k)\right|\psi_{0i}\right\rangle \approx \left\langle\psi_{0j}\left|\vec{\varepsilon}\cdot\sum_{k=1}^{N}\frac{q_k}{M_k}\hat{\vec{p}}_k\right|\psi_{0i}\right\rangle = \ldots$$

Trick: $\hat{\vec{p}}_k = -\frac{i}{\hbar}M_k[\hat{\vec{x}}_k, \hat{H}_0] \Rightarrow$ $\ldots = \frac{i}{\hbar}\underbrace{(E_{0j}-E_{0i})}_{\hbar\omega}\left\langle\psi_{0j}\left|\vec{\varepsilon}\cdot\underbrace{\sum_{k=1}^{N}q_k\hat{\vec{x}}_k}_{\hat{\vec{D}}}\right|\psi_{0i}\right\rangle$

where we introduced the operator of **electric dipole moment**:

$$\Rightarrow \sigma_{ji}^{\rm abs} \approx \frac{\pi\omega}{\epsilon_0 c}\left|\langle\psi_{0j}|\vec{\varepsilon}\cdot\hat{\vec{D}}|\psi_{0i}\rangle\right|^2 \delta(E_{0i}+\hbar\omega-E_{0j})$$

For $\vec{\varepsilon}=\vec{n}_x$: $\boxed{\sigma_{ji}^{\rm abs}\,d\omega \approx \frac{\pi\omega}{\epsilon_0\hbar c}\left|\langle\psi_{0j}|\hat{D}_x|\psi_{0i}\rangle\right|^2}$ $\neq 0$ *iff* $\boxed{|j_i-1| \leq j_j \leq (j_i+1)}$

▶ **Multipole expansion**

To go beyond the dipole approximation, it is appropriate to expand the incoming planar wave into spherical waves with increasing multipolarities. This is not quite trivial as one needs to correctly treat the wave polarization, which on the quantum level results from the **photon spin** (s=1).

Planar wave expansion (cf. p. 167): $e^{i\vec{k}\cdot\vec{x}} = 4\pi \sum_{l=0}^{\infty} \sum_{m=-l}^{+l} i^l\, j_l(kr)\, Y^*_{lm}\left(\frac{\vec{k}}{k}\right) Y_{lm}\left(\frac{\vec{x}}{x}\right)$

To include the polarization, we introduce **circular & linear polarization** bases in a general coordinate system: $\begin{cases} \vec{e}_{\pm} = \mp\frac{1}{\sqrt{2}}(\vec{n}_x \pm i\vec{n}_y) \\ \vec{e}_0 = \vec{n}_z \end{cases}$

Arbitrary lin. polarization vector $\vec{\varepsilon} \equiv (\varepsilon_x\vec{n}_x + \varepsilon_y\vec{n}_y + \varepsilon_z\vec{n}_z) = \sqrt{\frac{4\pi}{3}} \sum_{\nu=0,\pm1} Y^*_{1\nu}(\vec{\varepsilon})\, \vec{e}_\nu$

Note: the circular polarization vector $\vec{e}_0$ is present because the evaluation is done in a general system unrelated to the wave vector $\vec{k}$.

Introduce a "vector spherical function" with total angular momentum (multipolarity) λ: $\vec{\mathcal{Y}}_{l\lambda\mu}\left(\frac{\vec{x}}{x}\right) = \sum_{\nu,m} C^{\lambda\mu}_{1\nu lm} \vec{e}_\nu Y_{lm}\left(\frac{\vec{x}}{x}\right) \;\Leftrightarrow\; \vec{e}_\nu Y_{lm}\left(\frac{\vec{x}}{x}\right) = \sum_{\lambda,\mu} C^{\lambda\mu}_{1\nu lm} \vec{\mathcal{Y}}_{l\lambda\mu}\left(\frac{\vec{x}}{x}\right)$

$$\boxed{\vec{\varepsilon}\, e^{i\vec{k}\cdot\vec{x}} = \frac{(4\pi)^{\frac{3}{2}}}{3} \sum_{\lambda,\mu} \sum_{l,m} \sum_{\nu} i^l\, C^{\lambda\mu}_{1\nu lm}\, Y^*_{1\nu}(\vec{\varepsilon})\, Y^*_{lm}\left(\frac{\vec{k}}{k}\right) \underbrace{j_l(kr)\, \vec{\mathcal{Y}}_{l\lambda\mu}\left(\frac{\vec{x}}{x}\right)}_{\text{spatial dependence}}}$$

For each multipolarity λ it is possible to separate terms with both parities: **electric** (E) & **magnetic** (M) components. From the resulting expansion one can construct transition probabilities for Eλ & Mλ transitions. The above dipole approximation is identified as E1.

◂ **Historical remark**
1900's-10's: Multipole expansion of elmg. field elaborated within the classical theory
1940's-50's: Multipole expansion applied in QM (M.E. Rose *et al.*)

6. SCATTERING THEORY

Description of the processes induced by scattering of particles belongs to the most important application domains of quantum theory. Knowing the interaction Hamiltonian between the particles and the initial state, can one predict all outcomes & probabilities? And inversely: knowing the initial & final states, can one determine the interaction? This may resemble a task to analyze an internal structure of a watch by detecting tiny parts shot out when the thing is smashed on an anvil. In the quantum world, this is often the only research method available.

The scattering theory is a rather wide area, of which we are going to taste only a little bit. Here is a general typology of scattering processes:

(1) $a + A \to A + a$	**elastic scattering** (total kinetic energy conserved)
(2) $a + A \to A^* + a^*$	**inelastic scattering** (intrinsic excitations of particles involved, total kinetic energy not conserved)
(3) $a + A \to B + b + b' + \ldots$	more **complex reaction** (reconfiguration of the interacting particles, appearance of new objects)

6.1 Elementary description of elastic scattering

In a large part of this chapter we will deal with *elastic scattering*—the simplest scattering process which does not change the nature or internal structure of the scattered objects. First we focus on some basic concepts. The description of elastic scattering requires to solve the *stationary* Schrödinger equation with the specific interaction potential and an appropriate asymptotic behavior imposed on the wavefunction.

■ Scattering by fixed potential

Consider a spinless projectile particle moving in a fixed finite-range field. This corresponds to elastic scattering of the projectile on an infinite-mass target particle, the target-projectile interaction being assumed to have a limited reach.

▶ Formulation

Infinite-mass scattering center with *finite-range potential*, i.e., $V(\vec{x}) \approx 0$ for $|\vec{x}| > R$. Particle with scalar wavefunction scattered by the potential.

Initial state $\equiv$ momentum eigenstate of the scattered particle (assume $\vec{p} \propto \vec{n}_z$)

$\equiv$ **plane wave** e^{ikz} with $k = \frac{1}{\hbar}|\vec{p}|$, energy $\boxed{E = \frac{(\hbar k)^2}{2M}}$

For the solution of the scattering problem, we solve the stationary Schrödinger equation with the same energy E, which is in the **continuous spectrum** of the full Hamiltonian:
$$\left[-\frac{\hbar^2}{2M}\Delta + V(\vec{x})\right]\psi(\vec{x}) = \frac{(\hbar k)^2}{2M}\psi(\vec{x})$$

▶ Required **asymptotic form** of the wavefunction for $|\vec{x}| \gg R$

$$\boxed{\psi_k(\vec{x}) \;\propto\; e^{ikz} + f_k(\vartheta,\varphi)\frac{e^{ikr}}{r}}$$

$$\equiv \begin{pmatrix}\textbf{incoming}\\ \textbf{plane wave}\end{pmatrix} + \begin{pmatrix}\textbf{outgoing}\\ \textbf{spherical wave}\end{pmatrix}$$

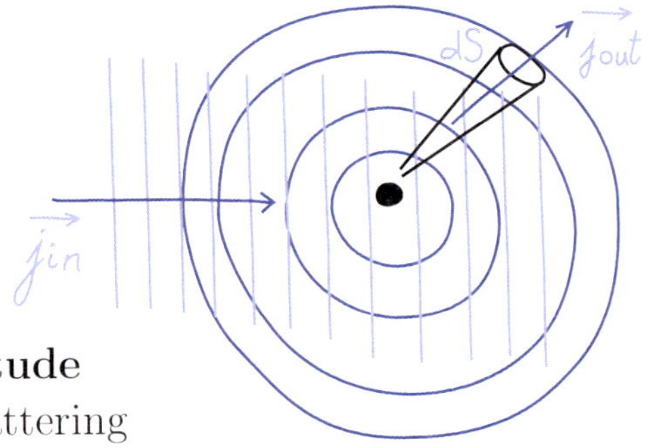

The function $f(\vartheta,\varphi) \equiv$ **scattering amplitude** contains all relevant information on the scattering of the incoming plane wave to various angles

▶ Cross section

Incoming flux: $\vec{j}_{\rm in} = \frac{\hbar k}{M}\vec{n}_z$ Outgoing flux: $\vec{j}_{\rm out}(r,\vartheta,\varphi) = \frac{|f_k(\vartheta,\varphi)|^2}{r^2}\frac{\hbar k}{M}\vec{n}_r$

So-called differential cross section is the flux to a an infinitesimal space angle $d\Omega$ around direction (ϑ,φ) normalized by the incoming flux:

$$d\sigma(\vartheta,\varphi) = \frac{\text{outgoing flux to space angle } d\Omega}{\text{incoming flux}} = \frac{|\vec{j}_{\rm out}(r,\vartheta,\varphi)|\overbrace{r^2 d\Omega}^{dS}}{|\vec{j}_{\rm in}|}$$

Differential cross section: $\boxed{\left(\frac{d\sigma}{d\Omega}\right)_k(\vartheta,\varphi) = |f_k(\vartheta,\varphi)|^2}$ (units of area)

■ Two-body problem & center-of-mass system

Elastic scattering of a projectile particle on a *finite-mass* target particle represents a genuine two-body problem. The familiar way of solving this problem proceeds via separating the relative target-projectile degree of freedom from that related the system's center of mass.

▶ Canonical transformation to relative & center-of-mass coordinates

2 particles with masses $\binom{M_1}{M_2}$. Position & momentum operators $\begin{pmatrix}\hat{\vec{x}}_1\\ \hat{\vec{x}}_2\end{pmatrix}$ & $\begin{pmatrix}\hat{\vec{p}}_1\\ \hat{\vec{p}}_2\end{pmatrix}$
New pair of canonically conjugate coordinates & momenta:

$$\boxed{\begin{aligned}\hat{\vec{x}}_{\rm c} &= \tfrac{M_1}{M_1+M_2}\hat{\vec{x}}_1 + \tfrac{M_2}{M_1+M_2}\hat{\vec{x}}_2\\ \hat{\vec{x}}_{\rm r} &= \hat{\vec{x}}_1 - \hat{\vec{x}}_2\end{aligned}} \Leftrightarrow \boxed{\begin{aligned}\hat{\vec{p}}_{\rm c} &= \hat{\vec{p}}_1 + \hat{\vec{p}}_2\\ \hat{\vec{p}}_{\rm r} &= \tfrac{M_2}{M_1+M_2}\hat{\vec{p}}_1 - \tfrac{M_1}{M_1+M_2}\hat{\vec{p}}_2\end{aligned}} \quad \begin{matrix}\text{center of mass}\\ \text{relative}\end{matrix}$$

Commutators: $[\hat{x}_{{\rm c}i},\hat{p}_{{\rm c}j}] = [\hat{x}_{{\rm r}i},\hat{p}_{{\rm r}j}] = i\hbar\delta_{ij}, \quad [\hat{x}_{{\rm c}i},\hat{p}_{{\rm r}j}] = [\hat{x}_{{\rm r}i},\hat{p}_{{\rm c}j}] = 0$

$\Rightarrow$ corresponding Poisson brackets $\Rightarrow$ the transformation is canonical

▶ Transformation of Hamiltonian

Kinetic energy of both particles: $\boxed{\hat{T} = \frac{\hat{\vec{p}}_1^{\,2}}{2M_1} + \frac{\hat{\vec{p}}_2^{\,2}}{2M_2} = \frac{\hat{\vec{p}}_{\rm c}^{\,2}}{2(M_1+M_2)} + \frac{\hat{\vec{p}}_{\rm r}^{\,2}}{2\frac{M_1M_2}{M_1+M_2}}}$

Define **reduced mass**: $\boxed{\mathcal{M} = \frac{M_1M_2}{M_1+M_2}}$

Potential depending on $\underbrace{\vec{x}_1-\vec{x}_2}_{\vec{x}_{\rm r}} \Rightarrow$ Hamiltonian $\boxed{\hat{H} = \overbrace{\frac{\hat{\vec{p}}_{\rm c}^{\,2}}{2M_{\rm tot}}}^{\hat{H}_{\rm c}} + \overbrace{\frac{\hat{\vec{p}}_{\rm r}^{\,2}}{2\mathcal{M}} + V(\hat{\vec{x}}_{\rm r})}^{\hat{H}_{\rm r}}}$

This represents the separation of center-of-mass and relative motions. Solution of the Schrödinger eq. with $\hat{H}_{\rm c}$ is a plane wave in center-of-mass coordinates. We need to solve the equation with $\hat{H}_{\rm r}$ in relative coordinates. This represents just the $\boxed{M \mapsto \mathcal{M}}$ change with respect to the fixed-potential problem.

▶ Transformation of scattering angles & cross section

Once the two-body problem is solved in the the center-of-mass (CM) system (as described above), one has to return back to the laboratory (LAB) system, in which the scattering angles and cross sections are measured.

Notation: particle 1 ≡ projectile, particle 2 ≡ target
$\vec{v}_1, \vec{v}_2,\ \vec{p}_1, \vec{p}_2,\ \vartheta, \varphi$ ≡ velocities & momenta & scattering angles in LAB
$\vec{v}_{\rm C1}, \vec{v}_{\rm C2},\ \vec{p}_{\rm C1}, \vec{p}_{\rm C2},\ \vartheta_{\rm C}, \varphi_{\rm C}$ ≡ velocities & momenta & scattering angles in CM

Center-of-mass speed in LAB:

$$\boxed{\vec{u} = \tfrac{M_1}{M_1+M_2}\vec{v}_1 + \tfrac{M_2}{M_1+M_2}\vec{v}_2}$$

= constant (along z)

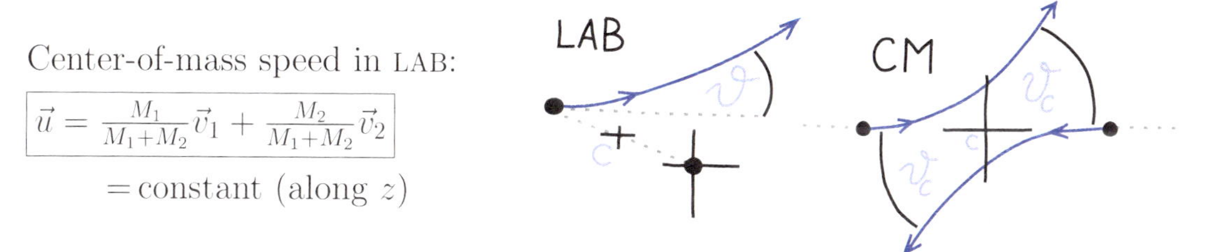

$\vec{p}_{C1} = M_1(\vec{v}_1 - \vec{u}) = \mathcal{M}(\vec{v}_1 - \vec{v}_2) \equiv +\vec{p}_C = \vec{p}_r \qquad \vec{p}_{C2} = M_2(\vec{v}_2 - \vec{u}) = \mathcal{M}(\vec{v}_2 - \vec{v}_1) = -\vec{p}_C = -\vec{p}_r$

$\boxed{\varphi = \varphi_C}$ assume $\varphi = 0 \quad \Rightarrow \quad p_1 \sin\vartheta = p_C \sin\vartheta_C \qquad p_1 \cos\vartheta - M_1 u = p_C \cos\vartheta_C$

$\Rightarrow \boxed{\tan\vartheta = \frac{p_C \sin\vartheta_C}{p_C \cos\vartheta_C + M_1 u}} \quad \boxed{\tan\vartheta_C = \frac{p_1 \sin\vartheta}{p_1 \cos\vartheta - M_1 u}}$

Outgoing fluxes in both LAB & CM systems must be *the same*!

$\left(\frac{d\sigma}{d\Omega}\right)_L d\Omega_L \overset{!}{=} \left(\frac{d\sigma}{d\Omega}\right)_C d\Omega_C \quad \Rightarrow \quad \left(\frac{d\sigma}{d\Omega}\right)_L = \left(\frac{d\sigma}{d\Omega}\right)_C \frac{d\Omega_C}{d\Omega_L} \qquad \begin{array}{l} d\Omega_L = \sin\vartheta\, d\vartheta\, d\varphi \\ d\Omega_C = \sin\vartheta_C\, d\vartheta_C\, d\varphi_C \end{array}$

$$\boxed{\left(\frac{d\sigma}{d\Omega}\right)_L(\vartheta, \varphi) = \left(\frac{d\sigma}{d\Omega}\right)_C(\vartheta_C, \varphi_C)\, \frac{\sin\vartheta_C}{\sin\vartheta} \frac{d\vartheta_C}{d\vartheta}}$$

This is the desired relation between LAB and CM differential cross sections (the derivative $\frac{d\vartheta_C}{d\vartheta}$ can be evaluated from the above relation $\vartheta_C \leftrightarrow \vartheta$).
From now on we will work in CM, *skipping the indices* "C" and "R".

■ Effect of particle indistinguishability in cross section

As the last pre-requisite of the scattering theory, let us discuss a rather important effect connected with quantum indistinguishability of identical particles. Depending on whether the scattered particles are identical bosons or fermions, the elastic cross section must be substantially modified with respect to the one for distinguishable particles.

▸ Asymptotic wavefunction in CM: $\psi(\vec{x}) \propto e^{ikz} + f_k(\vartheta, \varphi) \frac{e^{ikr}}{r}$

Exchange of particles in CM:

$\boxed{\vec{x} \mapsto -\vec{x} \Rightarrow \begin{cases} r \mapsto r \\ \vartheta \mapsto \pi - \vartheta \\ \varphi \mapsto \pi + \varphi \end{cases}}$

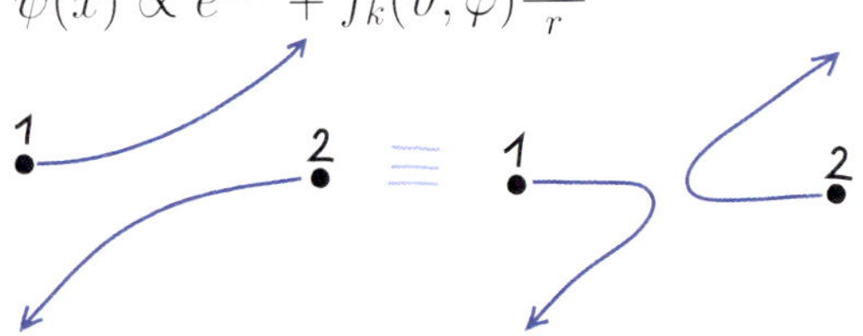

▸ **Symmetrized wavefunction** (for example ${}^4_2$He+${}^4_2$He scattering):

$$\psi(\vec{x}) \propto \left[e^{ikz} + e^{-ikz}\right] + \left[f_k(\vartheta, \varphi) + f_k(\pi - \vartheta, \pi + \varphi)\right] \frac{e^{ikr}}{r}$$

Cross section:

$$\boxed{\left(\frac{d\sigma}{d\Omega}\right)^+_k = \frac{1}{2}\left\{|f_k(\vartheta, \varphi)|^2 + |f_k(\pi - \vartheta, \pi + \varphi)|^2 + 2\mathrm{Re}[f_k(\vartheta, \varphi) f^*_k(\pi - \vartheta, \pi + \varphi)]\right\}}$$

where $\frac{1}{2}$ comes from the normalization of incoming flux

The same expression applies for 2 fermions in antisymmetric spin state (for example $e + e$ in spin singlet)

▸ **Antisymmetrized wavefunction** (for example $e + e$ in spin triplet):

$$\psi(\vec{x}) \propto \left[e^{ikz} - e^{-ikz}\right] + \left[f_k(\vartheta, \varphi) - f_k(\pi - \vartheta, \pi + \varphi)\right] \frac{e^{ikr}}{r}$$

Cross section:

$$\boxed{\left(\frac{d\sigma}{d\Omega}\right)^-_k = \frac{1}{2}\left\{|f_k(\vartheta, \varphi)|^2 + |f_k(\pi - \vartheta, \pi + \varphi)|^2 - 2\mathrm{Re}[f_k(\vartheta, \varphi) f^*_k(\pi - \vartheta, \pi + \varphi)]\right\}}$$

▶ **Example: unpolarized e+e scattering**

Probabilities for finding spin singlet & triplet states are $\frac{1}{4}$ & $\frac{3}{4}$ $\Rightarrow$

$$\left(\frac{d\sigma}{d\Omega}\right)_k = \frac{1}{4}\left(\frac{d\sigma}{d\Omega}\right)_k^+ + \frac{3}{4}\left(\frac{d\sigma}{d\Omega}\right)_k^- =$$
$$= \frac{1}{2}\left\{|f_k(\vartheta,\varphi)|^2 + |f_k(\pi-\vartheta,\pi+\varphi)|^2 - \mathrm{Re}[f_k(\vartheta,\varphi)f_k^*(\pi-\vartheta,\pi+\varphi)]\right\}$$

◀ **Historical remark**

1926: M. Born applies QM to scattering processes (probabilistic interpretation)
1930: N. Mott describes the effects of indistinguishability in Coulomb scattering

6.2 Perturbative approach to the scattering problem

There is a strong link of scattering theory to the nonstationary perturbation theory. Indeed, if the interaction between scattered particles is much smaller than the corresponding free Hamiltonian (energy), the scattering problem can be reformulated in terms of an equation which allows for iterative solutions.

■ Lippmann-Schwinger equation

The L.-S. equation is a clone of stationary Schrödinger equation tailored for general scattering problems. It results from the nonstationary formulation and leads to a suitable perturbative expansion.

▶ **Green operator** defining equation: $\left(i\hbar\frac{\partial}{\partial t} - \hat{H}\right)\hat{G}(t-t_0) = i\hbar\delta(t-t_0)$

Green operator of free particle: $\hat{G}_0(t-t_0)$. Green op. for $\hat{H}(t) = \hat{H}_0(t) + \hat{H}'(t)$ satisfies the equation: $\hat{G}(t-t_0) = \hat{G}_0(t-t_0) - \frac{i}{\hbar}\int\limits_{-\infty}^{+\infty}\hat{G}_0(t-t_1)\hat{H}'(t_1)\hat{G}(t_1-t_0)\,dt_1$

Equivalent expression: $$\boxed{|\psi(t)\rangle = |\phi(t)\rangle - \frac{i}{\hbar}\int\limits_{t_0}^{t}\hat{G}_0(t-t_1)\hat{H}'(t_1)|\psi(t_1)\rangle\,dt_1}$$

where $\left.\begin{matrix}|\psi(t)\rangle\\|\phi(t)\rangle\end{matrix}\right\}$ $\equiv$ states evolved from the same $t{=}t_0$ initial state by $\left\{\begin{matrix}\hat{H}(t)\\\hat{H}_0(t)\end{matrix}\right.$

▶ **Free and interaction Hamiltonians**

$\hat{H}_0 \equiv \frac{\hat{\vec{p}}_r^{\,2}}{2\mathcal{M}} + \hat{H}_{\rm int} \equiv$ free Hamiltonian, with $\hat{H}_{\rm int} \equiv$ intrinsic Hamiltonian of both scattering objects

$\hat{H}' \equiv$ interaction Hamiltonian, including potential term $V(\hat{\vec{x}}_1-\hat{\vec{x}}_2)$ as well as other terms affecting internal degrees of freedom of the objects

▶ **Transition from time-dependent to time-independent description**

Trick: Instead of $\hat{H} = \hat{H}_0 + \hat{H}'$ we use $\boxed{\hat{H}(t) = \hat{H}_0 + \underbrace{e^{\eta t}\hat{H}'}_{\hat{H}'(t)}}$ with $\boxed{\eta \gtrsim 0}$

Later we will apply the limit $\boxed{\eta \to 0_+}$

The initial state is prepared at $\boxed{t_0 \to -\infty}$ as an eigenstate $|\phi_{\vec{k}}\rangle$ of $\hat{H}_0$ with $\vec{p}_{\rm R}=\hbar\vec{k}$. Since $\hat{H}(t=-\infty)=\hat{H}_0$, the initial state is an eigenstate of the full Hamiltonian at that time. Due to the adiabatic onset of perturbation we may assume that the system at *any time* is in the eigenstate of $\hat{H}(t)$ with the same energy E (cf. the adiabatic theorem for discrete spectrum). This allows us to integrate over the time variable and obtain a time-independent equation.

Denote $\left.\begin{matrix}|\phi_{\vec{k}}(t)\rangle\\|\psi_{\vec{k}}(t)\rangle\end{matrix}\right\} \equiv$ state evolved to a finite time t from $|\phi_{\vec{k}}\rangle$ by $\left\{\begin{matrix}\hat{H}_0\\ \hat{H}(t)\end{matrix}\right.$

$$\underbrace{|\psi_{\vec{k}}(t)\rangle}_{e^{-\frac{i}{\hbar}Et}|\psi_{\vec{k}}\rangle} = \underbrace{|\phi_{\vec{k}}(t)\rangle}_{e^{-\frac{i}{\hbar}Et}|\phi_{\vec{k}}\rangle} - \frac{i}{\hbar}\int_{-\infty}^{t} \underbrace{\hat{G}_0(t-t_1)}_{\theta(t-t_1)e^{-\frac{i}{\hbar}\hat{H}_0(t-t_1)}}\ \underbrace{\hat{H}'(t_1)}_{e^{\eta t_1}\hat{H}'}\,\underbrace{|\psi_{\vec{k}}(t_1)\rangle}_{e^{-\frac{i}{\hbar}Et_1}|\psi_{\vec{k}}\rangle}\ dt_1 \qquad \text{subst. } \tau = t_1 - t$$

$$e^{-\frac{i}{\hbar}Et}|\psi_{\vec{k}}\rangle = e^{-\frac{i}{\hbar}Et}|\phi_{\vec{k}}\rangle - \frac{i}{\hbar}\,e^{\eta t-\frac{i}{\hbar}Et}\overbrace{\int_{-\infty}^{0} e^{-\frac{i}{\hbar}(E-\hat{H}_0+i\hbar\eta)\tau}\,\hat{H}'|\psi_{\vec{k}}\rangle\,d\tau}^{\frac{1}{-\frac{i}{\hbar}(E-\hat{H}_0+i\hbar\eta)}\hat{H}'|\psi_{\vec{k}}\rangle}$$

$$|\psi_{\vec{k}}\rangle = |\phi_{\vec{k}}\rangle + \frac{e^{\eta t}}{E-\hat{H}_0+i\hbar\eta}\hat{H}'|\psi_{\vec{k}}\rangle$$

▶ **Lippmann-Schwinger equation** the $\eta \to 0$ limit of the above eq. (the limit in the denominator cannot be performed by plain substitution)

$$\boxed{|\psi_{\vec{k}}\rangle = |\phi_{\vec{k}}\rangle + \frac{1}{E-\hat{H}_0+i\hbar\eta}\hat{H}'|\psi_{\vec{k}}\rangle}$$

Comments:

(a) The L.-S. eq. in the above form is general, valid for all types of processes

(b) The state $|\psi_{\vec{k}}\rangle$ represents the eigenstate of $\hat{H} = \hat{H}_0 + \hat{H}'$ which is the result of infinite-time evolution (by $\hat{H}$) from $|\phi_{\vec{k}}\rangle \equiv$ eigenstate of $\hat{H}_0$ $\Rightarrow$

$\boxed{\lim_{t\to\infty}\langle\phi_{\vec{k}'}|\hat{U}(+t,-t)|\phi_{\vec{k}}\rangle \propto \langle\phi_{\vec{k}'}|\psi_{\vec{k}}\rangle}$ $\equiv$ the $|\phi_{\vec{k}}\rangle\to|\phi_{\vec{k}'}\rangle$ **element of S-matrix**

(c) The L.-S. equation $\boxed{|\psi\rangle = |\phi\rangle + \frac{1}{E-\hat{H}_0\pm i\varepsilon}\hat{H}'|\psi\rangle}$ trivially holds for any states $|\phi\rangle, |\psi\rangle$ satisfying a pair of ordinary Schrödinger equations $\left\{\begin{matrix}\hat{H}_0|\phi\rangle=E|\phi\rangle\\(\hat{H}_0+\hat{H}')|\psi\rangle=E|\psi\rangle\end{matrix}\right\}$.

The above derivation shows, in addition, that the L.-S. eq. (with $+i\varepsilon$ sign and $|\phi\rangle, |\psi\rangle$ related in the above-described way) represents the correct transformation of a time-dependent problem to the corresponding **stationary problem**.

(d) Expression $\frac{1}{E-\hat{H}_0\pm i\varepsilon} = [(E\pm i\varepsilon)\hat{I} - \hat{H}_0]^{-1}$ stands for the *operator inverse* defined on the whole Hilbert space $\mathcal{H}$ (because of the $\pm i\varepsilon$ term). This expression represents the **Fourier transform** of the **Green operator** of the stationary Hamiltonian $\hat{H}_0$ (cf. p. 76):

$$\hat{G}_0(t) = \lim_{\varepsilon\to 0}\frac{i}{2\pi}\int_{-\infty}^{+\infty}\frac{e^{-\frac{i}{\hbar}Et}}{E-\hat{H}_0+i\varepsilon}dE \quad\Leftrightarrow\quad \boxed{\lim_{\varepsilon\to 0}\frac{1}{E-\hat{H}_0+i\varepsilon} = \frac{i}{\hbar}\int_{-\infty}^{\infty}\hat{G}_0(t)e^{+\frac{i}{\hbar}Et}dt = \hat{G}_0(E)}$$

▶ **Evaluation of Lippmann-Schwinger equation for elastic scattering** by a general **local potential** $V(\vec{x})$

L.-S. eq. in **x-representation:** $\psi_{\vec{k}}(\vec{x}) = \phi_{\vec{k}}(\vec{x}) + \int \langle \vec{x} \,|\, \frac{1}{E-\hat{H}_0+i\hbar\eta} \,|\, \vec{x}\,' \rangle \langle \vec{x}\,' | \hat{H}' | \psi_{\vec{k}} \rangle \, d\vec{x}\,'$

(a) $\langle \vec{x}\,' | \hat{H}' | \psi_{\vec{k}} \rangle = V(\vec{x}\,')\, \psi_{\vec{k}}(\vec{x}\,')$

(b) $$\boxed{\langle \vec{x} \,|\, \frac{1}{E-\hat{H}_0+i\hbar\eta} \,|\, \vec{x}\,' \rangle} = \iint \underbrace{\langle \vec{x} | \vec{p}\,' \rangle}_{\frac{1}{(2\pi\hbar)^{\frac{3}{2}}} e^{+\frac{i}{\hbar}\vec{p}\,'\cdot\vec{x}}} \underbrace{\langle \vec{p}\,' \,|\, \frac{1}{E-\hat{H}_0+i\hbar\eta} \,|\, \vec{p}\,'' \rangle}_{\frac{1}{E-\frac{1}{2\mathcal{M}}\vec{p}\,'^2+i\hbar\eta}\delta(\vec{p}\,'-\vec{p}\,'')} \underbrace{\langle \vec{p}\,'' | \vec{x}\,' \rangle}_{\frac{1}{(2\pi\hbar)^{\frac{3}{2}}} e^{-\frac{i}{\hbar}\vec{p}\,''\cdot\vec{x}\,'}} d\vec{p}\,' d\vec{p}\,'' = \ldots$$

$E \equiv \frac{(\hbar k)^2}{2\mathcal{M}} \qquad \frac{2\mathcal{M}\eta}{\hbar} \equiv \varepsilon \qquad \vec{p}\,' \equiv \hbar\vec{q} \qquad$ polar coordinates of $\vec{q}$ with $\vec{n}_z \propto (\vec{x}-\vec{x}\,')$

$$= \frac{1}{(2\pi\hbar)^3} \int \frac{e^{\frac{i}{\hbar}\vec{p}\,'\cdot(\vec{x}-\vec{x}\,')}}{E-\frac{1}{2\mathcal{M}}\vec{p}\,'^2+i\hbar\eta} d\vec{p}\,' = \frac{2\mathcal{M}}{\hbar^2(2\pi)^3} \int \frac{e^{i\vec{q}\cdot(\vec{x}-\vec{x}\,')}}{k^2-q^2+i\varepsilon} d\vec{q} = \frac{2\mathcal{M}}{\hbar^2(2\pi)^3} \int_0^\infty \int_0^{2\pi} \int_0^\pi \frac{e^{iq|\vec{x}-\vec{x}\,'|\cos\vartheta}}{k^2-q^2+i\varepsilon} q^2 \sin\vartheta \, d\varphi \, d\vartheta \, dq$$

$$= \frac{2\mathcal{M}}{(2\pi\hbar)^2} \int_0^\infty \left[-\frac{e^{iq|\vec{x}-\vec{x}\,'|\cos\vartheta}}{iq|\vec{x}-\vec{x}\,'|} \right]_{\vartheta=0}^{\vartheta=\pi} \frac{1}{k^2-q^2+i\varepsilon} q^2 dq = -\frac{2\mathcal{M}}{(2\pi\hbar)^2} \frac{1}{i|\vec{x}-\vec{x}\,'|} \int_0^\infty \frac{e^{+iq|\vec{x}-\vec{x}\,'|} - e^{-iq|\vec{x}-\vec{x}\,'|}}{q^2-k^2-i\varepsilon} q \, dq$$

Poles at $q = \pm\sqrt{k^2+i\varepsilon} \approx \pm\left(k + i\frac{\varepsilon}{2k}\right) \quad \Rightarrow \quad$ use the residuum theorem

$$= -\frac{2\mathcal{M}}{(2\pi\hbar)^2} \frac{1}{i|\vec{x}-\vec{x}\,'|} \frac{1}{2} \Bigg[\underbrace{\int_{-\infty}^{+\infty} \frac{e^{+iq|\vec{x}-\vec{x}\,'|}}{q^2-k^2-i\varepsilon} q \, dq}_{\to 2\pi i \frac{e^{ik|\vec{x}-\vec{x}\,'|}}{2k} k} - \underbrace{\int_{-\infty}^{+\infty} \frac{e^{-iq|\vec{x}-\vec{x}\,'|}}{q^2-k^2-i\varepsilon} q \, dq}_{\to -2\pi i \frac{e^{ik|\vec{x}-\vec{x}\,'|}}{2k} k} \Bigg] \xrightarrow{\varepsilon\to 0} \boxed{\underbrace{-\frac{2\mathcal{M}}{\hbar^2} \frac{1}{4\pi} \frac{e^{ik|\vec{x}-\vec{x}\,'|}}{|\vec{x}-\vec{x}\,'|}}_{G_k(\vec{x},\vec{x}\,')}}$$

Im q, Re q, 1), 2), $+\sqrt{k^2+i\varepsilon}$, $-\sqrt{k^2+i\varepsilon}$

The solution $\equiv$ Green function satisfying:

$$(\Delta + k^2) G_k(\vec{x}, \vec{x}\,') = \delta(\vec{x}-\vec{x}\,')$$

$\Rightarrow$ L.-S. eq. with local potential in x-representation:

$$\boxed{\psi_{\vec{k}}(\vec{x}) = \phi_{\vec{k}}(\vec{x}) - \frac{2\mathcal{M}}{\hbar^2} \frac{1}{4\pi} \int \frac{e^{ik|\vec{x}-\vec{x}\,'|}}{|\vec{x}-\vec{x}\,'|} V(\vec{x}\,')\, \psi_{\vec{k}}(\vec{x}\,')\, d\vec{x}\,'}$$

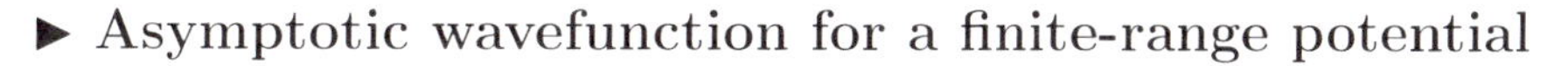

▶ **Asymptotic wavefunction for a finite-range potential**

Now we show that the L.-S. equation in x-representation yields automatically the wavefunction of the asymptotic form required in the elastic scattering ansatz. Assume $V(\vec{x}) \approx 0$ for $|\vec{x}| > R$ and consider $\psi_{\vec{k}}(\vec{x})$ for $\boxed{|\vec{x}| \gg R \gtrsim |\vec{x}\,'|}$

$$|\vec{x}-\vec{x}\,'| = \sqrt{r^2 + r'^2 - 2rr'\cos\alpha} \approx r - r'\cos\alpha \quad \Rightarrow \quad \frac{e^{ik|\vec{x}-\vec{x}\,'|}}{|\vec{x}-\vec{x}\,'|} \approx e^{-ikr'\cos\alpha} \frac{e^{ikr}}{r}$$

$$\boxed{\vec{k}\,' = k\frac{\vec{x}}{|\vec{x}|}}$$

scattering direction

$$\Rightarrow \quad \psi_{\vec{k}}(\vec{x}) = \overbrace{\phi_{\vec{k}}(\vec{x})}^{(2\pi)^{-\frac{3}{2}} e^{i\vec{k}\cdot\vec{x}}} + \underbrace{\left[-\frac{2\mathcal{M}}{\hbar^2} \frac{1}{4\pi} \overbrace{\int e^{-i\vec{k}\,'\cdot\vec{x}\,'} V(\vec{x}\,')\, \psi_{\vec{k}}(\vec{x}\,')\, d\vec{x}\,'}^{(2\pi)^{\frac{3}{2}} \langle \phi_{\vec{k}\,'} | \hat{V} | \psi_{\vec{k}} \rangle} \right]}_{(2\pi)^{-\frac{3}{2}} f_{\vec{k}}(\vec{k}\,')} \frac{e^{ikr}}{r}$$

L.-S. equation

$\Rightarrow$ explicit expression of the scattering amplitude from the exact solution $\psi_{\vec{k}}(\vec{x})$:

$$f_{\vec{k}}(\vartheta,\varphi) \equiv \boxed{f_{\vec{k}}(\vec{k}\,') = -\tfrac{4\pi^2\mathcal{M}}{\hbar^2}\langle\phi_{\vec{k}'}|\hat{V}|\psi_{\vec{k}}\rangle}$$

◀ **Historical remark**
1937-40's: J.A. Wheeler & W. Heisenberg introduce & investigate the scattering matrix (elaborate the asymptotic-time formulation of scattering)
1950: B.A. Lippmann & J. Schwinger derive the equation known by their names

■ Iterative expression of the scattering amplitude

The form of the Lippmann-Schwinger equation incites us to try an iterative solution. This leads to the Born series for the cross section. Individual terms of this series are closely related to expressions for transition amplitudes with increasing order, as obtained in the nonstationary perturbation theory.

▶ **Iterations in L.-S. equation**

$$\begin{aligned}|\psi_{\vec{k}}\rangle &= |\phi_{\vec{k}}\rangle + \tfrac{1}{E-\hat{H}_0+i\hbar\eta}\hat{H}'|\psi_{\vec{k}}\rangle \\ &= |\phi_{\vec{k}}\rangle + \tfrac{1}{E-\hat{H}_0+i\hbar\eta}\hat{H}'|\phi_{\vec{k}}\rangle + \tfrac{1}{E-\hat{H}_0+i\hbar\eta}\hat{H}'\tfrac{1}{E-\hat{H}_0+i\hbar\eta}\hat{H}'|\psi_{\vec{k}}\rangle \\ &= |\phi_{\vec{k}}\rangle + \tfrac{1}{E-\hat{H}_0+i\hbar\eta}\hat{H}'|\phi_{\vec{k}}\rangle + \tfrac{1}{E-\hat{H}_0+i\hbar\eta}\hat{H}'\tfrac{1}{E-\hat{H}_0+i\hbar\eta}\hat{H}'|\phi_{\vec{k}}\rangle + \cdots \\ &= \left(\hat{I} + \tfrac{1}{E-\hat{H}_0+i\hbar\eta}\hat{H}' + \tfrac{1}{E-\hat{H}_0+i\hbar\eta}\hat{H}'\tfrac{1}{E-\hat{H}_0+i\hbar\eta}\hat{H}' + \cdots\right)|\phi_{\vec{k}}\rangle\end{aligned}$$

▶ **T-operator**

The above iterative expression can be rewritten in terms of operator $\hat{T}$ ("transition matrix") defined via the equation: $\boxed{\hat{T}|\phi_{\vec{k}}\rangle = \hat{H}'|\psi_{\vec{k}}\rangle}$

$\hat{H}'\times$ Lippmann-Schwinger eq.: $\quad \underbrace{\hat{H}'|\psi_{\vec{k}}\rangle}_{\hat{T}|\phi_{\vec{k}}\rangle} = \hat{H}'|\phi_{\vec{k}}\rangle + \hat{H}'\tfrac{1}{E-\hat{H}_0+i\hbar\eta}\underbrace{\hat{H}'|\psi_{\vec{k}}\rangle}_{\hat{T}|\phi_{\vec{k}}\rangle}$

$\Rightarrow \quad \hat{T} = \hat{H}' + \hat{H}'\tfrac{1}{E-\hat{H}_0+i\hbar\eta}\hat{T}$

$$\boxed{\hat{T} = \hat{H}' + \hat{H}'\tfrac{1}{E-\hat{H}_0+i\hbar\eta}\hat{H}' + \hat{H}'\tfrac{1}{E-\hat{H}_0+i\hbar\eta}\hat{H}'\tfrac{1}{E-\hat{H}_0+i\hbar\eta}\hat{H}' + \cdots}$$

▶ **Born series**

The above iterative expressions yield an expansion of the scattering amplitude:

$$f_{\vec{k}}(\vec{k}\,') = -\tfrac{4\pi^2\mathcal{M}}{\hbar^2}\underbrace{\langle\phi_{\vec{k}'}|\hat{V}|\psi_{\vec{k}}\rangle}_{\langle\phi_{\vec{k}'}|\hat{T}|\phi_{\vec{k}}\rangle} = f^{(1)}_{\vec{k}}(\vec{k}\,') + f^{(2)}_{\vec{k}}(\vec{k}\,') + f^{(3)}_{\vec{k}}(\vec{k}\,') + \cdots$$

Interpretation through sequences of free evolutions & point interactions

m = 0 + 1 + 2 + 3 + ...

$$\boxed{\begin{array}{lcll} f^{(1)}_{\vec{k}}(\vec{k}\,') &=& -\tfrac{4\pi^2\mathcal{M}}{\hbar^2}\langle\phi_{\vec{k}'}|\hat{V}|\phi_{\vec{k}}\rangle & \text{1}^{\text{st}}\text{ Born approx.} \\ f^{(2)}_{\vec{k}}(\vec{k}\,') &=& -\tfrac{4\pi^2\mathcal{M}}{\hbar^2}\langle\phi_{\vec{k}'}|\hat{V}\tfrac{1}{E-\hat{H}_0+i\hbar\eta}\hat{V}|\phi_{\vec{k}}\rangle & \text{2}^{\text{nd}}\text{ Born approx.} \\ f^{(3)}_{\vec{k}}(\vec{k}\,') &=& -\tfrac{4\pi^2\mathcal{M}}{\hbar^2}\langle\phi_{\vec{k}'}|\hat{V}\tfrac{1}{E-\hat{H}_0+i\hbar\eta}\hat{V}\tfrac{1}{E-\hat{H}_0+i\hbar\eta}\hat{V}|\phi_{\vec{k}}\rangle & \text{3}^{\text{rd}}\text{ Born approx.} \\ \cdots && \cdots & \cdots \end{array}}$$

▶ Relation to non-stationary perturbation theory

Comparison of the 1st Born approximation with the Fermi golden rule

Transition rate $\boxed{W_{\vec{k}\to\vec{k}'} = \frac{2\pi}{\hbar}\,|\langle\tilde{\phi}_{\vec{k}'}|\hat{V}|\tilde{\phi}_{\vec{k}}\rangle|^2\,\rho_f(E) = |\vec{j}_{\rm in}|\,\left(\frac{d\sigma}{d\Omega}\right)_{\vec{k}}(\vec{k}')\,d\Omega}$

$\boxed{|\tilde{\phi}_{\vec{k}}\rangle \equiv \frac{1}{L^{\frac{3}{2}}}\,e^{i\vec{k}\cdot\vec{x}}}$ is a **plane wave in a box** of linear size L

$$\Rightarrow \vec{k} = \frac{2\pi}{L}\vec{n} \text{ with } \vec{n} = \begin{pmatrix} n_x \\ n_y \\ n_z \end{pmatrix} \text{ and } n_x, n_y, n_z = 0, 1, 2, \ldots$$

(a) $|\langle\tilde{\phi}_{\vec{k}'}|\hat{V}|\tilde{\phi}_{\vec{k}}\rangle|^2 = \frac{1}{L^6}\left|\int e^{i(\vec{k}-\vec{k}')\cdot\vec{x}'}V(\vec{x}')\,d\vec{x}'\right|^2$

(b) $\rho_f(E) = \frac{dN}{dE} = \frac{\left(\frac{L}{2\pi}\right)^3 k^2\,dk\,d\Omega}{\frac{\hbar^2 k}{\mathcal{M}}dk} = \left(\frac{L}{2\pi}\right)^3\,\frac{\mathcal{M}k}{\hbar^2}\,d\Omega \quad$ with $E = \frac{(\hbar k)^2}{2\mathcal{M}}$

(c) $|\vec{j}_{\rm in}| = \frac{\hbar k}{L^3\mathcal{M}}$

$$\Rightarrow \quad \left(\frac{d\sigma}{d\Omega}\right)_{\vec{k}}(\vec{k}') = \left(\frac{4\pi^2\mathcal{M}}{\hbar^2}\right)^2\left|\frac{1}{(2\pi)^3}\int e^{i(\vec{k}-\vec{k}')\cdot\vec{x}'}V(\vec{x}')\,d\vec{x}'\right|^2 \equiv |f^{(1)}_{\vec{k}}(\vec{k}')|^2$$

The 1st order of nonstationary perturbation theory yields the 1st Born approx.

▶ Convergence criteria

The Born series for scattering amplitude converges for finite-range potentials. For infinite-range potentials, the series may converge if the potential decreases "fast enough". For a given potential $V(\vec{x})$ there $\exists$ a function of energy $\lambda_{\rm max}(E)$ (convergence radius) such that the Born series of potential $V_\lambda(\vec{x}) \equiv \lambda V(\vec{x})$ converges for $\lambda \leq \lambda_{\rm max}(E)$.

▶ 1st Born approximation for spherically symmetric potentials

For potentials depending just on $r{=}|\vec{x}|$ the integration in each term of the Born series is reduced. For the first term, in particular, we proceed as follows:

$$f^{(1)}_{\vec{k}}(\vec{k}') = -\frac{4\pi^2\mathcal{M}}{\hbar^2}\,\frac{1}{(2\pi)^3}\int e^{i(\vec{k}-\vec{k}')\cdot\vec{x}'}V(|\vec{x}'|)\,d\vec{x}' \qquad \textbf{Fourier transform of } V$$

Transferred momentum $\boxed{\hbar\vec{q} = \hbar(\vec{k}'-\vec{k})}$

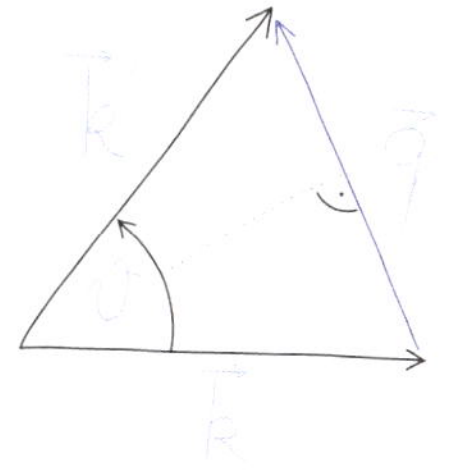

$$q = |\vec{k}'-\vec{k}| = \sqrt{k'^2 + k^2 - 2k'k\cos\vartheta} = \sqrt{2k^2(1-\cos\vartheta)} = 2k\sin\tfrac{\vartheta}{2}$$

Introduce local coord. system (x', y', z') with z' along $\vec{q}$ and then spherical coordinates (r', θ', ϕ'):

$$f^{(1)}_{\vec{k}}(\vec{k}') = -\frac{\mathcal{M}}{2\pi\hbar^2}\int_0^\infty\int_0^\pi\int_0^{2\pi} e^{-iqr'\cos\theta'}V(r')\,r'^2\sin\theta'\,d\phi'\,d\theta'\,dr' =$$

$$-\frac{\mathcal{M}}{\hbar^2}\int_0^\infty \underbrace{\left[\frac{e^{-iqr'\cos\theta'}}{-iqr'}\right]_0^\pi}_{-\frac{2\sin qr'}{qr'}} V(r')\,r'^2 dr' \Rightarrow \boxed{f^{(1)}_{\vec{k}}(\vec{k}') = \frac{\mathcal{M}}{\hbar^2 k\sin\frac{\vartheta}{2}}\int_0^\infty r'V(r')\sin\!\left(2kr'\sin\tfrac{\vartheta}{2}\right)dr'}$$

Scattering amplitude depends only on angle ϑ (not on φ). This is valid for all terms of the Born series. This can be seen directly from the axial symmetry of the problem with an isotropic potential around the incoming-particle direction.

► **Yukawa scattering**

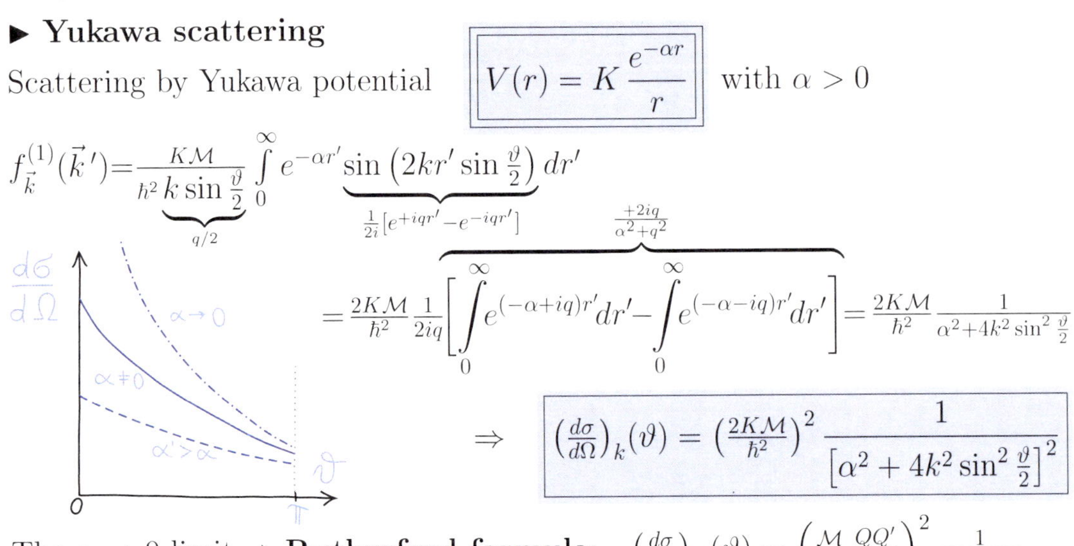

Scattering by Yukawa potential $\boxed{V(r) = K\,\frac{e^{-\alpha r}}{r}}$ with $\alpha > 0$

$$f^{(1)}_{\vec{k}}(\vec{k}\,') = \frac{K\mathcal{M}}{\hbar^2 \underbrace{k \sin\frac{\vartheta}{2}}_{q/2}} \int_0^\infty e^{-\alpha r'} \underbrace{\sin\left(2kr'\sin\frac{\vartheta}{2}\right)}_{\frac{1}{2i}[e^{+iqr'}-e^{-iqr'}]} dr'$$

$$= \frac{2K\mathcal{M}}{\hbar^2}\frac{1}{2iq}\overbrace{\left[\int_0^\infty e^{(-\alpha+iq)r'}dr' - \int_0^\infty e^{(-\alpha-iq)r'}dr'\right]}^{\frac{+2iq}{\alpha^2+q^2}} = \frac{2K\mathcal{M}}{\hbar^2}\frac{1}{\alpha^2+4k^2\sin^2\frac{\vartheta}{2}}$$

$$\Rightarrow \boxed{\left(\frac{d\sigma}{d\Omega}\right)_k(\vartheta) = \left(\frac{2K\mathcal{M}}{\hbar^2}\right)^2 \frac{1}{\left[\alpha^2 + 4k^2\sin^2\frac{\vartheta}{2}\right]^2}}$$

The $\alpha \to 0$ limit $\Rightarrow$ **Rutherford formula**: $\left(\frac{d\sigma}{d\Omega}\right)_p(\vartheta) = \left(\frac{\mathcal{M}}{2}\frac{QQ'}{4\pi\epsilon_0}\right)^2 \frac{1}{p^4\sin^4\frac{\vartheta}{2}}$

This formula can be obtained classically (it does not contain $\hbar$). However, **Coulomb scattering** cannot be described by the spherical-wave asymptotics used here, as this asymptotics is applicable only for finite-range or quickly decreasing potentials (cf. Sec. 6.3).

◄ **Historical remark**

1911: E. Rutherford derives classically the cross-section formula for Coulomb scattering to describe the 1909 experiment by H. Geiger & E. Marsden

1926: M. Born describes the scattering processes within QM; he derives explicitly the 1st approximation of a general scattering amplitude

1935: H. Yukawa introduces the potential for meson-mediated interaction of nucleons; this potential is now used to describe screened Coulomb interactions

6.3 Method of partial waves

We turn now to another method of analyzing scattering processes. It strictly relies on the assumption of *spherical symmetry*. The cross section is again expressed as an infinite series, but of a different type than in the perturbative approach.

■ Expression of elastic scattering in terms of spherical waves

The basic idea of the method is to express the scattered particle wavefunction in terms of states with good orbital angular momenta. This is always possible for spherically symmetric potentials.

► Asymptotic wavefunction $\psi_{\vec{k}}(\vec{x}) \approx \frac{1}{(2\pi)^{\frac{3}{2}}}\left[e^{ikz} + f_k(\vartheta)\frac{e^{ikr}}{r}\right]$

for a general **isotropic potential** is expanded in **orbital-momentum basis**: $\boxed{|klm\rangle \propto R_{kl}(r)Y_{lm}(\vartheta,\varphi)}$ with $l,m \equiv$ conserved quantum numbers.

Since z is associated with the direction of the incoming-particle linear momentum, the angular-momentum projection to z is $0 \Rightarrow$ only $\boxed{m=0}$ components $Y_{l0}(\vartheta,\varphi) \propto P_l(\cos\vartheta)$ contribute to the expansion:

► Asymptotic expansion of the **incoming plane wave** into spherical waves:

$$e^{ikz} = \sum_{l=0}^{\infty}(2l+1)i^l j_l(kr)P_l(\cos\vartheta) \approx \sum_{l=0}^{\infty}(2l+1)\frac{e^{+ikr}-e^{-i(kr-l\pi)}}{2ikr}P_l(\cos\vartheta)$$

where we used asymptotics of Bessel functions for $r \gg \frac{1}{k} = \frac{\hbar}{p} = \frac{\lambda}{2\pi}$:

$$j_l(kr) \sim \frac{\sin\left(kr - l\frac{\pi}{2}\right)}{kr} = \frac{e^{+i\left(kr-l\frac{\pi}{2}\right)}-e^{-i\left(kr-l\frac{\pi}{2}\right)}}{2ikr}$$

► Expansion of **scattering amplitude**: where $F_l(k) \equiv$ **partial-wave amplitude**

$$\boxed{f_k(\vartheta) = \sum_{l=0}^{\infty}(2l+1)F_l(k)P_l(\cos\vartheta)}$$

(general expression of any function of ϑ)

► Entire wavefunction

$$\boxed{\psi_{\vec{k}}(\vec{x}) \approx \frac{1}{(2\pi)^{\frac{3}{2}}}\sum_{l=0}^{\infty}(2l+1)\frac{1}{2ik}\left\{\underbrace{[1+2ikF_l(k)]}_{S_l(k)}\frac{e^{+ikr}}{r} - \frac{e^{-i(kr-l\pi)}}{r}\right\}P_l(\cos\vartheta)}$$

$\boxed{S_l(k) \equiv \langle +kl0|\hat{S}|+kl0\rangle}$ is the diagonal **S-matrix element** in basis $|+klm\rangle$ of outgoing (sign +) spherical waves with given l & k. This can be seen from the evolution: $e^{ikz} \equiv \sum_{l=0}^{\infty}(2l+1)\frac{e^{+ikr}-e^{-i(kr-l\pi)}}{2ikr}P_l(\cos\vartheta)$

$$\xrightarrow{t\to\infty} \quad \psi_{\vec{k}}(\vec{x}) \equiv \sum_{l=0}^{\infty}(2l+1)\frac{S_l(k)e^{+ikr}-e^{-i(kr-l\pi)}}{2ikr}P_l(\cos\vartheta)$$

► Continuity equation $\Rightarrow$ incoming flux = outgoing flux

$\Rightarrow$ coefficients for each l at $\frac{e^{+ikr}}{r}$ and $\frac{e^{-ikr}}{r}$ differ just by a phase $\Rightarrow |S_l(k)|=1$

$$\boxed{1+2ikF_l(k) = S_l(k) = e^{2i\delta_l(k)} \quad\Leftrightarrow\quad \boxed{F_l(k) = \frac{S_l(k)-1}{2ik} = e^{i\delta_l(k)}\frac{\sin\delta_l(k)}{k}}}$$

$\delta_l(k) \equiv$ relative **phase shift** of outgoing partial wave l

The above relation defines alternative **parametrizations** (but just parametrizations!) of the scattering amplitude & elastic cross section

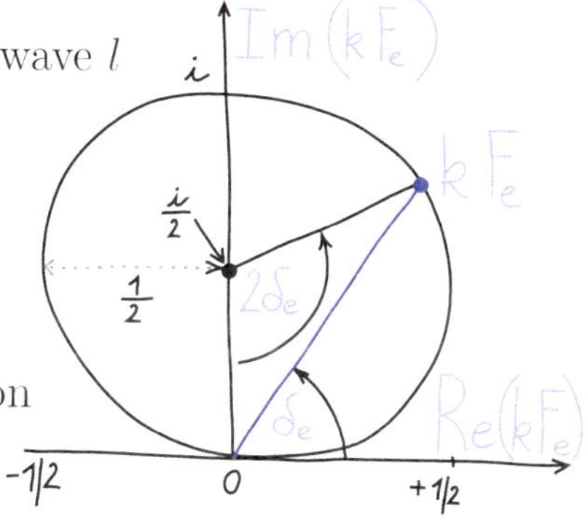

▶ **Expression of scattering amplitude through S-matrix & phase shifts**

$$f_k(\vartheta)=\frac{1}{2ik}\sum_{l=0}^{\infty}(2l+1)[S_l(k)-1]P_l(\cos\vartheta)=\frac{1}{k}\sum_{l=0}^{\infty}(2l+1)\,e^{i\delta_l(k)}\sin\delta_l(k)P_l(\cos\vartheta)$$

This allows one to express the differential cross section $\left(\frac{d\sigma}{d\Omega}\right)_k(\vartheta)=|f_k(\vartheta)|^2$

▶ **Integral cross section of elastic scattering**

Integrating the differential cross section over the full space angle we obtain the integral cross section:

$$\sigma^{\rm el}(k)=\int_0^{2\pi}\int_0^{\pi}|f_k(\vartheta)|^2\sin\vartheta\,d\varphi\,d\vartheta = 2\pi\sum_{l,l'}(2l+1)(2l'+1)F_l(k)F^*_{l'}(k)\overbrace{\int_0^{\pi}P_l(\cos\vartheta)P_{l'}(\cos\vartheta)\underbrace{\sin\vartheta\,d\vartheta}_{d(\cos\vartheta)}}^{\frac{2}{2l+1}\delta_{ll'}}$$

$$\sigma^{\rm el}(k)=4\pi\sum_{l=0}^{\infty}(2l+1)|F_l(k)|^2=\frac{\pi}{k^2}\sum_{l=0}^{\infty}(2l+1)|S_l(k)-1|^2=\frac{4\pi}{k^2}\sum_{l=0}^{\infty}(2l+1)\sin^2\delta_l(k)$$

$$=\sum_{l=0}^{\infty}\sigma_l^{\rm el}(k) \qquad \sigma_l^{\rm el}(k)=0 \text{ for } [\ F_l(k)=0 \ \Leftrightarrow\ \sin\delta_l(k)=0\ \Leftrightarrow\ S_l(k)=1\ ]$$

▶ **Classical calculation using impact factor**

The above expressions for integral cross sections can be easily interpreted in a classical language, making use of the so-called *impact factor* b, which is defined as the transverse projectile–target distance for $z\to-\infty$

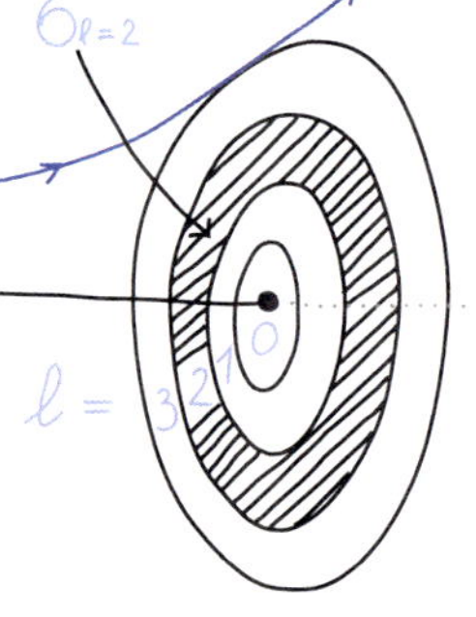

Orbital momentum $\underbrace{L}_{\sqrt{\hbar^2 l(l+1)}}=b\underbrace{p}_{\hbar k}$

$\Rightarrow$ for given l we have: $b_l(k)\approx\frac{\sqrt{l(l+1)}}{k}$

Estimated cross section of $l^{\rm th}$ part. wave:

$$\sigma_l^{\rm el}(k)=\pi(b^2_{l+\frac{1}{2}}-b^2_{l-\frac{1}{2}})=\tfrac{\pi}{k^2}(2l+1)$$

In the quantum calculation we obtained:

$$\sigma_l^{\rm el}(k)=\tfrac{\pi}{k^2}(2l+1)\overbrace{4\sin^2\delta_l(k)}^{\in[0,4]}$$

$\Rightarrow$ possibility of constructive/destructive interference for each term

▶ **Estimate of maximal angular momentum**

The classical impact-factor considerations make it possible to estimate the upper value of l where the cross-section series can be cut off. This is obtained from the maximal angular momentum for which the particle still hits the finite spatial region of nonzero potential:

We expect $\boxed{\sigma_l^{\rm el}(k)\approx 0 \text{ for } l>l_{\rm max}}$ where $\boxed{l_{\rm max}\approx kR}$ with $R\equiv$ range of V

In this way, all infinite sums become effectively **finite sums**:

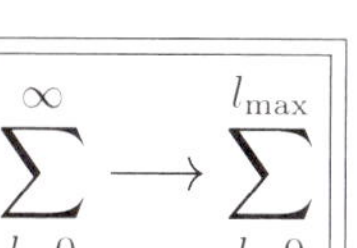

For instance: $f_k(\vartheta) \approx \frac{1}{k} \sum_{l=0}^{l_{\max}} (2l+1)\, e^{i\delta_l(k)} \sin\delta_l(k)\, P_l(\cos\vartheta)$

▶ Comparison with the asymptotics of a real solution

If we happen to know the *actual unbound solution* of the Schrödinger equation for the given potential (with the given energy in the continuous spectrum), we can directly calculate the required phase shifts & amplitudes of individual partial waves:

Radial Schrödinger equation $\boxed{\frac{d^2u_{kl}(r)}{dr^2} - \left[\frac{2\mathcal{M}}{\hbar^2} V(r) + \frac{l(l+1)}{r^2}\right] u_{kl}(r) + k^2 u_{kl}(r) = 0}$

Solution $\boxed{R_{kl}(r) = \frac{u_{kl}(r)}{r}}$ outside the range of the potential (for $r \geq R$):

$$= a_l\, j_l(kr) + b_l\, n_l(kr) = c_l^+\, h_l^+(kr) + c_l^-\, h_l^-(kr)$$

$j_l(kr),\ n_l(kr) \ \equiv$ **Bessel, Neumann functions** with asymptotics:	$h_l^{\pm}(kr) = j_l(kr) \pm i n_l(kr) \ \equiv$ **Hankel functions** with asymptotics:
$j_l(kr) \approx \frac{1}{kr}\sin\left(kr - l\frac{\pi}{2}\right) = \frac{1}{kr}\cos\left[kr - (l+1)\frac{\pi}{2}\right]$ $n_l(kr) \approx -\frac{1}{kr}\cos\left(kr - l\frac{\pi}{2}\right) = \frac{1}{kr}\sin\left[kr - (l+1)\frac{\pi}{2}\right]$	$h_l^+(kr) \approx \frac{1}{kr} e^{+i\left[kr - (l+1)\frac{\pi}{2}\right]}$ $h_l^-(kr) \approx \frac{1}{kr} e^{-i\left[kr - (l+1)\frac{\pi}{2}\right]}$

The general form of $r \geq R$ wavefunction yielding $\psi_{kl}(\vec{x}) = \frac{1}{(2\pi)^{3/2}} e^{ikz}$ for $V(r)=0$:

$$\psi_{kl}(\vec{x}) = \frac{1}{(2\pi)^{\frac{3}{2}}} \sum_{l=0}^{\infty} (2l+1) i^l \overbrace{\left[c_l^+ h_l^+(kr) + c_l^- h_l^-(kr)\right]}^{R_{kl}(r)} P_l(\cos\vartheta)$$

$$\approx \frac{1}{(2\pi)^{\frac{3}{2}}} \sum_{l=0}^{\infty} (2l+1) \frac{1}{ik} \left[c_l^+ \frac{e^{+ikr}}{r} - c_l^- \frac{e^{-i(kr - l\pi)}}{r}\right] P_l(\cos\vartheta)$$

This is compared with the required asymptotics:

$$\psi_{\vec{k}}(\vec{x}) = \frac{1}{(2\pi)^{\frac{3}{2}}} \sum_{l=0}^{\infty} (2l+1) \frac{1}{2ik} \left[e^{2i\delta_l(k)} \frac{e^{+ikr}}{r} - \frac{e^{-i(kr - l\pi)}}{r}\right] P_l(\cos\vartheta)$$

$\Rightarrow \quad r \geq R$ solution of radial Schrödinger eq. expressed in terms of $\delta_l(k)$:

$\boxed{c_l^+ = \frac{e^{2i\delta_l(k)}}{2}, \quad c_l^- = \frac{1}{2}} \quad \Rightarrow \quad R_{kl}(r) = \frac{e^{2i\delta_l(k)}}{2}[j_l(kr) + i n_l(kr)] + \frac{1}{2}[j_l(kr) - i n_l(kr)]$

$$\boxed{R_{kl}(r) = e^{i\delta_l(k)}\left[\cos\delta_l(k)\ j_l(kr) - \sin\delta_l(k)\ n_l(kr)\right]}$$

Note: Bessel functions $j_l(kr)$ are present in the incoming wave, while the Neumann functions $n_l(kr)$ are only in the outgoing wave (they disappear for $\delta_l \to 0$).

Conclusion: If one writes the actual asymptotic solution of the radial Schrödinger eq. in the above form, the phase shifts $\delta_l(k)$ for all partial waves are read out from that expression.

▶ Determination of phase shifts for a sharp potential

The above-described general method yields explicit results for potentials that *vanish identically* outside the range R: $V(r)\begin{cases}\neq 0 & \text{for } r\leq R \quad \text{(inside)}\\ =0 & \text{for } r>R \quad \text{(outside)}\end{cases}$

We require continuous connection of "inside-outside" **logarithmic derivative**

$$\beta_{kl}(R)\equiv R\tfrac{d}{dr}\ln R_{kl}(r)\Big|_{r=R}=R\,\frac{R'_{kl}(r)}{R_{kl}(r)}\Big|_{r=R}\qquad \forall\, l=0,1,2\ldots$$

$$\beta_{kl}(R)=kR\frac{\cos\delta_l(k)\frac{dj_l}{dr}(kR)-\sin\delta_l(k)\frac{dn_l}{dr}(kR)}{\cos\delta_l(k)\,j_l(kR)-\sin\delta_l(k)\,n_l(kR)}\Leftrightarrow \boxed{\tan\delta_l(k)=\frac{kR\frac{dj_l}{dr}(kR)-\beta_{kl}(R)\,j_l(kR)}{kR\frac{dn_l}{dr}(kR)-\beta_{kl}(R)\,n_l(kR)}}$$

Values of $\beta_{kl}(R)$ calculated from the **inside solution** $\Rightarrow$ we determine $\delta_l(k)$

▶ **Hard-sphere scattering**

$$V(r)=\begin{cases}\infty & \text{for } r\leq R\\ 0 & \text{for } r>R\end{cases}$$

R

$$R_{kl}(R)=e^{i\delta_l(k)}\big[\cos\delta_l(k)\,j_l(kR)-\sin\delta_l(k)\,n_l(kR)\big]=0\quad\Rightarrow \boxed{\tan\delta_l(k)=\frac{j_l(kR)}{n_l(kR)}}$$

l=0: $j_0(kR)=\frac{\sin kR}{kR},\ n_0(kR)=-\frac{\cos kR}{kR}\quad\Rightarrow\ \boxed{\delta_0(k)=-kR}$

(a) High-energy case $(kR\gg 1)$

$$\boxed{l\ll kR\quad\Rightarrow\quad j_l(kR)\approx\tfrac{1}{kR}\sin(kR-l\tfrac{\pi}{2}),\quad n_l(kR)\approx-\tfrac{1}{kR}\cos(kR-l\tfrac{\pi}{2})}$$

$\Rightarrow \tan\delta_l(k)=-\tan\big(kR-l\frac{\pi}{2}\big)\Rightarrow$ the $l^{\rm th}$ and $(l{+}1)^{\rm th}$ phase shifts differ by $\frac{\pi}{2}$

$\Rightarrow$ their contrib. to $\sigma^{\rm el}$ is $\frac{4\pi}{k^2}\big[(2l{+}1)\sin^2\delta_l(k)+(2l{+}3)\cos^2\delta_l(k)\big]\approx\frac{4\pi}{k^2}(2l{+}2)$

$\Rightarrow$ each l-term of the series contributes by $\approx\frac{4\pi}{k^2}\frac{2l+2}{2}$

$$\boxed{l\gg kR\quad\Rightarrow\quad j_l(kR)\approx\frac{(kR)^l}{(2l{+}1)!!},\quad n_l(kR)\approx-\frac{(2l{-}1)!!}{(kR)^{l+1}}}$$

$$\Rightarrow\tan\delta_l(k)\approx-\frac{(kR)^{2l+1}}{(2l{+}1)!!(2l{-}1)!!}\Rightarrow\tan\delta_{l+1}(k)\approx\underbrace{\Big(\frac{kR}{2l}\Big)^2}_{\lll 1}\tan\delta_l(k)\Rightarrow\text{decrease with } l$$

Assume $\boxed{l_{\rm max}\approx kR}$

$$\sigma^{\rm el}(k)\approx\frac{4\pi}{k^2}\sum_{l=0}^{l_{\rm max}}(2l{+}1)\sin^2\delta_l(k)\approx\frac{4\pi}{k^2}\overbrace{\sum_{l=0}^{kR}\frac{2l+2}{2}}^{\approx\frac{(kR)^2}{2}}\approx\boxed{2\pi R^2\approx\sigma^{\rm el}}\qquad\ldots 2\times\ \pi R^2$$

(b) Low-energy case $(kR\ll 1)$

Only the l=0 term works: $\delta_0(k)=-kR\approx\sin\delta_0(k)$

$$\sigma^{\rm el}(k)\approx\frac{4\pi}{k^2}\sin^2\delta_0(k)\approx\boxed{4\pi R^2\approx\sigma^{\rm el}}\qquad\ldots 4\times\ \pi R^2$$

In no case the classical geometrical cross section $\boxed{\sigma_{\rm clas}=\pi R^2}$ was obtained. The reason for low energy is a quantum interference phenomenon, but why is it so for high energy, when one would expect the classical behavior?

▸ Shadow scattering

Answer to the above question concerning the geometric cross section in high-E case: For $\sigma^{\rm el}{=}0$ the wavefunction would be $\psi(\vec{x})\propto e^{ikz}$, which is $\neq 0$ everywhere, including the region behind the sphere. Just to generate $\psi(\vec{x})$ vanishing in the region behind the sphere, the cross section must be $\sigma^{\rm el}\approx\pi R^2$. The reflected part of $\psi(\vec{x})$ produces another contribution $\sigma^{\rm el}\approx\pi R^2$. Together: $\sigma^{\rm el}\approx 2\pi R^2$

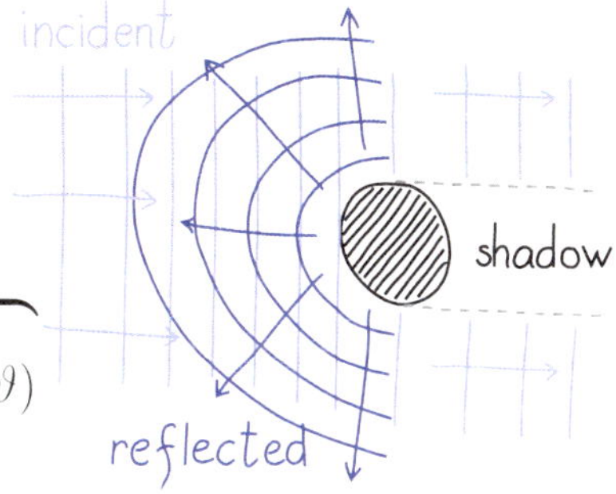

"Reflected" & "shadow" parts identified in:

$$f(\vartheta)=\sum_{l=0}^{\infty}(2l+1)\underbrace{\frac{e^{2i\delta_l(k)}-1}{2ik}}_{F_l(k)}P_l(\cos\vartheta)=$$

$$=\overbrace{\frac{1}{2ik}\sum_{l=0}^{\infty}(2l{+}1)e^{2i\delta_l(k)}P_l(\cos\vartheta)}^{f_{\rm refl}(\vartheta)}\overbrace{-\frac{1}{2ik}\sum_{l=0}^{\infty}(2l{+}1)P_l(\cos\vartheta)}^{f_{\rm shad}(\vartheta)}$$

$$\sigma_{\rm refl}=\iint|f_{\rm refl}(\vartheta)|^2\sin\vartheta d\varphi d\vartheta=$$

$$=\frac{1}{4k^2}\sum_{l,l'}(2l{+}1)(2l'{+}1)e^{i[\delta_l(k)-\delta_{l'}(k)]}\frac{4\pi}{2l{+}1}\delta_{ll'}=\frac{\pi}{k^2}\sum_{l=0}^{l_{\max}}(2l{+}1)\approx\pi R^2$$

$$\sigma_{\rm shad}=\iint|f_{\rm shad}(\vartheta)|^2\sin\vartheta d\varphi d\vartheta=\cdots\cdots\qquad\cdots\cdots\approx\pi R^2$$

$$\sigma_{\rm interf}=\iint 2{\rm Re}[f_{\rm refl}(\vartheta)f^*_{\rm shad}(\vartheta)]\sin\vartheta d\varphi d\vartheta=\cdots=\frac{2\pi}{k^2}\sum_{l=0}^{l_{\max}}(2l{+}1)\cos[2\delta_l(k)]\approx 0$$

▸ Coulomb scattering

Coulomb potential is a **long-range** one ⇒ **special treatment** needed. Here we just outline the method of solution without performing all calculations.

Consider the **repulsive case**:

$$\left[-\frac{\hbar^2}{2\mathcal{M}}\Delta+\frac{QQ'}{4\pi\epsilon_0}\frac{1}{r}-\frac{(\hbar k)^2}{2\mathcal{M}}\right]\psi_k(\vec{x})=0\quad\Leftrightarrow\quad\left[\Delta+k^2-\frac{2\gamma k}{r}\right]\psi_k(\vec{x})=0$$

$$\gamma=\frac{QQ'\mathcal{M}}{4\pi\epsilon_0\hbar^2 k}=\underbrace{\frac{e^2}{4\pi\epsilon_0(\hbar c)}}_{\alpha\doteq\frac{1}{137}}\underbrace{\frac{c\mathcal{M}}{\hbar k}}_{\left(\frac{v}{c}\right)^{-1}}ZZ'\qquad\left[\frac{d^2}{dr^2}+k^2-\frac{2\gamma k}{r}-\frac{l(l+1)}{r^2}\right]u_{kl}(r)=0$$

Schrödinger eq. is solved analytically in terms of hypergeometric functions. This yields the following **asymptotic solution**:

$$\psi_k(\vec{x})\overset{r\to\infty}{\propto}e^{i[kz-\gamma\ln k(r-z)]}+f_k(\vartheta)\frac{e^{i(kr-\gamma\ln 2kr)}}{r}$$

$$\propto\frac{1}{2ik}\sum_{l=0}^{\infty}(2l+1)P_l(\cos\vartheta)\left[e^{2i\delta_l(k)}\frac{e^{i(kr-\gamma\ln 2kr)}}{r}-\frac{e^{-i(kr-\gamma\ln 2kr-l\pi)}}{r}\right]$$

with known amplitude $f_k(\vartheta)=-\gamma\frac{e^{-i\left[\gamma\ln\left(\sin^2\frac{\vartheta}{2}\right)-2\delta_0(k)\right]}}{2k\sin^2\frac{\vartheta}{2}}$ and phase shifts $\delta_l(k)$

$$\vec{j}_{\rm in}\propto-\frac{\hbar\gamma}{\mathcal{M}}\frac{x}{r(r-z)}\vec{n}_x-\frac{\hbar\gamma}{\mathcal{M}}\frac{y}{r(r-z)}\vec{n}_y+\left(\frac{\hbar k}{\mathcal{M}}-\frac{\hbar\gamma}{\mathcal{M}}\frac{1}{r}\right)\vec{n}_z\quad\overset{r\to\infty}{\longrightarrow}\quad\frac{\hbar k}{\mathcal{M}}\vec{n}_z$$

$$\vec{j}_{\rm out}\propto\frac{|f_k(\vartheta)|^2}{r^2}\left(\frac{\hbar k}{\mathcal{M}}-\frac{\hbar\gamma}{\mathcal{M}}\frac{1}{r}\right)\vec{n}_r\quad\overset{r\to\infty}{\longrightarrow}\quad|f_k(\vartheta)|^2\frac{\hbar k}{\mathcal{M}r^2}\vec{n}_r$$

Cross section:

$$\left(\frac{d\sigma}{d\Omega}\right)_k(\vartheta)=|f_k(\vartheta)|^2=\left|\frac{1}{k}\sum_{l=0}^{\infty}(2l+1)e^{i\delta_l(k)}\sin\delta_l(k)P_l(\cos\vartheta)\right|^2=\alpha^2\frac{ZZ'}{16}\left(\frac{\hbar c}{E}\right)^2\frac{1}{\sin^4\frac{\vartheta}{2}}$$

Superposition Coulomb potential plus a finite-range potential $\Rightarrow$ the same asymptotics is used, in which $\delta_l(k)$ must be determined numerically

Inclusion of inelastic scattering

The method of partial waves makes it easy to include into the description the presence of inelastic scattering. More precisely, the inelastic scattering is included only through its *influence on elastic scattering*, the method providing nothing more but just a convenient phenomenological *parametrization*. A microscopic description requires to keep under control all the segments of the full Hilbert space where products of various inelastic channels appear, which is a hard problem. Nevertheless, even with these limitations, the parametrization provided by the partial-wave method has rather important consequences.

▶ Elastic scattering in presence of inelastic channels

The S-matrix is no more a complex unity but satisfies: $|S_l(k)|\in[0,1]$:

$$S_l(k)=\underbrace{\eta_l(k)}_{\in[0,1]}e^{2i\delta_l(k)}\quad\Rightarrow\quad \begin{aligned}F_l(k)&=\frac{S_l(k)-1}{2ik}\\&=\frac{1}{2k}\Big\{\eta_l(k)\sin 2\delta_l(k)+i\left[1-\eta_l(k)\cos 2\delta_l(k)\right]\Big\}\end{aligned}$$

$f_k(\vartheta)=\sum_{l=0}^{\infty}(2l+1)F_l(k)P_l(\cos\vartheta)\quad\Rightarrow$ the same expressions for $\sigma^{\rm el}(k)$ as before:

$$\sigma^{\rm el}(k)=\frac{\pi}{k^2}\sum_{l=0}^{\infty}(2l+1)\,|S_l(k)-1|^2=\frac{\pi}{k^2}\sum_{l=0}^{\infty}(2l+1)\left[1+\eta_l^2(k)-2\eta_l(k)\cos 2\delta_l(k)\right]$$

▶ Integral cross section of inelastic processes

The integral (but not differential!) inelastic cross section can be calculated through the balance of the overall incoming & outgoing flows derived from asymp. wavefunction: $\psi_{\vec{k}}(\vec{x})\approx\frac{1}{(2\pi)^{\frac{3}{2}}}\sum_{l=0}^{\infty}(2l+1)\frac{1}{2ik}\left\{S_l(k)\,\frac{e^{+ikr}}{r}-\frac{e^{-i(kr-l\pi)}}{r}\right\}P_l(\cos\vartheta)$

Radial flow: $\vec{j}_r(\vec{x})=\frac{1}{\mathcal{M}}\,{\rm Re}\left(\psi^*_{\vec{k}}(\vec{x})\,\underbrace{\left[-i\hbar\frac{\partial}{\partial r}-\frac{i\hbar}{r}\right]}_{\hat{p}_r\ \text{rad. momentum}}\psi_{\vec{k}}(\vec{x})\right)\vec{n}_r=$

$$=\frac{1}{\mathcal{M}}\frac{1}{(2\pi)^3}\vec{n}_r\sum_{l,l'}(2l+1)(2l'+1)P_l(\cos\vartheta)P_{l'}(\cos\vartheta)\times$$
$$\times\,{\rm Re}\frac{1}{-2ik}\left\{S^*_l(k)\,\frac{e^{-ikr}}{r}-\frac{e^{+i(kr-l\pi)}}{r}\right\}\frac{\hbar k}{2ik}\left\{S_l(k)\,\frac{e^{+ikr}}{r}+\frac{e^{-i(kr-l\pi)}}{r}\right\}$$
$$=\frac{1}{\mathcal{M}}\frac{1}{(2\pi)^3}\vec{n}_r\sum_{l,l'}(2l+1)(2l'+1)P_l(\cos\vartheta)P_{l'}(\cos\vartheta)\frac{\hbar}{4kr^2}\left[|S_l(k)|^2-1\right]$$

$$J^{\rm el}(k)=\iint j_r(r,\vartheta,\varphi)r^2\sin\vartheta d\varphi d\vartheta=-\frac{1}{(2\pi)^3}\frac{\pi\hbar}{\mathcal{M}k}\sum_{l=0}^{\infty}(2l+1)\left[1-|S_l(k)|^2\right]\leq 0$$

This is the total incoming flow which is not compensated by the outgoing flow because of inelastic processes. The integral cross section of inelastic process:

$$\sigma^{\rm inel}(k)=\frac{J^{\rm inel}(k)}{j_{\rm in}(k)}=\frac{-J^{\rm el}(k)}{\frac{1}{(2\pi)^3}\frac{\hbar k}{\mathcal{M}}}\quad\Rightarrow\quad\boxed{\sigma^{\rm inel}(k)=\frac{\pi}{k^2}\sum_{l=0}^{\infty}(2l+1)\left[1-|S_l(k)|^2\right]}$$

▶ **Total cross section**

$$\sigma^{\rm tot}(k)=\sigma^{\rm el}(k)+\sigma^{\rm inel}(k)=\frac{\pi}{k^2}\sum_{l=0}^{\infty}(2l+1)\left\{|S_l(k)-1|^2+\left[1-|S_l(k)|^2\right]\right\}$$

$$\boxed{\sigma^{\rm tot}(k)=\frac{2\pi}{k^2}\sum_{l=0}^{\infty}(2l+1)\big[1-\underbrace{{\rm Re}\,S_l(k)}_{\eta_l(k)\cos 2\delta_l(k)}\big]}$$

▶ **Relation between elastic and inelastic cross sections**

$$\text{Define}\begin{cases}x_l(k)\equiv\frac{\sigma_l^{\rm inel}(k)}{\frac{\pi}{k^2}(2l+1)}=1-\eta_l^2(k) & \in[0,1]\\ y_l(k)\equiv\frac{\sigma_l^{\rm el}(k)}{\frac{\pi}{k^2}(2l+1)}=1+\eta_l^2(k)-2\eta_l(k)\cos 2\delta_l(k) & \in[0,4]\end{cases}$$

$$\Rightarrow\quad\boxed{y_l(k)=2-x_l(k)-2\sqrt{1-x_l(k)}\cos 2\delta_l(k)}$$

Considering $-1\leq\cos 2\delta_l(k)\leq+1$ we obtain:

$$2-x_l(k)-2\sqrt{1-x_l(k)}\leq y_l(k)\leq 2-x_l(k)+2\sqrt{1-x_l(k)}$$

This represents an important **constraint** upon the possible values of elastic & inelastic integral cross sections for a given partial wave.

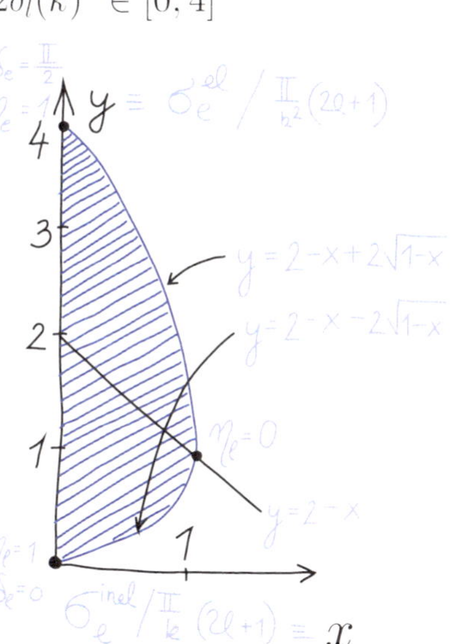

▶ **Optical theorem**

Statement: Imaginary part of the **elastic forward scattering amplitude** $\propto$ **total cross section** including all processes: $\boxed{{\rm Im}\,f_k^{\rm el}(\vartheta{=}0)=\frac{k}{4\pi}\sigma^{\rm tot}(k)}$

Proof for **isotropic potentials**:

$${\rm Im}f_k^{\rm el}(\vartheta{=}0)=\sum_{l=0}^{\infty}(2l+1)\underbrace{{\rm Im}\,F_l(k)}_{\frac{1}{2k}[1-\eta_l(k)\cos 2\delta_l(k)]}\overbrace{P_l(1)}^{1}=\frac{1}{2k}\sum_{l=0}^{\infty}(2l+1)\big[1-\underbrace{\eta_l(k)\cos 2\delta_l(k)}_{{\rm Re}S_l(k)}\big]=\frac{k}{4\pi}\sigma^{\rm tot}(k)$$

This relation is valid in the **most general case**, i.e., also for anisotropic potentials (beyond the method of partial waves).

For the elastic scattering, it can be proven from the L.-S. equation that:

$$\underbrace{-\tfrac{4\pi^2\mathcal{M}}{\hbar^2}\,\mathrm{Im}\langle\phi_{\vec{k}}|\hat{T}|\phi_{\vec{k}}\rangle}_{\mathrm{Im}f_{\vec{k}}(\vec{k})} = \tfrac{k}{4\pi}\underbrace{\left(\tfrac{4\pi^2\mathcal{M}}{\hbar^2}\right)^2\int|\langle\phi_{\vec{k}'}|\hat{T}|\phi_{\vec{k}}\rangle|^2\,\delta\big(k'-\sqrt{\tfrac{2\mathcal{M}E}{\hbar^2}}\,\big)\;k'^2dk'd\Omega'}_{\sigma^{\rm el}(k)}$$

A close analogue of the optical theorem can be formulated within the (non)stationary perturbation theory: The amplitude of the initial unperturbed state in the final state, obtained through the perturbation, is determined just from the normalization condition. This depends on the total admixture of all other unperturbed states in the final state.

■ Low-energy & resonance scattering

We conclude this section by sketching two additional topics: The low-energy scattering, which is a tool to determine basic properties of interaction, and resonance scattering, which indicates the existence of metastable states. Both these topic became much expanded in more advanced courses of QM.

▶ Low-energy limit of scattering amplitude

For $k\to 0$, only the $l=0$ partial wave active $\Rightarrow$ if no inelastic scattering present, there is just 1 real parameter determining this limit:

scattering length $\boxed{a\equiv\lim_{k\to 0}\left[-\dfrac{\sin\delta_0(k)}{k}\right]}$

$$\lim_{k\to 0}\sigma^{\rm el}(k)=\lim_{k\to 0}\tfrac{4\pi}{k^2}\sin^2\delta_0(k)=\boxed{4\pi a^2=\sigma^{\rm el}(E\approx 0)}$$

The visual meaning of scattering length is derived from the wavefunction form at $r>R$:

$$R_{k,l=0}(r)=e^{i\delta_0(k)}\Big[\cos\delta_0(k)\overbrace{j_0(kr)}^{\frac{\sin kr}{kr}}-\sin\delta_0(k)\overbrace{n_0(kr)}^{-\frac{\cos kr}{kr}}\Big]=\tfrac{e^{i\delta_0(k)}}{kr}\sin\left[kr+\delta_0(k)\right]$$

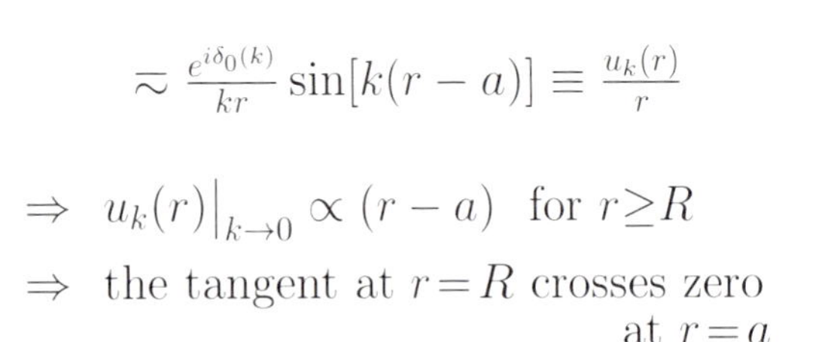

$$\approx\tfrac{e^{i\delta_0(k)}}{kr}\sin[k(r-a)]\equiv\tfrac{u_k(r)}{r}$$

$\Rightarrow\ u_k(r)\big|_{k\to 0}\propto(r-a)$ for $r\geq R$

$\Rightarrow$ the tangent at $r=R$ crosses zero at $r=a$

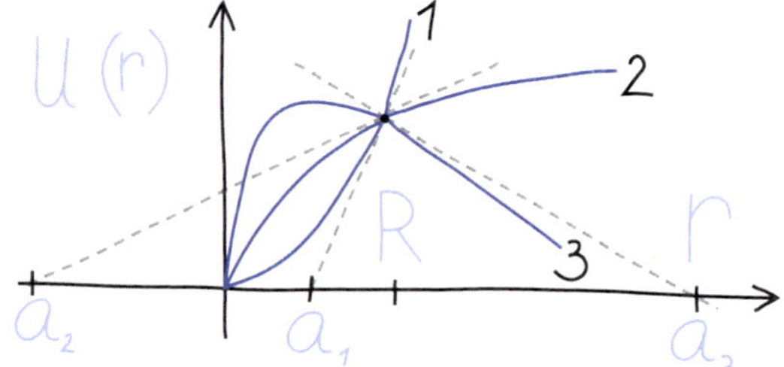

The value of a indicates some basic properties of the potential, although the information it gives is usually not unique:

$$\left.\begin{array}{c} 0<a<R \\ a\lesssim R \\ \hline -\infty<a<+\infty \\ a<0 \\ R\ll a \end{array}\right\}\Leftrightarrow\left\{\begin{array}{l} \text{repulsive potential (convex}\to\text{concave wf.)} \\ \text{strong repulsive potential}(a=R\text{ for hard sphere)} \\ \hline \text{attractive potential} \\ \text{shallow attractive potential} \\ \text{attractive potential with weakly bound state }E\lesssim 0 \end{array}\right.$$

▶ **Isolated resonance**

Assume that the **S-matrix extended to complex plane** $k \in \mathbb{C}$ has a **simple pole** at $k = k_{\rm R}$ with $\boxed{\frac{(\hbar k_{\rm R})^2}{2\mathcal{M}} = E_0 - i\frac{\Gamma}{2} \equiv E}$ $\Rightarrow$ $\boxed{S_l^{\rm R}(k) = \frac{k - k_{\rm R}^*}{k - k_{\rm R}}}$

$$k_{\rm R} = \sqrt{\frac{2\mathcal{M}E_0}{\hbar^2}}\sqrt{1 - i\frac{\Gamma}{2E_0}} \approx \overbrace{\sqrt{\frac{2\mathcal{M}E_0}{\hbar^2}}}^{k_0} - i\overbrace{\sqrt{\frac{\mathcal{M}\Gamma^2}{8\hbar^2 E_0}}}^{\kappa} \quad \text{for } \Gamma \ll E_0$$

$|S_l^{\rm R}(k)| = 1$ for $k \in \mathbb{R}$

For energy $E \in \mathbb{C}$ the evolution is not unitary $\Rightarrow |\langle\psi_{\rm R}(t)|\psi_{\rm R}(t)\rangle|^2 \equiv \left|e^{-\frac{i}{\hbar}\left(E_0 - i\frac{\Gamma}{2}\right)t}\right|^2$ $= e^{-\frac{\Gamma}{\hbar}t} \equiv P_0(t) \equiv$ survival probability $\Rightarrow$ **quasistationary state**, lifetime $\boxed{\tau = \frac{\hbar}{\Gamma}}$

Only the outgoing wave is present at $k = k_{\rm R}$ since $S_l^{\rm R}(k_{\rm R}) = \infty$

$\Rightarrow$ scattering wavefunction: $R(r) \propto \frac{e^{+ik_{\rm R}r}}{r} \approx \frac{e^{+ik_0 r}}{r}\, e^{+\kappa r}$

Approximation of the cross section for $E - E_0 \ll E_0$:

$$\sigma_l^{\rm R}(k) = \frac{\pi}{k^2}(2l+1)\left|S_l^{\rm R}(k) - 1\right|^2 = \frac{\pi}{k^2}(2l+1)\left|\frac{k_{\rm R} - k_{\rm R}^*}{k - k_{\rm R}}\right|^2 \approx \frac{\pi}{k^2}(2l+1)\overbrace{\left|\frac{-2i\kappa}{(k-k_0)+i\kappa}\right|^2}^{\frac{4\kappa^2}{(k-k_0)^2+\kappa^2}} =$$

$$\frac{4\pi}{k^2}(2l+1)\frac{\frac{\mathcal{M}}{2\hbar^2}\left(\frac{\Gamma}{2}\right)^2}{\frac{2\mathcal{M}}{\hbar^2}E_0\left(\sqrt{E} - \sqrt{E_0}\right)^2 + \frac{\mathcal{M}}{2\hbar^2}\left(\frac{\Gamma}{2}\right)^2} \approx \frac{4\pi}{k^2}(2l+1)\frac{\left(\frac{\Gamma}{2}\right)^2}{(E-E_0)^2 + \left(\frac{\Gamma}{2}\right)^2}$$

with $\left(\sqrt{E} - \sqrt{E_0}\right)^2 \approx \frac{(E-E_0)^2}{4E_0}$

Breit-Wigner resonance

$$\boxed{\sigma_l^{\rm R}(k) \approx \frac{4\pi\hbar^2}{2\mathcal{M}E}(2l+1)\frac{\left(\frac{\Gamma}{2}\right)^2}{(E - E_0)^2 + \left(\frac{\Gamma}{2}\right)^2}}$$

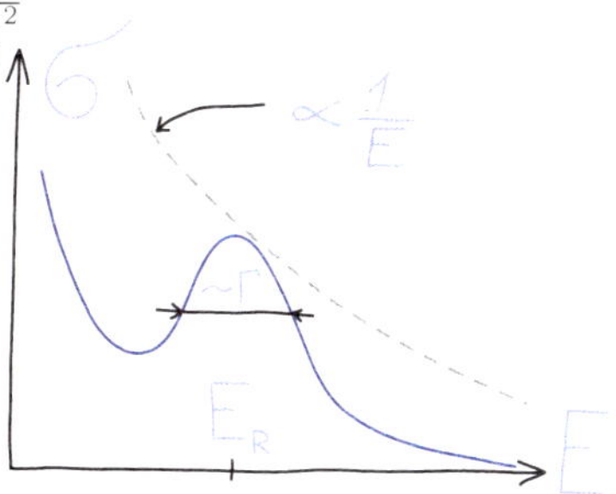

◀ **Historical remark**

1870-90's: Lord Rayleigh develops the scattering theory for elmg. and sound waves (he derives the "optical theorem" and elaborates the partial-wave expansion)
1927: H. Faxen & J. Holtsmark apply the partial-wave expansion in QM
1929: G. Breit & E. Wigner describe resonant states via the B.-W. distribution
1939: N. Bohr, R. Peierls, G. Placzek apply the Rayleigh optical relation in QM

7. MANY-BODY SYSTEMS

It this Chapter, we will deal with systems consisting of a number (fixed or variable) of mutually interacting particles. In the main focus will be the systems of *indistinguishable particles*, either bosons or fermions. The concept of indistinguishability and its various consequences were already discussed at several places of this course, starting from Chapter 1. Now we intend to formulate a general language describing all non-relativistic many-particle systems like atoms, nuclei, molecules, condensates etc. As a by product, we will learn how to quantize the electromagnetic field.

7.1 Formalism of particle creation/annihilation operators

We introduce the principal mathematical gear of many-body physics: the operators that can create or annihilate particles in a given state. These operators make it possible to generate a basis of the whole Fock space and to express any physical operator in this space. Moreover, basic algebraic properties of the creation/annihilation operators capture elegantly the difference between bosons and fermions.

■ Hilbert space of bosons & fermions

Let us first recall the relevant properties of bosonic and fermionic subspaces of a general many-particle Hilbert space. We introduce a so-called representation of occupation numbers in these subspaces, which is a natural starting point for creation/annihilation operators.

▸ Indistinguishable particles

N identical distinguishable particles: Hilbert space $\mathcal{H}^{(N)}=\mathcal{H}_1\otimes\mathcal{H}_2\otimes\cdots\otimes\mathcal{H}_N$, where all single-particle spaces $\mathcal{H}_k$ are the same

Projection to bosonic/fermionic spaces $\mathcal{H}^{(N)}_{\pm}$ expressed via sums over **particle permutations** $\boxed{(1,2,\dots N)\ \rightarrow\ (k_1^\pi,k_2^\pi,\dots k_N^\pi)\qquad \pi=1,2,\dots N!}$

$$\boxed{\textbf{bosons: } \hat{P}_+=\frac{1}{N!}\sum_{\pi=1}^{N!}\hat{\mathcal{E}}_\pi}\qquad\qquad \boxed{\textbf{fermions: } \hat{P}_-=\frac{1}{N!}\sum_{\pi=1}^{N!}\underbrace{\sigma_\pi}_{\substack{\pm1\\ \text{permutation sign}}}\hat{\mathcal{E}}_\pi}$$

(factor $\frac{1}{N!}\Rightarrow$ projector property $\hat{P}_\pm^2=\hat{P}_\pm$)

▸ Bases in the bosonic & fermionic spaces

Separable (non-entangled) basis in $\mathcal{H}^{(N)}$: $|\Phi_{i_1i_2\dots i_N}\rangle\equiv|\phi_{i_1}\rangle_1|\phi_{i_2}\rangle_2\dots|\phi_{i_N}\rangle_N$

where $|\phi_i\rangle_k\equiv i^{\text{th}}$ basis state in the kth single-particle space

Simplified notation: $|\Phi_{i_1i_2\dots i_N}\rangle\rightarrow\boxed{|\Phi_{12\dots N}\rangle\equiv|\phi_1\rangle_1|\phi_2\rangle_2\dots|\phi_N\rangle_N}$ $\begin{cases}\text{with}\\ |\phi_k\rangle_k\equiv|\phi_{i_k}\rangle_k\end{cases}$

$\Rightarrow|\phi_k\rangle_k$ is *any* (not the kth) basis state of the k^{th}particle

Action of permutation operators: $\hat{\mathcal{E}}_\pi|\Phi_{12\dots N}\rangle\equiv|\phi_{k_1^\pi}\rangle_1|\phi_{k_2^\pi}\rangle_2\dots|\phi_{k_N^\pi}\rangle_N$

Projections of the separable basis $\{|\Phi_{12\dots N}\rangle\}$ to $\mathcal{H}^{(N)}_\pm\quad\Rightarrow$

(a) basis in boson space: $\mathcal{N}_+\hat{P}_+|\Phi_{12\dots N}\rangle=\frac{\mathcal{N}_+}{N!}\sum_{\pi=1}^{N!}|\phi_{k_1^\pi}\rangle_1|\phi_{k_2^\pi}\rangle_2\dots|\phi_{k_N^\pi}\rangle_N$

(b) basis in fermion space: $\mathcal{N}_-\hat{P}_-|\Phi_{12\dots N}\rangle=\frac{\mathcal{N}_-}{N!}\underbrace{\sum_{\pi=1}^{N!}\sigma_\pi|\phi_{k_1^\pi}\rangle_1|\phi_{k_2^\pi}\rangle_2\dots|\phi_{k_N^\pi}\rangle_N}$

$$\text{Det}\begin{pmatrix}|\phi_1\rangle_1 & |\phi_1\rangle_2 & \cdots & |\phi_1\rangle_N\\ |\phi_2\rangle_1 & |\phi_2\rangle_2 & \cdots & |\phi_2\rangle_N\\ \vdots & & & \vdots\\ |\phi_N\rangle_1 & |\phi_N\rangle_2 & \cdots & |\phi_N\rangle_N\end{pmatrix}\quad \textbf{Slater determinant}$$

antisymmetry of the determinant under any exchange of columns or rows

Normalization coefficients
$$\boxed{\mathcal{N}_+ = \sqrt{\frac{N!}{n_1!\,n_2!\,n_3!\dots}} \qquad \mathcal{N}_- = \sqrt{N!}}$$

$n_k \equiv$ number of repetitions of the state $|\phi_k\rangle$ in the ensemble, i.e., number of particles in the state $|\phi_k\rangle$ (bosons: n_k=0,1,2,3..., fermions: n_k=0,1)

$$\Rightarrow \quad \boxed{n_1 + n_2 + n_3 + \dots = N}$$

Reasoning:

$$\mathcal{N}_+\hat{P}_+|\Phi_{12\dots N}\rangle \equiv \mathcal{N}_+\tfrac{1}{N!} \times \left[\text{sum of } N! \text{ states, partly identical}\right]$$
$$= \underbrace{\mathcal{N}_+\frac{n_1!\,n_2!\,n_3!\dots}{N!}}_{\sqrt{\frac{n_1!\,n_2!\,n_3!\dots}{N!}}} \times \left[\text{sum of } \tfrac{N!}{n_1!\,n_2!\,n_3!\dots} \text{ orthogonal terms}\right]$$
$$\Rightarrow \quad \mathcal{N}_+ = \sqrt{\frac{N!}{n_1!\,n_2!\,n_3!\dots}}$$

▶ **Representation of occupation numbers**

$$\boxed{\mathcal{N}_\pm\hat{P}_\pm|\Phi_{12\dots N}\rangle \equiv |n_1, n_2, n_3, \cdots, n_k, \cdots\rangle_\pm}$$

$$\text{with } n_k \equiv \left\{\begin{array}{l}\text{occupation number of the}\\ \text{basis state } |\phi_k\rangle \text{ (with } k{=}1,2,3,\dots\infty)\end{array}\right\} = \left\{\begin{array}{ll}0,1,2,3\dots & \text{for bosons}\\ 0,1 & \text{for fermions}\end{array}\right.$$

These vectors form a basis in the space of indistinguishable particles (bosons or fermions) $\equiv$ representation of occupation numbers

Bosonic & fermionic creation/annihilation operators

Creation and annihilation operators, respectively, increase and decrease the number of particles in a given single-particle state by one, forming a system of "ladder" operators in the Fock space. Their repeated application enables one to generate any basis state in the occupation-number representation from a unique state called *vacuum*. Mutual permutations of these operators obey simple commutation or anticommutation rules, depending on the bosonic or fermionic nature of the particles involved.

▶ **Fock space** (Hilbert space with indefinite number of particles N)

$$\mathcal{H}_\pm = \mathcal{H}_\pm^{(0)} \oplus \mathcal{H}_\pm^{(1)} \oplus \mathcal{H}_\pm^{(2)} \oplus \cdots \mathcal{H}_\pm^{(N-1)} \oplus \mathcal{H}_\pm^{(N)} \oplus \mathcal{H}_\pm^{(N+1)} \dots$$

Particle creation operators: $\mathcal{H}_\pm^{(N)} \mapsto \mathcal{H}_\pm^{(N+1)}$

Particle annihilation operators: $\mathcal{H}_\pm^{(N)} \mapsto \mathcal{H}_\pm^{(N-1)}$ (for N=0: $\mathcal{H}_\pm^{(N)} \mapsto 0$)

▶ **Creation operators**

$$\text{Bosons:} \quad \hat{b}_k^\dagger|n_1,..n_k,...\rangle_+ = \sqrt{n_k{+}1}\,|n_1,..(n_k{+}1),...\rangle_+$$

$$\text{Fermions:} \quad \hat{a}_k^\dagger|n_1,..n_k,...\rangle_- = \left\{\begin{array}{ll}\overbrace{\sqrt{n_k{+}1}}^{1}\,|n_1,..(n_k{+}1),...\rangle_- & \text{for } n_k = 0\\ 0 & \text{for } n_k = 1\end{array}\right.$$

Square-root coefficients included into these definitions ensure simple algebraic properties; see below (cf. the harmonic-oscillator ladder operators, Sec. 2.5).

▶ Annihilation operators

Definition:
$$\boxed{\begin{array}{ll}\text{Bosons}: & \hat{b}_k|n_1,..n_k,..\rangle_+ = \sqrt{n_k}\,|n_1,..(n_k{-}1),...\rangle_+ \\ \text{Fermions}: & \hat{a}_k|n_1,..n_k,..\rangle_- = \sqrt{n_k}\,|n_1,..(n_k{-}1),...\rangle_-\end{array}}$$

Defined in this way, the annihilation operators are Hermitian conjugates of creation operators:

$$\underbrace{{}_+\langle n'_1,..n'_k,..|\hat{b}_k|n_1,..n_k,..\rangle_+}_{\sqrt{n_k}\ \delta_{n'_1n_1}\cdots\delta_{n'_k(n_k-1)}\cdots} = \underbrace{{}_+\langle n_1,..n_k,..|\hat{b}^\dagger_k|n'_1,..n'_k,..\rangle^*_+}_{\sqrt{n'_k+1}\ \delta_{n_1n'_1}\cdots\delta_{n_k(n'_k+1)}\cdots}$$

$$\overbrace{{}_-\langle n'_1,..n'_k,..|\hat{a}_k|n_1,..n_k,..\rangle_-} = \overbrace{{}_-\langle n_1,..n_k,..|\hat{a}^\dagger_k|n'_1,..n'_k,..\rangle^*_-}$$

▶ Commutation relations for boson operators

$$\boxed{[\hat{b}^\dagger_k,\hat{b}^\dagger_l]=0=[\hat{b}_k,\hat{b}_l]}$$ (order of creation/annihilation of 2 bosons is irrelevant)

Proof for $k{=}l$ trivial, for $k{\neq}l$ below:

$$\hat{b}^\dagger_k\hat{b}^\dagger_l|..n_k..n_l...\rangle_+ = \hat{b}^\dagger_l\hat{b}^\dagger_k|..n_k..n_l...\rangle_+ = \sqrt{(n_k{+}1)(n_l{+}1)}\,|..(n_k{+}1)..(n_l{+}1)...\rangle_+$$

(the relation for annihilation operators obtained by the Hermitian conjugation)

$$\boxed{[\hat{b}_k,\hat{b}^\dagger_l]=\delta_{kl}}$$ (do not commute for $k{=}l$)

Proof for $k{=}l$:
$$\Big(\hat{b}_k\hat{b}^\dagger_k-\hat{b}^\dagger_k\hat{b}_k\Big)\,|..n_k...\rangle_+ = \overbrace{\Big(\sqrt{(n_k{+}1)^2}-\sqrt{n_k^2}\,\Big)}^{1}\,|..n_k...\rangle_+$$

For $k{\neq}l$:
$$\hat{b}_k\hat{b}^\dagger_l|..n_k..n_l...\rangle_+ = \hat{b}^\dagger_l\hat{b}_k|..n_k..n_l...\rangle_+ = \sqrt{n_k(n_l{+}1)}\,|..(n_k{-}1)..(n_l{+}1)...\rangle_+$$

▶ Anticommutation relations for fermion operators

All relations for fermions are expressed through the **anticommutator**:
$$\Big\{\hat{A},\hat{B}\Big\}\equiv\hat{A}\hat{B}+\hat{B}\hat{A}$$

Pauli principle $\Rightarrow\ \hat{a}^\dagger_k\hat{a}^\dagger_k|..n_k...\rangle_- = 0 = \hat{a}_k\hat{a}_k|..n_k...\rangle_-$

$$\Rightarrow\ \hat{a}^\dagger_k\hat{a}^\dagger_k=0=\hat{a}_k\hat{a}_k \quad\Rightarrow\quad \boxed{\{\hat{a}^\dagger_k,\hat{a}^\dagger_k\}=0=\{\hat{a}_k,\hat{a}_k\}}$$

$$\hat{a}_k\hat{a}^\dagger_k|..n_k...\rangle_- = \begin{cases}0 & \text{for } n_k{=}1\\ |..n_k...\rangle_- & \text{for } n_k{=}0\end{cases} \qquad \hat{a}^\dagger_k\hat{a}_k|..n_k...\rangle_- = \begin{cases}|..n_k...\rangle_- & \text{for } n_k{=}1\\ 0 & \text{for } n_k{=}0\end{cases}$$

$$\Rightarrow\ \underbrace{(\hat{a}_k\hat{a}^\dagger_k+\hat{a}^\dagger_k\hat{a}_k)}_{\{\hat{a}_k,\hat{a}^\dagger_k\}}|..n_k...\rangle_- = |..n_k...\rangle_- \quad\Rightarrow\quad \boxed{\{\hat{a}_k,\hat{a}^\dagger_k\}=\hat{I}}$$

We require more general relations:
$$\boxed{\{\hat{a}^\dagger_k,\hat{a}^\dagger_l\}=0=\{\hat{a}_k,\hat{a}_l\}}\quad\boxed{\{\hat{a}_k,\hat{a}^\dagger_l\}=\delta_{kl}}$$

Their validity for $k{=}l$ was just proven. For $k{\neq}l$ these relations represent some satisfiable requirements upon the *phases*, namely:

$$\hat{a}^\dagger_k\hat{a}^\dagger_l\,|..\overbrace{n_k}^{0}..\overbrace{n_l}^{0}...\rangle_- = -\hat{a}^\dagger_l\hat{a}^\dagger_k\,|..\overbrace{n_k}^{0}..\overbrace{n_l}^{0}...\rangle_-$$
$$\hat{a}_k\hat{a}^\dagger_l\,|..\underbrace{n_k}_{1}..\underbrace{n_l}_{0}...\rangle_- = -\hat{a}^\dagger_l\hat{a}_k\,|..\underbrace{n_k}_{1}..\underbrace{n_l}_{0}...\rangle_-$$

In this way, the fermionic creation/annihilation operators are fully analogous to the bosonic ones except that the commutators are replaced by anticommutators

▶ Particle number operators

Number of particles in the single-particle state $|\phi_k\rangle$:

bosons $\boxed{\hat{N}_k = \hat{b}_k^\dagger \hat{b}_k}$ fermions $\boxed{\hat{N}_k = \hat{a}_k^\dagger \hat{a}_k}$

$$\hat{b}_k^\dagger \hat{b}_k |..n_k...\rangle_+ = \underbrace{\sqrt{n_k^2}}_{n_k} |..n_k...\rangle_+ \qquad \hat{a}_k^\dagger \hat{a}_k |..n_k...\rangle_- = \underbrace{\sqrt{n_k^2}}_{n_k=0,1} |..n_k...\rangle_-$$

⇒ **total number of particles**:

bosons $\boxed{\hat{N} = \sum_k \hat{b}_k^\dagger \hat{b}_k}$ fermions $\boxed{\hat{N} = \sum_k \hat{a}_k^\dagger \hat{a}_k}$

We identify standard commutation relations of ladder operators (Sec. 2.4):

$$\begin{cases} \left[\hat{N}_k, \hat{b}_l^\dagger\right] = \delta_{kl}\hat{b}_l^\dagger & \left[\hat{N}_k, \hat{b}_l\right] = -\delta_{kl}\hat{b}_l \\ \left[\hat{N}, \hat{b}_l^\dagger\right] = +\hat{b}_l^\dagger & \left[\hat{N}, \hat{b}_l\right] = -\hat{b}_l \\ \left[\hat{N}_k, \hat{a}_l^\dagger\right] = \delta_{kl}\hat{a}_l^\dagger & \left[\hat{N}_k, \hat{a}_l\right] = -\delta_{kl}\hat{a}_l \\ \left[\hat{N}, \hat{a}_l^\dagger\right] = +\hat{a}_l^\dagger & \left[\hat{N}, \hat{a}_l\right] = -\hat{a}_l \end{cases}$$

▶ Creation of basis states from the vacuum

Consecutive creation of individual particles into the occupied single-particle states:

$$\boxed{|n_1, n_2, n_3...\rangle_\pm = \begin{cases} \frac{1}{\sqrt{n_1!n_2!n_3!...}}(\hat{b}_1^\dagger)^{n_1}(\hat{b}_2^\dagger)^{n_2}(\hat{b}_3^\dagger)^{n_3}\cdots|0\rangle & \text{for bosons} \\ (\hat{a}_1^\dagger)^{n_1}(\hat{a}_2^\dagger)^{n_2}(\hat{a}_3^\dagger)^{n_3}\cdots|0\rangle & \text{for fermions} \end{cases}}$$

$|0\rangle \equiv$ **vacuum state** ($\in \mathcal{H}_\pm^{(0)}$, no particle present) satisfying: $\hat{b}_k|0\rangle = 0 = \hat{a}_k|0\rangle$

▶ Relation between spin and statistics

Theorem: elementary particles belong to the families of bosons and fermions according to their spins:

Particles with $\boldsymbol{s}$ = **half-integer** are **fermions**: electron, muon, tauon, all neutrinos, all quarks (leptons & hadrons ≡ matter particles)

Particles with $\boldsymbol{s}$ = **integer** are **bosons**: photon, W, Z, gluon, graviton?, Higgs (interaction mediators & an "auxiliary" particle)

▶ Bifermions vs. bosons

Bifermion ≡ a **pair of fermions**. Example: meson (quark-antiquark). Any bifermion must have an integer spin. Question: is it a real boson?

Exchange of 2 bifermions ⇒ 2× change of sign ⇒ boson-like behavior

However, consider the creation/annihilation operators of a general bifermion:

$$\left.\begin{array}{c}\hat{A}^\dagger=\sum_{k,l}\alpha_{kl}\hat{a}_k^\dagger\hat{a}_l^\dagger\\ \text{creation}\end{array}\right\}\Leftrightarrow\left\{\begin{array}{c}\hat{A}=\sum_{k,l}\alpha_{kl}^*\hat{a}_l\hat{a}_k\\ \text{annihilation}\end{array}\right.$$

Antisymmetry : $\alpha_{kl}=-\alpha_{lk}$
Normalization : $\sum_{k,l}|\alpha_{kl}|^2=\frac{1}{2}$

Normalization: $1=\langle 0|\hat{A}\hat{A}^\dagger|0\rangle=\sum_{k,l}\sum_{k',l'}\alpha_{k'l'}^*\alpha_{kl}\langle 0|\hat{a}_{l'}\hat{a}_{k'}\hat{a}_k^\dagger\hat{a}_l^\dagger|0\rangle=2\sum_{k,l}|\alpha_{kl}|^2$

Commutator:

$$[\hat{A},\hat{A}^\dagger]=\sum_{k,l}\sum_{k',l'}\alpha_{k'l'}^*\alpha_{kl}[\hat{a}_{l'}\hat{a}_{k'},\hat{a}_k^\dagger\hat{a}_l^\dagger]=\sum_{k,l}\sum_{k',l'}\alpha_{k'l'}^*\alpha_{kl}\Big(-\delta_{kk'}\hat{a}_l^\dagger\hat{a}_{l'}+\delta_{kl'}\hat{a}_l^\dagger\hat{a}_{k'}$$
$$+\delta_{lk'}\hat{a}_k^\dagger\hat{a}_{l'}-\delta_{ll'}\hat{a}_k^\dagger\hat{a}_{k'}+\delta_{kk'}\delta_{ll'}-\delta_{kl'}\delta_{lk'}\Big)=\hat{I}+\underbrace{4\sum_{l,l'}\Big(\sum_k\alpha_{l'k}^*\alpha_{kl}\Big)\hat{a}_l^\dagger\hat{a}_{l'}}_{\hat{\Delta}}$$

$\hat{\Delta}\equiv$ **correction** to the boson-type of commutator:
Its impact depends on a concrete state $|\Psi\rangle$ of the many-body system. In general, $\langle\Psi|\hat{\Delta}|\Psi\rangle\approx 0$ for many-body states $|\Psi\rangle$ in which the single-fermion states present in the bifermion operator $\hat{A}^\dagger$ are "far enough" from the states occupied by the rest of the system. Example: a pair of mesons far from each other. $\langle\Psi|\hat{\Delta}|\Psi\rangle\neq 0$ for states $|\Psi\rangle$ in which the states contained in $\hat{A}^\dagger$ are partly occupied by the rest of the system. Example: a pair of nucleons in the nucleus.

▶ Transformations of creation/annihilation operators

Consider 2 single-particle bases: $\{|\phi_j\rangle\}_j\overset{\hat{U}}{\leftrightarrow}\{|\tilde{\phi}_i\rangle\}_i\quad\Leftrightarrow\quad|\tilde{\phi}_i\rangle=\sum_{i'}\underbrace{\langle\phi_{i'}|\tilde{\phi}_i\rangle}_{U_{ii'}}|\phi_{i'}\rangle$

$\hat{U}$ represents a unitary operator relating the two bases, which also constitutes the transformation between boson & fermion creation/annihilation operators:

$$\hat{\tilde{b}}_i^\dagger\equiv\sum_{i'}\langle\phi_{i'}|\tilde{\phi}_i\rangle\hat{b}_{i'}^\dagger\qquad\hat{\tilde{b}}_j\equiv\sum_{j'}\langle\tilde{\phi}_j|\phi_{j'}\rangle\hat{b}_{j'}\qquad\hat{\tilde{a}}_i^\dagger\equiv\sum_{i'}\langle\phi_{i'}|\tilde{\phi}_i\rangle\hat{a}_{i'}^\dagger\qquad\hat{\tilde{a}}_j\equiv\sum_{j'}\langle\tilde{\phi}_j|\phi_{j'}\rangle\hat{a}_{j'}$$

$$\Big[\hat{\tilde{b}}_j,\hat{\tilde{b}}_i^\dagger\Big]=\sum_{j',i'}\langle\tilde{\phi}_j|\phi_{j'}\rangle\langle\phi_{i'}|\tilde{\phi}_i\rangle\overbrace{\Big[\hat{b}_{j'},\hat{b}_{i'}^\dagger\Big]}^{\delta_{i'j'}}=\delta_{ij}\qquad\Big\{\hat{\tilde{a}}_j,\hat{\tilde{a}}_i^\dagger\Big\}=\sum_{j',i'}\langle\tilde{\phi}_j|\phi_{j'}\rangle\langle\phi_{i'}|\tilde{\phi}_i\rangle\overbrace{\Big\{\hat{a}_{j'},\hat{a}_{i'}^\dagger\Big\}}^{\delta_{i'j'}}=\delta_{ij}$$

$\Rightarrow$ commutation/anticommutation relations remain the same

▶ "Second quantization"

A transformation of creation/annihilation operators for general particles to the **coordinate & spin eigenbasis** $\{|\tilde{\phi}_{\vec{x},m_s}\rangle\}$

$$\hat{\tilde{b}}_{\vec{x},m_s}^\dagger\equiv\hat{\psi}_+^\dagger(\vec{x},m_s)=\sum_i\overbrace{\langle\phi_i|\tilde{\phi}_{\vec{x},m_s}\rangle}^{\phi_i^*(\vec{x},m_s)}\hat{b}_i^\dagger\qquad\hat{\tilde{a}}_{\vec{x},m_s}^\dagger\equiv\hat{\psi}_-^\dagger(\vec{x},m_s)=\sum_i\overbrace{\langle\phi_i|\tilde{\phi}_{\vec{x},m_s}\rangle}^{\phi_i^*(\vec{x},m_s)}\hat{a}_i^\dagger$$

$$\hat{\tilde{b}}_{\vec{x},m_s}\equiv\hat{\psi}_+(\vec{x},m_s)=\sum_j\underbrace{\langle\tilde{\phi}_{\vec{x},m_s}|\phi_j\rangle}_{\phi_j(\vec{x},m_s)}\hat{b}_j\qquad\hat{\tilde{a}}_{\vec{x},m_s}\equiv\hat{\psi}_-(\vec{x},m_s)=\sum_j\underbrace{\langle\tilde{\phi}_{\vec{x},m_s}|\phi_j\rangle}_{\phi_j(\vec{x},m_s)}\hat{a}_j$$

The new single-particle basis is not discrete (countable) $\Rightarrow$ commutation/ anticommutation relations will contain the δ-function instead of Kronecker δ:

Commutation relations (bosons) Anticommutation relations (fermions)

$$\begin{aligned}
\left[\hat\psi_+^\dagger(\vec x, m_s), \hat\psi_+^\dagger(\vec x\,', m_s')\right] &= 0 = \left\{\hat\psi_-^\dagger(\vec x, m_s), \hat\psi_-^\dagger(\vec x\,', m_s')\right\} \\
\left[\hat\psi_+(\vec x, m_s), \hat\psi_+(\vec x\,', m_s')\right] &= 0 = \left\{\hat\psi_-(\vec x, m_s), \hat\psi_-(\vec x\,', m_s')\right\} \\
\left[\hat\psi_+(\vec x, m_s), \hat\psi_+^\dagger(\vec x\,', m_s')\right] &= \delta(\vec x - \vec x\,')\delta_{m_s m_s'} = \left\{\hat\psi_-(\vec x, m_s), \hat\psi_-^\dagger(\vec x\,', m_s')\right\}
\end{aligned}$$

Proof of the last line:

$$\left[\hat\psi_+(\vec x, m_s), \hat\psi_+^\dagger(\vec x\,', m_s')\right] = \sum_{i,j} \langle\tilde\phi_{\vec x, m_s}|\phi_j\rangle\langle\phi_i|\tilde\phi_{\vec x\,', m_s'}\rangle \overbrace{[\hat b_j, \hat b_i^\dagger]}^{\delta_{ij}} = \overbrace{\langle\tilde\phi_{\vec x, m_s}|\tilde\phi_{\vec x\,', m_s'}\rangle}^{\delta(\vec x - \vec x')\delta_{m_s m_s'}}$$

$$\left\{\hat\psi_-(\vec x, m_s), \hat\psi_-^\dagger(\vec x\,', m_s')\right\} = \sum_{i,j} \langle\tilde\phi_{\vec x, m_s}|\phi_j\rangle\langle\phi_i|\tilde\phi_{\vec x\,', m_s'}\rangle \underbrace{\{\hat a_j, \hat a_i^\dagger\}}_{\delta_{ij}} = \underbrace{\langle\tilde\phi_{\vec x, m_s}|\tilde\phi_{\vec x\,', m_s'}\rangle}_{\delta(\vec x - \vec x')\delta_{m_s m_s'}}$$

Particle number operator: $$\hat N_\pm = \sum_{m_s} \int \underbrace{\hat\psi_\pm^\dagger(\vec x, m_s)\hat\psi_\pm(\vec x, m_s)}_{\hat n_\pm(\vec x, m_s)\text{ particle density}} d\vec x$$

The above procedure is often referred to as the "second quantization" (in analogy to the "first quantization", in which physical quantities became operators) since it induces the transition: wavefunction $\left.\begin{matrix}\psi^*(\vec x, m_s)\\ \psi(\vec x, m_s)\end{matrix}\right\} \mapsto \left\{\begin{matrix}\hat\psi_\pm^\dagger(\vec x, m_s)\\ \hat\psi_\pm(\vec x, m_s)\end{matrix}\right.$ operator

■ Operators in bosonic & fermionic N-particle spaces

Creation/annihilation operators enable one to express *any* operator acting in the whole Fock space! In particular, the operators that conserve the total number of particles (those keeping the subspaces $\mathcal{H}_\pm^{(N)}$ invariant) can be written through products containing the same number of creation and annihilation operators. This results in an important classification of such operators according to the number of particles ($n = 1, 2, 3\ldots$) they influence in a single "action". We talk about n-body operators.

▶ General operator expressed via creation/annihilation operators

Creation/annihilation operators of bosons or fermions **unified notation**: $\left\{\begin{matrix}\hat c_k^\dagger \equiv \hat b_k^\dagger \text{ or } \hat a_k^\dagger \\ \hat c_k \equiv \hat b_k \text{ or } \hat a_k\end{matrix}\right.$

Consider operator $\hat O$ conserving the particle number $\Rightarrow$ $[\hat O, \hat N] = 0$

$\hat O$ acts within any N-particle subspace $\mathcal{H}^{(N)}$, where it can be expressed as:

$$\hat O = \sum_{i_1 \ldots i_N} \sum_{i_1' \ldots i_N'} \langle\phi_{i_1}..\phi_{i_N}|\hat O|\phi_{i_1'}..\phi_{i_N'}\rangle |\phi_{i_1}..\phi_{i_N}\rangle\langle\phi_{i_1'}..\phi_{i_N'}|$$

Assume that the operator (observable) $\hat O$ is physical for *indistinguishable particles* $\Rightarrow$ it acts inside $\mathcal{H}_\pm^{(N)}$ $\Rightarrow$ $[\hat O, \hat P_\pm] = 0$

$$\hat{O}\hat{P}_\pm = \hat{P}_\pm\hat{O}\hat{P}_\pm = \sum_{i_1,..i_N}\sum_{i'_1,..i'_N} \langle\phi_{i_1}..\phi_{i_N}|\hat{O}|\phi_{i'_1}..\phi_{i'_N}\rangle \underbrace{\hat{P}_\pm|\phi_{i_1}..\phi_{i_N}\rangle}_{\sqrt{\frac{n_1!n_2!..}{N!}}|n_1,n_2,..\rangle} \underbrace{\langle\phi_{i'_1}..\phi_{i'_N}|\hat{P}_\pm}_{\langle n'_1,n'_2,..|\sqrt{\frac{n'_1!n'_2!..}{N!}}}$$

$$= \frac{1}{N!}\sum_{i_1,..i_N}\sum_{i'_1,..i'_N} \langle\phi_{i_1}..\phi_{i_N}|\hat{O}|\phi_{i'_1}..\phi_{i'_N}\rangle\, \hat{c}^\dagger_{i_1}\hat{c}^\dagger_{i_2}..\hat{c}^\dagger_{i_N} \underbrace{|0\rangle\langle 0|}_{\hat{P}^{(0)}} \hat{c}_{i'_N}..\hat{c}_{i'_2}\hat{c}_{i'_1}$$

within the space $\mathcal{H}^{(N)}$... $\hat{P}^{(0)}$ ...can be removed

$$\boxed{\hat{P}_\pm\hat{O}\hat{P}_\pm = \frac{1}{N!}\sum_{i_1,...i_N}\sum_{i'_1,...i'_N} \langle\phi_{i_1}..\phi_{i_N}|\hat{O}|\phi_{i'_1}..\phi_{i'_N}\rangle\, \hat{c}^\dagger_{i_1}\hat{c}^\dagger_{i_2}..\hat{c}^\dagger_{i_N}\hat{c}_{i'_N}..\hat{c}_{i'_2}\hat{c}_{i'_1}}$$

This is the most general expression in the N-particle subspace of an operator respecting particle indistinguishability and conserving particle number. However, as shown below, for some classes of operators this can be further simplified.

▶ **One-body operators**

Operator defined in the $N{=}1$ subspace through: $\boxed{\hat{T}|\psi\rangle_k = \sum_{i_k}\langle\phi_{i_k}|\hat{T}|\psi\rangle|\phi_{i_k}\rangle_k}$

The action of $\hat{T}$ is extended to all $N{>}1$ subspaces via summation over all particles:

$$\boxed{\hat{O}^{(1)} = \sum_{k=1}^{N}(\hat{T})_k} \equiv \sum_{k=1}^{N}\big(\hat{I}_1\otimes\cdots\hat{I}_{k-1}\otimes \underbrace{\hat{T}}_{k^{\text{th}}\text{place}} \otimes\hat{I}_{k+1}\cdots\otimes\hat{I}_N\big)$$

⇒ defining property of 1-body operator:

$$\boxed{\hat{O}^{(1)} \underbrace{\hat{P}_\pm|\phi_1\cdots\phi_k\cdots\phi_N\rangle}_{\frac{1}{\sqrt{N!}}\hat{c}^\dagger_1\cdots\hat{c}^\dagger_k\cdots\hat{c}^\dagger_N|0\rangle} = \sum_{k=1}^{N}\sum_{i_k}\langle\phi_{i_k}|\hat{T}|\phi_k\rangle \underbrace{\hat{P}_\pm|\phi_1\cdots\phi_{i_k}\cdots\phi_N\rangle}_{\frac{1}{\sqrt{N!}}\hat{c}^\dagger_1\cdots\hat{c}^\dagger_{i_k}\cdots\hat{c}^\dagger_N|0\rangle}}$$

We consider an operator defined as $\hat{O}^{(1)} \equiv \sum_{i,i'}\langle\phi_i|\hat{T}|\phi_{i'}\rangle\,\hat{c}^\dagger_i\hat{c}_{i'}$ and show that it satisfies the above property:

Note that: $[\hat{O}^{(1)},\hat{c}^\dagger_k] = \sum_{i,i'}\langle\phi_i|\hat{T}|\phi_{i'}\rangle \overbrace{[\hat{c}^\dagger_i\hat{c}_{i'},\hat{c}^\dagger_k]}^{\delta_{i'k}\hat{c}^\dagger_i} = \sum_i\langle\phi_i|\hat{T}|\phi_k\rangle\hat{c}^\dagger_i$

$$\hat{O}^{(1)}\hat{c}^\dagger_1..\hat{c}^\dagger_k..\hat{c}^\dagger_N|0\rangle = \Big\{\big(\underbrace{[\hat{O}^{(1)},\hat{c}^\dagger_1]}_{\sum_{i_1}\langle\phi_{i_1}|\hat{T}|\phi_1\rangle\hat{c}^\dagger_{i_1}}\hat{c}^\dagger_2..\hat{c}^\dagger_N\big)+\cdots+\big(\hat{c}^\dagger_1..\hat{c}^\dagger_{k-1}\underbrace{[\hat{O}^{(1)},\hat{c}^\dagger_k]}_{\sum_{i_k}\langle\phi_{i_k}|\hat{T}|\phi_k\rangle\hat{c}^\dagger_{i_k}}..\hat{c}^\dagger_N\big)$$

$$+\cdots+\big(\hat{c}^\dagger_1..\hat{c}^\dagger_{N-1}\overbrace{[\hat{O}^{(1)},\hat{c}^\dagger_N]}^{\sum_{i_N}\langle\phi_{i_N}|\hat{T}|\phi_N\rangle\hat{c}^\dagger_{i_N}}\big)\Big\}|0\rangle = \sum_{k=1}^{N}\sum_{i_k}\langle\phi_{i_k}|\hat{T}|\phi_k\rangle\hat{c}^\dagger_1..\hat{c}^\dagger_{i_k}..\hat{c}^\dagger_N|0\rangle$$

The above def. property is verified, so any 1-body operator can be expressed as:

$$\boxed{\hat{O}^{(1)} \equiv \sum_{i,i'}\langle\phi_i|\hat{T}|\phi_{i'}\rangle\,\hat{c}^\dagger_i\hat{c}_{i'}}$$

Graphical representation of this expression:

▶ Two-body operators

Operator defined in the N=2 subspace through:

$$\boxed{(\hat{V})_{kl}|\psi\rangle_{kl} = \sum_{i_k,i_l}\langle\phi_{i_k}\phi_{i_l}|\hat{V}|\psi\rangle_{kl}|\phi_{i_k}\phi_{i_l}\rangle_{kl}}$$

With respect to the exchange symmetry: $\langle\phi_i\phi_j|\hat{V}|\phi_{i'}\phi_{j'}\rangle = \langle\phi_j\phi_i|\hat{V}|\phi_{j'}\phi_{i'}\rangle$

The action of $\hat{V}$ is extended to all N>2 subspaces via summation over all particle pairs:

$$\boxed{\hat{O}^{(2)} = \sum_{k=1}^{N}\sum_{l=k+1}^{N}(\hat{V})_{kl}} = \sum_{\substack{k\\ l>k}\}=1}^{N} \hat{I}_1\otimes\cdot\cdot\hat{I}_{k-1}\otimes\hat{I}_{k+1}\otimes\cdot\cdot\hat{I}_{l-1}\otimes\hat{I}_{l+1}\otimes\cdot\cdot\hat{I}_N\otimes(\hat{V})_{kl}$$

⇒ defining property of a general 2-body operator:

$$\hat{O}^{(2)}\hat{c}_1^\dagger..\hat{c}_k^\dagger..\hat{c}_l^\dagger..\hat{c}_N^\dagger|0\rangle = \sum_{\substack{k\\ l>k}\}=1}^{N}\sum_{i_k,j_l}\langle\phi_{i_k}\phi_{j_l}|\hat{V}|\phi_k\phi_l\rangle\ \hat{c}_1^\dagger..\hat{c}_{i_k}^\dagger..\hat{c}_{j_l}^\dagger..\hat{c}_N^\dagger|0\rangle$$

We consider an operator defined as $\hat{O}^{(2)} \equiv \frac{1}{2}\sum_{i,i'}\sum_{j,j'}\langle\phi_i\phi_j|\hat{V}|\phi_{i'}\phi_{j'}\rangle\ \hat{c}_i^\dagger\hat{c}_j^\dagger\hat{c}_{j'}\hat{c}_{i'}$ and show that it satisfies the above property:

First note that: $$[\hat{O}^{(2)},\hat{c}_k^\dagger] = \frac{1}{2}\sum_{i,i'}\sum_{j,j'}\langle\phi_i\phi_j|\hat{V}|\phi_{i'}\phi_{j'}\rangle\ \overbrace{[\hat{c}_i^\dagger\hat{c}_j^\dagger\hat{c}_{j'}\hat{c}_{i'},\hat{c}_k^\dagger]}^{\delta_{i'k}\hat{c}_i^\dagger\hat{c}_j^\dagger\hat{c}_{j'}\pm\delta_{j'k}\hat{c}_i^\dagger\hat{c}_j^\dagger\hat{c}_{i'}} =$$

$$= \frac{1}{2}\sum_{i,j,j'}\langle\phi_i\phi_j|\hat{V}|\phi_k\phi_{j'}\rangle\hat{c}_i^\dagger\hat{c}_j^\dagger\hat{c}_{j'} + \frac{1}{2}\sum_{i,i',j}\langle\phi_j\phi_i|\hat{V}|\phi_k\phi_{i'}\rangle\hat{c}_j^\dagger\hat{c}_i^\dagger\hat{c}_{i'} = \sum_{i,j,l}\langle\phi_i\phi_j|\hat{V}|\phi_k\phi_l\rangle\hat{c}_i^\dagger\hat{c}_j^\dagger\hat{c}_l$$

$$\hat{O}^{(2)}\hat{c}_1^\dagger..\hat{c}_k^\dagger..\hat{c}_l^\dagger..\hat{c}_N^\dagger|0\rangle = \Big\{\Big(\underbrace{[\hat{O}^{(2)},\hat{c}_1^\dagger]}_{\sum_{i_1,j_1,l_1}\langle\phi_{i_1}\phi_{j_1}|\hat{V}|\phi_1\phi_{l_1}\rangle\hat{c}_{i_1}^\dagger\hat{c}_{j_1}^\dagger\hat{c}_{l_1}}\ \hat{c}_2^\dagger..\hat{c}_N^\dagger\Big) + \cdots + \Big(\hat{c}_1^\dagger..\hat{c}_{k-1}^\dagger\underbrace{[\hat{O}^{(2)},\hat{c}_k^\dagger]}_{\sum_{i_k,j_k,l_k}\langle\phi_{i_k}\phi_{j_k}|\hat{V}|\phi_k\phi_{l_k}\rangle\hat{c}_{i_k}^\dagger\hat{c}_{j_k}^\dagger\hat{c}_{l_k}}..\hat{c}_N^\dagger\Big)$$

$$+\cdots+\Big(\hat{c}_1^\dagger..\hat{c}_{N-1}^\dagger\ \underbrace{[\hat{O}^{(2)},\hat{c}_N^\dagger]}_{\sum_{i_N,j_N,l_N}\langle\phi_{i_N}\phi_{j_N}|\hat{V}|\phi_N\phi_{l_N}\rangle\hat{c}_{i_N}^\dagger\hat{c}_{j_N}^\dagger\hat{c}_{l_N}}\Big)\Big\}|0\rangle = \sum_{\substack{k\\ l>k}\}=1}^{N}\sum_{i_k,j_l}\langle\phi_{i_k}\phi_{j_l}|\hat{V}|\phi_k\phi_l\rangle\ \hat{c}_1^\dagger..\hat{c}_{i_k}^\dagger..\hat{c}_{j_l}^\dagger..\hat{c}_N^\dagger|0\rangle$$

The last equality results from the fact that $\hat{c}_{l_k}$ in the commutator expressions can only annihilate a state already created (otherwise the result=0) ⇒ l_k=(k+1) or (k+2) or … N. The pair $\hat{c}_{j_k}^\dagger\hat{c}_{l_k}$ commutes to the right to the position of the $\hat{c}_{l_k}^\dagger$ and the whole combination $\hat{c}_{j_k}^\dagger\hat{c}_{l_k}\hat{c}_{l_k}^\dagger$ is replaced by $\hat{c}_{j_k}^\dagger$. The last expression verifies the above property of 2-body operators ⇒

General 2-body operator reads as:

$$\boxed{\hat{O}^{(2)} \equiv \frac{1}{2}\sum_{i,i'}\sum_{j,j'}\langle\phi_i\phi_j|\hat{V}|\phi_{i'}\phi_{j'}\rangle\ \hat{c}_i^\dagger\hat{c}_j^\dagger\hat{c}_{j'}\hat{c}_{i'}}$$

Graphical representation of this expression:

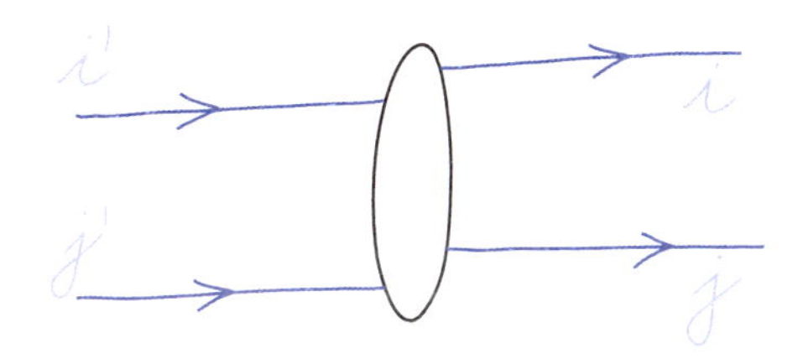

▶ **Higher-order operators**

An analogous procedure can be applied (though with increasing intricacy) to any n-body operator.

Example, **three-body**:

$$\hat{O}^{(3)} = \sum_{k=1}^{N}\sum_{l=k+1}^{N}\sum_{m=l+1}^{N}(\hat{W})_{klm} = \frac{1}{3!}\sum_{i,i'}\sum_{j,j'}\sum_{k,k'}\langle\phi_i\phi_j\phi_k|\hat{W}|\phi_{i'}\phi_{j'}\phi_{k'}\rangle\,\hat{c}_i^\dagger\hat{c}_j^\dagger\hat{c}_k^\dagger\hat{c}_{k'}\hat{c}_{j'}\hat{c}_{i'}$$

Graphical representation of 3- & n-body operators:

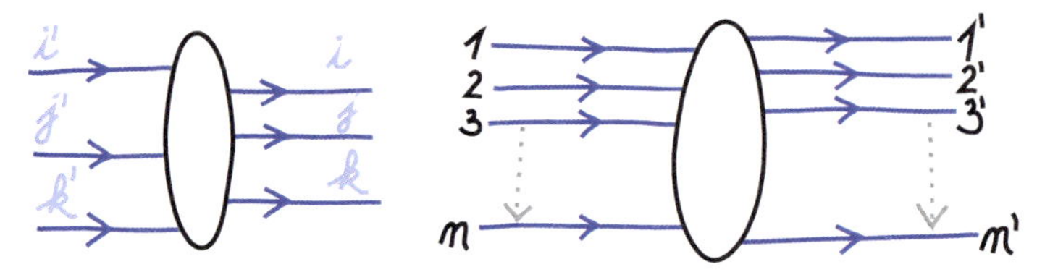

▶ **Normal ordering** of the products of creation/annihilation operators

Matrix elements of an n-body operator in the N-body space are expressed in terms of the following vacuum expectation values:

$$\langle 0|\underbrace{\hat{c}_{j_1}\cdots\hat{c}_{j_N}}_{N\times}\underbrace{\hat{c}_{k_1}^\dagger\cdots\hat{c}_{k_n}^\dagger}_{n\times}\underbrace{\hat{c}_{l_1}\cdots\hat{c}_{l_n}}_{n\times}\underbrace{\hat{c}_{i_N}^\dagger\cdots\hat{c}_{i_1}^\dagger}_{N\times}|0\rangle$$

The product inside is standardly rewritten in the normal-ordered form:

$$:\underbrace{\hat{c}_{i_1}^\dagger\cdots\hat{c}_{j_1}\cdots\hat{c}_{i_k}^\dagger\cdots\hat{c}_{j_l}\cdots\hat{c}_{j_m}\cdots\hat{c}_{i_n}^\dagger}_{\text{unsorted product of } n\,\times\,\hat{c}_\bullet^\dagger \text{ and } m\,\times\,\hat{c}_\circ}: = \underbrace{\sigma}_{\pm}\underbrace{\hat{c}_{i_1}^\dagger\cdots\hat{c}_{i_k}^\dagger\cdots\hat{c}_{i_n}^\dagger}_{n\,\times\,\hat{c}_\bullet^\dagger}\underbrace{\hat{c}_{j_1}\cdots\hat{c}_{j_l}\cdots\hat{c}_{j_m}}_{m\,\times\,\hat{c}_\circ}$$

$$\sigma = \begin{cases} +1 & \text{bosons} \\ \pm 1 = \sigma_\pi & \text{fermions: sign of permut. } (i_1..j_1..i_k..j_l..j_m..i_n)\to(i_1..i_k..i_n j_1..j_l..j_m) \end{cases}$$

Examples:
$$\begin{cases} :\hat{b}_i\hat{b}_j^\dagger: = \hat{b}_j^\dagger\hat{b}_i & :\hat{a}_i\hat{a}_j^\dagger: = -\hat{a}_j^\dagger\hat{a}_i \\ :\hat{b}_i^\dagger\hat{b}_j\hat{b}_k^\dagger: = \hat{b}_i^\dagger\hat{b}_k^\dagger\hat{b}_j & :\hat{a}_i^\dagger\hat{a}_j\hat{a}_k^\dagger: = -\hat{a}_i^\dagger\hat{a}_k^\dagger\hat{a}_j \\ \qquad = \hat{b}_k^\dagger\hat{b}_i^\dagger\hat{b}_j & \qquad = +\hat{a}_k^\dagger\hat{a}_i^\dagger\hat{a}_j \end{cases}$$

▶ **Wick theorem**

There exists a systematic way how a product of creation/annihilation operators can be transformed into the normal-ordered form. It makes use of so-called **contraction**, which for an operator product $\hat{A}\hat{B}$ is defined as the vacuum expectation value $\langle\hat{A}\hat{B}\rangle_0 = \langle 0|\hat{A}\hat{B}|0\rangle$

Examples:
$$\begin{cases} \langle\hat{b}_i\hat{b}_j^\dagger\rangle_0 = \delta_{ij} = \langle\hat{a}_i\hat{a}_j^\dagger\rangle_0 \\ \langle\hat{b}_j^\dagger\hat{b}_i\rangle_0 = 0 = \langle\hat{a}_j^\dagger\hat{a}_i\rangle_0 \\ \langle\hat{b}_j^\dagger\hat{b}_i^\dagger\rangle_0 = \langle\hat{b}_j\hat{b}_i\rangle_0 = 0 = \langle\hat{a}_j\hat{a}_i\rangle_0 = \langle\hat{a}_j^\dagger\hat{a}_i^\dagger\rangle_0 \end{cases}$$

Statement:

Product of creation & annihilation operators =

$$\sum_{k=0,1,2,\dots}\underbrace{\sigma}_{\pm}\;\begin{matrix}(\text{normal ordered product with } k \text{ pairs removed}) \equiv :\bullet_k: \\ \times\,(\text{product of } k \text{ contractions of the removed pairs}) \equiv c_k\end{matrix}$$

The proof not given here, instead we give some examples

Examples : **bosons| |fermions**

$\hat{b}_i\hat{b}_j^\dagger = \underbrace{:\hat{b}_i\hat{b}_j^\dagger:}_{\hat{b}_j^\dagger\hat{b}_i} + \underbrace{\langle\hat{b}_i\hat{b}_j^\dagger\rangle_0}_{\delta_{ij}}$ $\qquad$ $\hat{a}_i\hat{a}_j^\dagger = \underbrace{:\hat{a}_i\hat{a}_j^\dagger:}_{-\hat{a}_j^\dagger\hat{a}_i} + \underbrace{\langle\hat{a}_i\hat{a}_j^\dagger\rangle_0}_{\delta_{ij}}$

$\hat{b}_i^\dagger\hat{b}_j\hat{b}_k\hat{b}_l^\dagger = \hat{b}_i^\dagger\hat{b}_l^\dagger\hat{b}_j\hat{b}_k + \delta_{kl}\hat{b}_i^\dagger\hat{b}_j + \delta_{jl}\hat{b}_i^\dagger\hat{b}_k$ $\qquad$ $\hat{a}_i^\dagger\hat{a}_j\hat{a}_k\hat{a}_l^\dagger = \hat{a}_i^\dagger\hat{a}_l^\dagger\hat{a}_j\hat{a}_k + \delta_{kl}\hat{a}_i^\dagger\hat{a}_j - \delta_{jl}\hat{a}_i^\dagger\hat{a}_k$

General product :

$$\begin{aligned}\hat{A}\hat{B}\hat{C}\hat{D} = {} & :\hat{A}\hat{B}\hat{C}\hat{D}: + \langle\hat{A}\hat{B}\rangle_0 :\hat{C}\hat{D}: \\ & + \langle\hat{A}\hat{C}\rangle_0 :\hat{B}\hat{D}: + \langle\hat{A}\hat{D}\rangle_0 :\hat{B}\hat{C}: \\ & + \langle\hat{B}\hat{C}\rangle_0 :\hat{A}\hat{D}: + \langle\hat{B}\hat{D}\rangle_0 :\hat{A}\hat{C}: \\ & + \langle\hat{C}\hat{D}\rangle_0 :\hat{A}\hat{B}: + \langle\hat{A}\hat{B}\rangle_0\langle\hat{C}\hat{D}\rangle_0 \\ & + \langle\hat{A}\hat{C}\rangle_0\langle\hat{B}\hat{D}\rangle_0 + \langle\hat{A}\hat{D}\rangle_0\langle\hat{B}\hat{C}\rangle_0\end{aligned} \qquad \begin{aligned}\hat{A}\hat{B}\hat{C}\hat{D} = {} & :\hat{A}\hat{B}\hat{C}\hat{D}: + \langle\hat{A}\hat{B}\rangle_0 :\hat{C}\hat{D}: \\ & - \langle\hat{A}\hat{C}\rangle_0 :\hat{B}\hat{D}: + \langle\hat{A}\hat{D}\rangle_0 :\hat{B}\hat{C}: \\ & + \langle\hat{B}\hat{C}\rangle_0 :\hat{A}\hat{D}: - \langle\hat{B}\hat{D}\rangle_0 :\hat{A}\hat{C}: \\ & + \langle\hat{C}\hat{D}\rangle_0 :\hat{A}\hat{B}: + \langle\hat{A}\hat{B}\rangle_0\langle\hat{C}\hat{D}\rangle_0 \\ & - \langle\hat{A}\hat{C}\rangle_0\langle\hat{B}\hat{D}\rangle_0 + \langle\hat{A}\hat{D}\rangle_0\langle\hat{B}\hat{C}\rangle_0\end{aligned}$$

If the **vacuum expectation value** of an operator product is to be evaluated, one makes use of the obvious fact that $\boxed{\langle 0| :\bullet_k: |0\rangle = 0}$ Only the terms composed solely of contractions (if $\neq 0$) may contribute to the result.

▸ Two-state correlations

The N-body state $|\Psi\rangle$ contains complete information on the system, including information on statistical properties of all occupation numbers n_i associated with individual single-particle states $|\phi_i\rangle$. For any particular $|\Psi\rangle$, these properties can be described by means of the following general quantities:

(a) Average: $\qquad \langle n_i\rangle_\Psi = \langle\Psi|\hat{c}_i^\dagger\hat{c}_i|\Psi\rangle$

(b) Dispersion: $\qquad \langle\langle n_i^2\rangle\rangle_\Psi = \langle n_i^2\rangle_\Psi - \langle n_i\rangle_\Psi^2 = \overbrace{\underbrace{\langle\Psi|\hat{c}_i^\dagger\hat{c}_i\hat{c}_i^\dagger\hat{c}_i|\Psi\rangle}}^{\langle\Psi|\hat{b}_i^\dagger\hat{b}_i^\dagger\hat{b}_i\hat{b}_i|\Psi\rangle + \langle\Psi|\hat{b}_i^\dagger\hat{b}_i|\Psi\rangle \ \text{(bosons)} \qquad \langle\Psi|\hat{a}_i^\dagger\hat{a}_i|\Psi\rangle \ \text{(fermions)}} - \langle\Psi|\hat{c}_i^\dagger\hat{c}_i|\Psi\rangle^2$

(c) Correlation between occupation numbers of states $|\phi_i\rangle, |\phi_j\rangle$ (for $i \neq j$):

$$\boxed{\langle\langle n_i n_j\rangle\rangle_\Psi = \underbrace{\langle n_i n_j\rangle_\Psi - \langle n_i\rangle_\Psi\langle n_j\rangle_\Psi}_{\left\langle(n_i - \langle n_i\rangle_\Psi)(n_j - \langle n_j\rangle_\Psi)\right\rangle_\Psi}} = \underbrace{\langle\Psi|\hat{c}_i^\dagger\hat{c}_i\hat{c}_j^\dagger\hat{c}_j|\Psi\rangle}_{\langle\Psi|\hat{c}_i^\dagger\hat{c}_j^\dagger\hat{c}_j\hat{c}_i|\Psi\rangle} - \langle\Psi|\hat{c}_i^\dagger\hat{c}_i|\Psi\rangle\langle\Psi|\hat{c}_j^\dagger\hat{c}_j|\Psi\rangle$$

Normalized correlation coefficient: $\boxed{C_{ij}(\Psi) \equiv \dfrac{\langle\langle n_i n_j\rangle\rangle_\Psi}{\sqrt{\langle\langle n_i^2\rangle\rangle_\Psi\langle\langle n_j^2\rangle\rangle_\Psi}} \in [-1, +1]}$

$$C_{ij}(\Psi) = \left\{\begin{array}{rl} +1 & \text{for perfect correlation} \\ 0 & \text{for null correlation} \\ -1 & \text{for perfect anticorrelation}\end{array}\right\} \text{of } (n_i - \langle n_i\rangle_\Psi) \text{ and } (n_j - \langle n_j\rangle_\Psi)$$

▸ Many-body Hamiltonian

General expression of a Hamiltonian with **1-body terms** (kinetic energies of individual particles + potential energies in an external potential field) and

2-particle interactions:

$$\hat{H} = \sum_{i,i'} \varepsilon_{ii'}\, \hat{c}_i^\dagger \hat{c}_{i'} + \tfrac{1}{2} \sum_{\substack{i,i'\\ j,j'}} \nu_{iji'j'}\, \hat{c}_i^\dagger \hat{c}_j^\dagger \hat{c}_{j'} \hat{c}_{i'}$$

where $\varepsilon_{ii'} = \langle\phi_i|\hat{T}|\phi_{i'}\rangle$ and $\nu_{iji'j'} = \langle\phi_i\phi_j|\hat{V}|\phi_{i'}\phi_{j'}\rangle$ are matrix elements in the space of *distinguishable* particles. The 3-particle and higher interactions can also be included by the respective n-body expressions.

▶ Coordinate form of Hamiltonian

If the many-body Hamiltonian is expressed in terms of coordinates $\hat{\vec{x}}_k$ and spin projections $\hat{s}_{zk}$ of individual particles (k=1,...N), it is useful to utilize the coordinate form of creation & annihilation operators.

$$\hat{H} = \underbrace{\sum_{k=1}^{N} \overbrace{\left(-\tfrac{\hbar^2}{2M}\Delta_k\right)}^{(\hat{T})_k}}_{\text{kinetic term } \hat{O}^{(1)}_{\text{kin}}} + \underbrace{\sum_{k=1}^{N} \overbrace{U(\hat{\vec{x}}_k, \hat{s}_{zk})}^{(\hat{U})_k}}_{\text{external potential } \hat{O}^{(1)}_{\text{pot}}} + \underbrace{\sum_{\substack{k\\ l>k}=1}^{N} \overbrace{V(\hat{\vec{x}}_k, \hat{s}_{zk}; \hat{\vec{x}}_l, \hat{s}_{zl})}^{(\hat{V})_{kl}}}_{\text{interaction } \hat{O}^{(2)}_{\text{int}}}$$

$$\hat{O}^{(1)}_{\text{kin}} + \hat{O}^{(1)}_{\text{pot}} = \sum_{i,i'} \langle\phi_i|(\hat{T}+\hat{U})|\phi_{i'}\rangle \hat{c}_i^\dagger \hat{c}_{i'}$$
$$= \sum_{i,i'} \left\{ \sum_{m_s} \int \phi_i^*(\vec{x}, m_s) \left[-\tfrac{\hbar^2}{2M}\Delta + U(\vec{x}, m_s)\right] \phi_{i'}(\vec{x}, m_s) d\vec{x} \right\} \hat{c}_i^\dagger \hat{c}_{i'}$$
$$= \sum_{m_s} \int \underbrace{\left[\sum_i \phi_i^*(\vec{x}, m_s)\hat{c}_i^\dagger\right]}_{\hat{\psi}^\dagger_\pm(\vec{x}, m_s)} \left[-\tfrac{\hbar^2}{2M}\Delta + U(\vec{x}, m_s)\right] \underbrace{\left[\sum_{i'} \phi_{i'}(\vec{x}, m_s)\hat{c}_{i'}\right]}_{\hat{\psi}_\pm(\vec{x}, m_s)} d\vec{x}$$

$$\hat{O}^{(2)}_{\text{int}} = \tfrac{1}{2} \sum_{i,i',j,j'} \langle\phi_i\phi_j|\hat{V}|\phi_{i'}\phi_{j'}\rangle \hat{c}_i^\dagger \hat{c}_j^\dagger \hat{c}_{j'} \hat{c}_{i'} = \tfrac{1}{2}\times$$
$$\sum_{\substack{i,i'\\ j,j'}} \left\{ \sum_{\substack{m_s\\ m_s'}} \iint \phi_i^*(\vec{x}, m_s)\phi_j^*(\vec{x}', m_s') V(\vec{x}, m_s; \vec{x},' m_s') \phi_{i'}(\vec{x}, m_s)\phi_{j'}(\vec{x}', m_s') d\vec{x} d\vec{x}' \right\} \hat{c}_i^\dagger \hat{c}_j^\dagger \hat{c}_{j'} \hat{c}_{i'}$$
$$= \tfrac{1}{2} \sum_{\substack{m_s\\ m_s'}} \iint \hat{\psi}^\dagger_\pm(\vec{x}, m_s) \hat{\psi}^\dagger_\pm(\vec{x}', m_s') V(\vec{x}, m_s; \vec{x},' m_s') \hat{\psi}_\pm(\vec{x}', m_s') \hat{\psi}_\pm(\vec{x}, m_s) d\vec{x} d\vec{x}'$$

The final expression is of the **field-theory type**:

$$\hat{H} = \sum_{m_s} \int \hat{\psi}^\dagger_\pm(\vec{x}, m_s)\left[-\tfrac{\hbar^2}{2M}\Delta + U(\vec{x}, m_s)\right]\hat{\psi}_\pm(\vec{x}, m_s)\, d\vec{x}$$
$$+ \tfrac{1}{2} \sum_{\substack{m_s\\ m_s'}} \iint \hat{\psi}^\dagger_\pm(\vec{x}, m_s) \hat{\psi}^\dagger_\pm(\vec{x}', m_s') V(\vec{x}, m_s; \vec{x},' m_s') \hat{\psi}_\pm(\vec{x}', m_s') \hat{\psi}_\pm(\vec{x}, m_s)\, d\vec{x}\, d\vec{x}'$$

■ Quantization of electromagnetic field

The formalism built up in the above paragraphs will now be applied in a concrete task to quantize the electromagnetic field. We know that elmg. quanta, the photons, have spin s=1, so they are bosons. The quantized field enables one to describe all processes connected with elmg. interaction of matter, including, e.g., spontaneous elmg. decays of many-body systems (photon emissions).

▶ Photon creation/annihilation operators

The general solution of the wave equation $\vec{\nabla}^2\vec{A}-\frac{1}{c^2}\frac{\partial^2\vec{A}}{\partial t^2}=0$ for the elmg. vector potential $\vec{A}(\vec{x},t)$ in vacuum ($c=\frac{1}{\sqrt{\epsilon_0\mu_0}}$) is a superposition of planar waves:

$$\vec{A}(\vec{x},t)=\sum_{\nu=\pm}\int\mathcal{N}_{Vk}\Big\{\underbrace{\alpha_{\vec{k}\nu}}_{\mapsto\hat{b}_{\vec{k}\nu}}\ \vec{e}_{\vec{k}\nu}e^{+i(\vec{k}\cdot\vec{x}-\omega_k t)}+\underbrace{\alpha^*_{\vec{k}\nu}}_{\mapsto\hat{b}^\dagger_{\vec{k}\nu}}\ \vec{e}^{\,*}_{\vec{k}\nu}e^{-i(\vec{k}\cdot\vec{x}-\omega_k t)}\Big\}d\vec{k}$$

with $\omega_k=c|\vec{k}|$ and:

(a) $\mathcal{N}_{Vk}\equiv$ a scaling factor for each mode which will be determined later

(b) $\vec{e}_{\vec{k}\pm}=\mp\frac{1}{\sqrt{2}}\left[\vec{e}_{\vec{k}x}\pm i\vec{e}_{\vec{k}y}\right]$ ≡ **circular polarization** vectors composed of unit vectors of linear polarization satisfying the Coulomb gauge condition: $\vec{e}_{\vec{k}x}\cdot\vec{k}=0=\vec{e}_{\vec{k}y}\cdot\vec{k}\ \Rightarrow\ \vec{e}^{\,*}_{\vec{k}\nu}\cdot\vec{e}_{\vec{k}\nu'}=\delta_{\nu\nu'}$

(c) $\alpha_{\vec{k}\nu}\equiv$ arbitrary coefficients

Field quantization:

function $\vec{A}(\vec{x},t)\in\mathbb{R}\mapsto$ operator $\hat{\vec{A}}(\vec{x},t)=\hat{\vec{A}}^\dagger(\vec{x},t)\ \Leftrightarrow\ \begin{cases}\alpha_{\vec{k}\nu}\mapsto\hat{b}_{\vec{k}\nu}\\ \alpha^*_{\vec{k}\nu}\mapsto\hat{b}^\dagger_{\vec{k}\nu}\end{cases}$

Operators $\hat{b}^\dagger_{\vec{k}\nu}$ and $\hat{b}_{\vec{k}\nu}$, respectively, create and annihilate **photons** with **momentum** $\vec{p}_\gamma=\hbar\vec{k}$ and **spin projection** $s_{\vec{k}}=\nu\hbar=\pm\hbar$ to the flight direction $\vec{k}/k$:

$$\hat{b}^\dagger_{\vec{k}\nu}|0\rangle_\gamma=|\vec{k}\nu\rangle_\gamma \qquad \hat{b}_{\vec{k}\nu}|\vec{k}\nu\rangle_\gamma=|0\rangle_\gamma$$

▶ Energy of elmg. field

Classical expression for energy: $\mathcal{E}=\frac{1}{2}\int_V\big[\epsilon_0|\overbrace{\vec{E}(\vec{x},t)}^{-\frac{\partial\vec{A}}{\partial t}}|^2+\mu_0^{-1}|\overbrace{\vec{B}(\vec{x},t)}^{\vec{\nabla}\times\vec{A}}|^2\big]d\vec{x}$

$$-\frac{\partial}{\partial t}\vec{A}=\sum_{\nu=\pm}\int\mathcal{N}_{Vk}\big\{i\alpha_{\vec{k}\nu}\omega_k\vec{e}_{\vec{k}\nu}e^{+i(\vec{k}\cdot\vec{x}-\omega_k t)}-i\alpha^*_{\vec{k}\nu}\omega_k\vec{e}^{\,*}_{\vec{k}\nu}e^{-i(\vec{k}\cdot\vec{x}-\omega_k t)}\big\}d\vec{k}$$

$$c[\vec{\nabla}\times\vec{A}]=\sum_{\nu=\pm}\int\mathcal{N}_{Vk}\big\{i\alpha_{\vec{k}\nu}\overbrace{[c\vec{k}\times\vec{e}_{\vec{k}\nu}]}^{i\nu\omega_k\vec{e}_{\vec{k}\nu}}e^{+i(\vec{k}\cdot\vec{x}-\omega_k t)}-i\alpha^*_{\vec{k}\nu}\overbrace{[c\vec{k}\times\vec{e}^{\,*}_{\vec{k}\nu}]}^{-i\nu\omega_k\vec{e}^{\,*}_{\vec{k}\nu}}e^{-i(\vec{k}\cdot\vec{x}-\omega_k t)}\big\}d\vec{k}$$

For $V\to\infty$ the spatial integration yields: $\int_V e^{i(\vec{k}\pm\vec{k}')\cdot\vec{x}}d\vec{x}\approx V\delta_{\vec{k},\mp\vec{k}'}$

The resulting expression for energy: $\mathcal{E}=V\epsilon_0\sum_\nu\int(\mathcal{N}_{Vk}\omega_k)^2\big(\alpha^*_{\vec{k}\nu}\alpha_{\vec{k}\nu}+\alpha_{\vec{k}\nu}\alpha^*_{\vec{k}\nu}\big)d\vec{k}$

This after the quantization, with the choice of $\mathcal{N}_{Vk}=\sqrt{\frac{\hbar}{2V\epsilon_0\omega_k}}$, leads to:

$$\hat{\mathcal{E}}=\sum_{\nu=\pm}\int\hbar\omega_k\left(\hat{b}^\dagger_{\vec{k}\nu}\hat{b}_{\vec{k}\nu}+\tfrac{1}{2}\right)d\vec{k}$$

Interpretation: This expression is equivalent to an ensemble of harmonic oscillators, each one associated with a single field mode

Note: The term associated with zero-point motion yields diverging contribution and must be artificially removed in the field theory

▶ Photon emission & absorption

In Sec. 5.3, we outlined the theory of transitions stimulated by classical elmg. waves in systems of charged particles. Now this theory can be extended to describe interactions of matter with general, also *non-classical* field states.
Example: Any field state $|\Psi_\gamma\rangle$ with a definite photon number N_γ is non-classical as it yields *vanishing averages* of field intensities: $\langle\Psi_\gamma|\hat{\vec{E}}(\vec{x},t)|\Psi_\gamma\rangle = 0 = \langle\Psi_\gamma|\hat{\vec{B}}(\vec{x},t)|\Psi_\gamma\rangle$ (the terms of $\hat{\vec{E}}$ & $\hat{\vec{B}}$ contain either $\hat{b}^\dagger_{\vec{k}\nu}$ or $\hat{b}_{\vec{k}\nu}$ $\Rightarrow$ change N_γ), but the dispersions of field intensities are nonzero.
In particular, such general theory applies to the processes of single-photon absorption and spontaneous single-photon emission.

Consider a system composed of N particles with charges q_k and masses M_k. The **matter-field interaction Hamiltonian** (cf. Sec. 5.3):

$$\hat{H}'(t) = \sum_{k=1}^{N} \tfrac{q_k}{M_k}\left[\hat{\vec{A}}(\hat{\vec{x}}_k,t)\cdot\hat{\vec{p}}_k\right]$$

where $\hat{\vec{A}}(\hat{\vec{x}}_k,t)$ is taken from the above general expression with the $\left\{\begin{array}{l}\alpha_{\vec{k}\nu}\mapsto \hat{b}_{\vec{k}\nu}\\ \alpha^*_{\vec{k}\nu}\mapsto \hat{b}^\dagger_{\vec{k}\nu}\end{array}\right\}$ substitutions

Transition probabilities for single-photon absorptions & emissions are calculated with the aid of the Fermi golden rule:

Process	Initial state $\|\psi_{0i}\rangle$		Final state $\|\psi_{0j}\rangle$	Active term in $\hat{H}'(t)$
emission	$\|E_{0i}\rangle_{\rm a}\|0\rangle_\gamma$	$\longrightarrow$	$\|E_{0j}\rangle_{\rm a}\|\vec{k}\nu\rangle_\gamma$	one with $\hat{b}^\dagger_{\vec{k}\nu}$
absorption	$\|E_{0i}\rangle_{\rm a}\|\vec{k}\nu\rangle_\gamma$	$\longrightarrow$	$\|E_{0j}\rangle_{\rm a}\|0\rangle_\gamma$	one with $\hat{b}_{\vec{k}\nu}$

From this point on, the calculation of transition amplitudes is rather analogous to that presented in Sec. 5.3 (using either the dipole approximation or the complete multipole expansion). For spontaneous emissions one needs to include into the density of final states also the emitted-photon state density $\rho_\gamma(E_\gamma)$ (calculated as the number of modes per unit energy in a finite box of volume V; for details see elsewhere).

◀ Historical remark

1927: Paul Dirac shows the equivalence of an ensemble of non-interacting bosons with indefinite particle number (elmg. field) with a system of harmonic oscillators (the use of occupation number representation & creation/annihilation operators)
1928: Pascual Jordan & Eugene Wigner generalize Dirac's results to fermions (the use of anticommutators) & ensembles of interacting particles
1932: Vladimir Fock introduces the Hilbert space for q. fields/ many-body systems
1939,40: Markus Fierz and Wolfgang Pauli formulate the spin-statistics theorem
1950: G.-C. Wick provides a method for evaluating products of creat./annih. opers.

7.2 Many-body techniques

We are ready now to apply the above-derived general formalism in some sophisticated approximation methods, which are extremely useful for the description of various quantum many-body systems—atoms, molecules, nuclei, clusters etc.

■ Fermionic mean field & Hartree-Fock method

Atoms & nuclei represent genuine many-body systems since all their constituent particles (fermions) interact with each other. Nevertheless, it turns out—at least as far as the ground-state properties are considered—that one can transform this difficult problem into a much simpler problem of individual particles moving in a *single-particle mean field*. This field can be seen as a kind of averaged influence of all other particles on any selected particle.

► Hartree-Fock ansatz for the ground-state wavefunction

Fermionic **Hamiltonian** with **one + two body** terms written in arbitrary basis:

$$\hat{H} = \sum_{k,k'} \varepsilon_{kk'} \hat{a}^\dagger_k \hat{a}_{k'} + \frac{1}{2} \sum_{\substack{k,k'\\ l,l'}} \nu_{klk'l'} \hat{a}^\dagger_k \hat{a}^\dagger_l \hat{a}_{l'} \hat{a}_{k'}$$

The ground state of an N-particle system is searched as a **Slater-determinant** wavefunction

$$|\Psi_{\rm HF}\rangle = \hat{a}^\dagger_N \cdots \hat{a}^\dagger_2 \hat{a}^\dagger_1 |0\rangle$$

where $\hat{a}^\dagger_N, \ldots, \hat{a}^\dagger_2, \hat{a}^\dagger_1$ create some orthonormal single-particle states interpreted as the lowest eigenstates of an unknown **one-body Hamiltonian = mean field**
⇒ the ground state ≡ "**Fermi sea**"
(N lowest levels of the mean-field Hamiltonian occupied, higher levels empty)

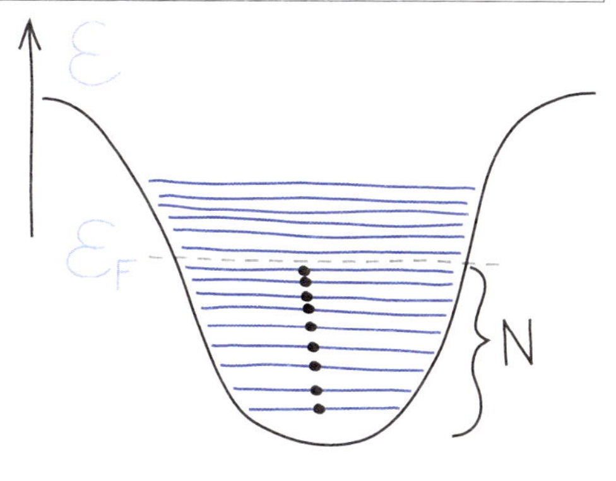

► Variation of the HF state

The unknown mean-field states $|\phi_i\rangle = \hat{a}^\dagger_i|0\rangle$ entering into $|\Psi_{\rm HF}\rangle$ will be determined by the **stationary variational method**:
Infinitesimal unitary variation $|\phi_i\rangle \mapsto |\phi_i\rangle + |\delta\phi_i\rangle$

$$|\phi_i\rangle \mapsto \underbrace{\sum_j u_{ij}|\phi_j\rangle}_{e^{i\hat{\epsilon}}|\phi_i\rangle} \approx |\phi_i\rangle + \underbrace{i\sum_j \epsilon_{ij}|\phi_j\rangle}_{|\delta\phi_i\rangle} \quad \Rightarrow \quad \boxed{\hat{a}^\dagger_i \mapsto \hat{a}^\dagger_i + \underbrace{i\sum_j \epsilon_{ij}\hat{a}^\dagger_j}_{\delta\hat{a}^\dagger_i}} \quad \epsilon_{ij} = \epsilon^*_{ji}$$

$$|\Psi_{\rm HF}\rangle = \hat{a}^\dagger_N \cdots \hat{a}^\dagger_2 \hat{a}^\dagger_1|0\rangle \quad \mapsto \quad \left(\hat{a}^\dagger_N + \delta\hat{a}^\dagger_N\right) \cdots \left(\hat{a}^\dagger_2 + \delta\hat{a}^\dagger_2\right)\left(\hat{a}^\dagger_1 + \delta\hat{a}^\dagger_1\right)|0\rangle$$

$$\approx |\Psi_{\rm HF}\rangle + (\underbrace{\delta\hat{a}^\dagger_N}_{i\sum_j \epsilon_{Nj}\hat{a}^\dagger_j} \cdots \hat{a}^\dagger_2 \hat{a}^\dagger_1)|0\rangle + \cdots\cdots + (\hat{a}^\dagger_N \cdots \underbrace{\delta\hat{a}^\dagger_2}_{i\sum_j \epsilon_{2j}\hat{a}^\dagger_j} \hat{a}^\dagger_1)|0\rangle + (\hat{a}^\dagger_N \cdots \hat{a}^\dagger_2 \underbrace{\delta\hat{a}^\dagger_1}_{i\sum_j \epsilon_{1j}\hat{a}^\dagger_j})|0\rangle$$

Ket variation:

$$|\Psi_{\rm HF}\rangle \mapsto |\Psi_{\rm HF}\rangle + \underbrace{i\sum_{i=1}^{N}\sum_{j=N+1}^{\infty}\epsilon_{ij}\hat{a}_j^{\dagger}\hat{a}_i|\Psi_{\rm HF}\rangle}_{|\delta\Psi_{\rm HF}\rangle}$$

Bra variation (independ. coeffs. $\epsilon'_{ij}=\epsilon'^{*}_{ji}$):

$$\langle\Psi'_{\rm HF}| \mapsto \langle\Psi_{\rm HF}| \underbrace{-i\sum_{i=1}^{N}\sum_{j=N+1}^{\infty}\epsilon'_{ji}\langle\Psi_{\rm HF}|\hat{a}_i^{\dagger}\hat{a}_j}_{\langle\delta\Psi'_{\rm HF}|}$$

▶ Variational condition

The condition for $|\Psi_{\rm HF}\rangle$ reads as follows:

$$\langle\Psi_{\rm HF}|\hat{H}|\delta\Psi_{\rm HF}\rangle + \langle\delta\Psi'_{\rm HF}|\hat{H}|\Psi_{\rm HF}\rangle = i\sum_{i\leq N}\sum_{j>N}\langle\Psi_{\rm HF}|\epsilon_{ij}\hat{H}\hat{a}_j^{\dagger}\hat{a}_i - \epsilon'_{ji}\hat{a}_i^{\dagger}\hat{a}_j\hat{H}|\Psi_{\rm HF}\rangle \stackrel{!}{=} 0 \ \ \forall\left\{\begin{smallmatrix}\epsilon_{ij}\\ \epsilon'_{ji}\end{smallmatrix}\right.$$

$$\Rightarrow \quad \boxed{\langle\Psi_{\rm HF}|\hat{H}\hat{a}_j^{\dagger}\hat{a}_i|\Psi_{\rm HF}\rangle \stackrel{!}{=} 0 \quad \forall\left\{\begin{smallmatrix}i\leq N\\ j>N\end{smallmatrix}\right.}$$

Assuming that $\hat{H}$ is written in terms of the single-particle states involved in the HF state we get:

$$\left\langle\Psi_{\rm HF}\middle|\left(\sum_{k,k'}\varepsilon_{kk'}\hat{a}_k^{\dagger}\hat{a}_{k'} + \frac{1}{2}\sum_{\substack{k,k'\\ l,l'}}\nu_{klk'l'}\hat{a}_k^{\dagger}\hat{a}_l^{\dagger}\hat{a}_{l'}\hat{a}_{k'}\right)\hat{a}_j^{\dagger}\hat{a}_i\middle|\Psi_{\rm HF}\right\rangle = 0 \quad \forall\left\{\begin{array}{l} i\leq N\\ j>N\end{array}\right.$$

Evaluation of both terms: anticommutation of $\hat{a}_j^{\dagger}$ or $\hat{a}_i$ to the leftmost position (the resulting matrix element =0 since $j>N$ and $i\leq N$)

(a) One-body term: $\sum_{k,k'}\varepsilon_{kk'}\langle\Psi_{\rm HF}|\hat{a}_k^{\dagger}\hat{a}_{k'}\hat{a}_j^{\dagger}\hat{a}_i|\Psi_{\rm HF}\rangle =$

$$= \sum_{k,k'}\varepsilon_{kk'}\Big(\underbrace{\langle\Psi_{\rm HF}|\hat{a}_j^{\dagger}\hat{a}_k^{\dagger}\hat{a}_{k'}\hat{a}_i|\Psi_{\rm HF}\rangle}_{0} + \delta_{jk'}\underbrace{\langle\Psi_{\rm HF}|\hat{a}_k^{\dagger}\hat{a}_i|\Psi_{\rm HF}\rangle}_{\langle\Psi_{\rm HF}|\hat{a}_i\hat{a}_k^{\dagger}|\Psi_{\rm HF}\rangle+\delta_{ik}\langle\Psi_{\rm HF}|\Psi_{\rm HF}\rangle}\Big) = \sum_{k,k'}\varepsilon_{kk'}\delta_{jk'}\delta_{ik} = \varepsilon_{ij}$$

(b) Two-body term: $\frac{1}{2}\sum_{\substack{k,k'\\ l,l'}}\nu_{klk'l'}\langle\Psi_{\rm HF}|\hat{a}_k^{\dagger}\hat{a}_l^{\dagger}\hat{a}_{l'}\hat{a}_{k'}\hat{a}_j^{\dagger}\hat{a}_i|\Psi_{\rm HF}\rangle =$

$$= \frac{1}{2}\sum_{\substack{k,k'\\ l,l'}}\nu_{klk'l'}\Big[\delta_{jk'}\delta_{ik}\underbrace{\langle\Psi_{\rm HF}|\hat{a}_l^{\dagger}\hat{a}_{l'}|\Psi_{\rm HF}\rangle}_{\substack{\delta_{ll'}\text{ for } l\leq N\\ 0\text{ for } l,l'>N}} + \delta_{jl'}\delta_{il}\underbrace{\langle\Psi_{\rm HF}|\hat{a}_k^{\dagger}\hat{a}_{k'}|\Psi_{\rm HF}\rangle}_{\substack{\delta_{kk'}\text{ for } k\leq N\\ 0\text{ for } k,k'>N}}$$

$$-\delta_{jk'}\delta_{il}\overbrace{\langle\Psi_{\rm HF}|\hat{a}_k^{\dagger}\hat{a}_{l'}|\Psi_{\rm HF}\rangle}^{\substack{\delta_{kl'}\text{ for } k\leq N\\ 0\text{ for } k,k'>N}} - \delta_{jl'}\delta_{ik}\overbrace{\langle\Psi_{\rm HF}|\hat{a}_l^{\dagger}\hat{a}_{k'}|\Psi_{\rm HF}\rangle}^{\substack{\delta_{lk'}\text{ for } l\leq N\\ 0\text{ for } l,l'>N}}\Big] =$$

$$= \frac{1}{2}\Big[\sum_{k\leq N}\underbrace{(\nu_{ikjk}+\nu_{kikj})}_{2\nu_{kikj}} - \sum_{k\leq N}\underbrace{(\nu_{ikkj}+\nu_{kijk})}_{2\nu_{ikkj}}\Big] = \sum_{k\leq N}(\nu_{kikj}-\nu_{ikkj})$$

Together:

$$\boxed{\varepsilon_{ij} + \sum_{k\leq N}(\nu_{kikj}-\nu_{ikkj}) = 0 \qquad \forall\left\{\begin{array}{l} i\leq N\\ j>N\end{array}\right.}$$

This represents a coupled set of conditions for the Hamiltonian matrix elements in the HF basis which must be satisfied to minimize the energy functional

▶ Mean-field equation

We know that $\varepsilon_{ij} \equiv \langle\phi_i|\hat{T}|\phi_j\rangle$. The above set of equations can be formally solved by introducing another one-body operator $\hat{V}_{\rm HF}$, which is defined through its matrix elements in the HF basis as follows: $\boxed{\langle\phi_i|\hat{V}_{\rm HF}|\phi_j\rangle \equiv \sum_{k\leq N}(\nu_{kikj}-\nu_{ikkj})}$

It represents the **Hartree-Fock mean field**

$\Rightarrow$ the above variational condition reads: $\langle\phi_i|(\hat{T}+\hat{V}_{\rm HF})|\phi_j\rangle = 0$ for $\left\{ {i\leq N \atop j>N} \right.$

This can be replaced by a stronger condition that $(\hat{T}+\hat{V}_{\rm HF})$ is diagonal in the basis $\{|\phi_n\rangle\}$, i.e: $\boxed{(\hat{T}+\hat{V}_{\rm HF})|\phi_n\rangle = \varepsilon_n|\phi_n\rangle}$ **one-body eigenvalue equation**

$\Rightarrow$ Many-body ground state approximated with the aid of eigensolutions of a one-body problem. However, the HF mean field is expressed via the eigensolutions *that we want to determine*:

$$\hat{V}_{\rm HF}|\phi_n\rangle = \sum_m \langle\phi_m|\hat{V}_{\rm HF}|\phi_n\rangle|\phi_m\rangle = \sum_m \left[\sum_{k\leq N}\Big(\langle\phi_k\phi_m|\hat{V}|\phi_k\phi_n\rangle - \langle\phi_m\phi_k|\hat{V}|\phi_k\phi_n\rangle\Big)\right]|\phi_m\rangle$$

$\Rightarrow$ **selfconsistent problem**

Solution searched in an **iterative procedure**: basis $\{|\phi_n^{(0)}\rangle\} \Rightarrow$ mean field $\hat{V}_{\rm HF}^{(0)}$ $\Rightarrow$ basis $\{|\phi_n^{(1)}\rangle\} \Rightarrow$ mean field $\hat{V}_{\rm HF}^{(1)} \Rightarrow$ basis $\{|\phi_n^{(2)}\rangle\} \Rightarrow$ mean field $\hat{V}_{\rm HF}^{(2)} \Rightarrow \ldots\ldots$

We may hope in a fast convergence.

▶ Coordinate representation of the mean field

Meaning of the above-defined mean field operator becomes more intuitive in the coordinate representation. The action of $\hat{V}_{\rm HF}$ on the HF single-particle basis read as: $\hat{V}_{\rm HF}\,\phi_n(\vec{x},\mu) =$

$$\sum_m\Big[\sum_{k\leq N}\sum_{\mu_1\mu_2}\iint\phi_k^*(\vec{x}_1,\mu_1)\phi_m^*(\vec{x}_2,\mu_2)V(\vec{x}_1,\vec{x}_2)\phi_k(\vec{x}_1,\mu_1)\phi_n(\vec{x}_2,\mu_2)d\vec{x}_1d\vec{x}_2\Big]\phi_m(\vec{x},\mu)$$

$$-\sum_m\Big[\sum_{k\leq N}\sum_{\mu_1\mu_2}\iint\phi_m^*(\vec{x}_1,\mu_1)\phi_k^*(\vec{x}_2,\mu_2)V(\vec{x}_1,\vec{x}_2)\phi_k(\vec{x}_1,\mu_1)\phi_n(\vec{x}_2,\mu_2)d\vec{x}_1d\vec{x}_2\Big]\phi_m(\vec{x},\mu)$$

Using $\sum_m \phi_m^*(\vec{x}_\bullet,\mu_\bullet)\phi_m(\vec{x},\mu) = \delta(\vec{x}_\bullet-\vec{x})\delta_{\mu_\bullet\mu}$ (with $\bullet$=1,2) we obtain:

$$\hat{V}_{\rm HF}\,\phi_n(\vec{x},\mu) = \underbrace{\left[\int\sum_{k\leq N}\sum_{\mu_1}|\phi_k(\vec{x}_1,\mu_1)|^2V(\vec{x}_1,\vec{x})d\vec{x}_1\right]}_{V_{\rm HF}(\vec{x})\ \textbf{local potential}}\phi_n(\vec{x},\mu) +$$

$$\overbrace{+\int\sum_{\mu_2}\underbrace{\left[\sum_{k\leq N}\phi_k^*(\vec{x}_2,\mu_2)V(\vec{x},\vec{x}_2)\phi_k(\vec{x},\mu)\right]}_{W_{\rm HF}(\vec{x},\mu,\vec{x}_2,\mu_2)\ \text{transformation kernel}}\phi_n(\vec{x}_2,\mu_2)d\vec{x}_2}^{\textbf{nonlocal potential}\ \int\sum_{\mu_2}W_{\rm HF}(\vec{x},\mu,\vec{x}_2,\mu_2)\phi_n(\vec{x}_2,\mu_2)d\vec{x}_2}$$

The **local potential** contains averaging of the value $V(\vec{x}_1,\vec{x})$ from all the remaining particles in occupied states weighted by the respective probability densities $|\phi_k(\vec{x}_1,\mu_1)|^2$. The **nonlocal potential** (also called **exchange term**) results from the antisymmetrization of 2-body wavefunctions.

▶ **Ground-state energy**

Estimate of the g.s. energy from the HF wavefunction:

$$E_0 \approx \langle \Psi_{\rm HF}|\hat{H}|\Psi_{\rm HF}\rangle = \cdots\cdots = \sum_{k\leq N}\varepsilon_{kk} + \tfrac{1}{2}\sum_{k\leq N}\sum_{l\leq N}(\nu_{klkl}-\nu_{lkkl})$$

Sum of single-particle energies of the occupied mean-field states:

$$\sum_{k\leq N}\varepsilon_k = \sum_{k\leq N}\langle\phi_k|(\hat{T}+\hat{V}_{\rm HF})|\phi_k\rangle = \sum_{k\leq N}\varepsilon_{kk} + \sum_{k\leq N}\sum_{l\leq N}(\nu_{klkl}-\nu_{lkkl})$$

Comparison of the above expressions:

$$\boxed{E_0 \approx \sum_{k\leq N}\Big[\varepsilon_k - \tfrac{1}{2}\underbrace{\sum_{l\leq N}(\nu_{klkl}-\nu_{lkkl})}_{\langle\phi_k|\hat{V}_{\rm HF}|\phi_k\rangle}\Big]}$$

The correction $\Delta\varepsilon_k = \frac{1}{2}\langle\phi_k|\hat{V}_{\rm HF}|\phi_k\rangle$ of energy ε_k, present in the last formula, compensates the double counting of particle interaction energies (e.g., the sum $\varepsilon_1+\varepsilon_2$ contains all interaction between particles $1\leftrightarrow k$ and $2\leftrightarrow k$, so that the $1\leftrightarrow 2$ interaction is counted twice)

◀ **Historical remark**

1927: D.R. Hartree introduces a self-consistent method to solve many-body Sch. eq.
1930: V. Fock and J.C. Slater modify the Hartree method to respect antisymmetry
1935: D.R. Hartree reformulates the method in a way suitable for computations

■ Bosonic condensates & Hartree-Bose method

The Hartree-Fock method has its bosonic counterpart, called after Hartree and Bose. It relies on the same principle, but is much simpler technically since bosons do not obey the Pauli exclusion law.

▶ **Bosonic condensate**

Bosonic Hamiltonian with one + two body terms:

$$\boxed{\hat{H} = \sum_{k,k'}\varepsilon_{kk'}\hat{b}_k^\dagger\hat{b}_{k'} + \tfrac{1}{2}\sum_{\substack{k,k'\\ l,l'}}\nu_{klk'l'}\hat{b}_k^\dagger\hat{b}_l^\dagger\hat{b}_{l'}\hat{b}_{k'}}$$

Ground state of the N-particle system searched in the form of the **condensate** type of wavefunction: $\boxed{|\Psi_{\rm HB}\rangle = \frac{1}{\sqrt{N!}}(\hat{B}^\dagger)^N|0\rangle}$

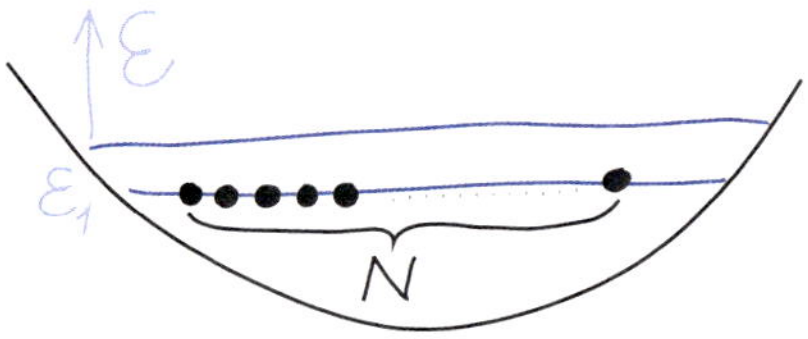

with $\boxed{\hat{B}^\dagger \equiv \sum_k \beta_k\hat{b}_k^\dagger}$ creating the boson into

a general single-particle state $|\psi_B\rangle = \sum_k \beta_k|\phi_k\rangle$

with unknown coefficients subject to normalization: $\sum_k |\beta_k|^2 = 1$

▶ Energy functional

To perform the variational procedure, we need to express the energy functional $\langle\Psi_{\rm HB}|\hat{H}|\Psi_{\rm HB}\rangle$ as a function of coefficients $\{\beta_k\}$. First we evaluate commutators:

$$\left.\begin{array}{l}\underbrace{[\hat{b}_k,\hat{B}^\dagger]}_{\hat{C}_1}=\beta_k\\ \underbrace{[\hat{b}_k,(\hat{B}^\dagger)^N]}_{\hat{C}_N}=\underbrace{[\hat{b}_k,\hat{B}^\dagger]}_{\beta_k}(\hat{B}^\dagger)^{N-1}+\hat{B}^\dagger\underbrace{[\hat{b}_k,(\hat{B}^\dagger)^{N-1}]}_{\hat{C}_{N-1}}\end{array}\right\}\Rightarrow\left\{\begin{array}{l}[\hat{b}_k,(\hat{B}^\dagger)^N]=N\beta_k(\hat{B}^\dagger)^{N-1}\\ [(\hat{B})^N,\hat{b}_k^\dagger]=N\beta_k^*(\hat{B})^{N-1}\end{array}\right.$$

From these relations we calculate the following averages:

$$\langle\Psi_{\rm HB}|\hat{b}_k^\dagger\hat{b}_{k'}|\Psi_{\rm HB}\rangle=\tfrac{1}{N!}\langle 0|(\hat{B})^N\hat{b}_k^\dagger\hat{b}_{k'}(\hat{B}^\dagger)^N|0\rangle=\beta_k^*\beta_{k'}\tfrac{N^2}{N!}\langle 0|(\hat{B})^{N-1}(\hat{B}^\dagger)^{N-1}|0\rangle=N\beta_k^*\beta_{k'}$$

$$\begin{aligned}\langle\Psi_{\rm HB}|\hat{b}_k^\dagger\hat{b}_l^\dagger\hat{b}_{l'}\hat{b}_{k'}|\Psi_{\rm HB}\rangle&=\tfrac{1}{N!}\langle 0|(\hat{B})^N\hat{b}_k^\dagger\hat{b}_l^\dagger\hat{b}_{l'}\hat{b}_{k'}(\hat{B}^\dagger)^N|0\rangle\\&=\beta_k^*\beta_{k'}\tfrac{N^2}{N!}\langle 0|(\hat{B})^{N-1}\hat{b}_l^\dagger\hat{b}_{l'}(\hat{B}^\dagger)^{N-1}|0\rangle=N(N-1)\beta_k^*\beta_l^*\beta_{k'}\beta_{l'}\end{aligned}$$

The energy average (energy functional) in the space of condensate states:

$$\boxed{\langle\Psi_{\rm HB}|\hat{H}|\Psi_{\rm HB}\rangle=N\sum_{k,k'}\varepsilon_{kk'}\beta_k^*\beta_{k'}+\frac{N(N-1)}{2}\sum_{\substack{k,k'\\l,l'}}\nu_{klk'l'}\beta_k^*\beta_l^*\beta_{k'}\beta_{l'}\equiv\mathcal{E}(\{\beta_k\})}$$

To find parameters $\{\beta_k\}$ of the condensate state, the function $\mathcal{E}(\{\beta_k\})$ must be **minimized**, respecting the normalization condition $\sum_k|\beta_k|^2=1$.

Alternatively, one can skip the normalization constraint and minimize the expression:
$$\tilde{\mathcal{E}}(\{\beta\})=\frac{\langle\Psi_{\rm HB}|\hat{H}|\Psi_{\rm HB}\rangle}{\langle\Psi_{\rm HB}|\Psi_{\rm HB}\rangle}$$

◀ Historical remark

1924-5: A. Einstein & S.N. Bose predict that systems of bosons at $T\to 0$ form a condensate state with unusual properties (the first laboratory preparation in 1995)
1938: F. London relates boson condensation to **superfluidity** & superconductivity

■ Pairing & BCS method

The Hartree-Fock method does not work well for the fermionic systems whose valence shell (or valence band) of single-particle states is filled up approximately to the middle. Indeed, the existence of a number of partly occupied valence orbits with nearly degenerate spectrum makes the HF method unstable (it has many almost equivalent solutions). In this situation, an attractive short-range type of interaction produces a new effect beyond the mean field—pairing of particles in conjugate states related by the time reversal. It turns out that at low temperatures, the systems with pairing exhibit **superconductivity**, a phenomenon partly analogous to the superfluidity of some Bose systems. The basic many-body theory which takes the fermionic pairing into account is abbreviated after its inventors Bardeen, Cooper, and Schrieffer.

▶ Pairing interaction

Consider an approx. contact interaction given by: $\boxed{V(\vec{x}_1-\vec{x}_2)\approx -V_0\,\delta(\vec{x}_1-\vec{x}_2)}$

Matrix element $\langle\phi_i\phi_j|\hat{V}|\phi_{i'}\phi_{j'}\rangle\approx$

$$-V_0\sum_{\mu_1,\mu_2}\iint\phi_i^*(\vec{x}_1,\mu_1)\phi_j^*(\vec{x}_2,\mu_2)\delta(\vec{x}_1-\vec{x}_2)\phi_{i'}(\vec{x}_1,\mu_1)\phi_{j'}(\vec{x}_2,\mu_2)\,d\vec{x}_1d\vec{x}_2$$
$$=-V_0\int\Big[\sum_{\mu_1}\phi_i^*(\vec{x},\mu_1)\phi_{i'}(\vec{x},\mu_1)\Big]\Big[\sum_{\mu_2}\phi_j^*(\vec{x},\mu_2)\phi_{j'}(\vec{x},\mu_2)\Big]d\vec{x}$$

For $\left\{\begin{matrix}\phi_i(\vec{x},\mu)=\phi_j^*(\vec{x},-\mu)\equiv\hat{\mathcal{T}}\phi_j(\vec{x},\mu)\\ \phi_{i'}(\vec{x},\mu)=\phi_{j'}^*(\vec{x},-\mu)\equiv\hat{\mathcal{T}}\phi_{j'}(\vec{x},\mu)\end{matrix}\right\}$ we get $\left\{\begin{matrix}\langle\phi_i\phi_j|\hat{V}|\phi_{i'}\phi_{j'}\rangle\approx\\ -V_0\int\Big|\sum_{\mu}\phi_i^*(\vec{x},\mu)\phi_{i'}(\vec{x},\mu)\Big|^2d\vec{x}\end{matrix}\right.$

(a particularly strong interaction element)

We may approximate this situation by assuming that $\hat{V}$ acts only between couples of states $\boxed{\underbrace{|\phi_k\rangle}_{\hat{a}_k^\dagger|0\rangle}\leftrightarrow\underbrace{|\phi_{\bar{k}}\rangle}_{\hat{a}_{\bar{k}}^\dagger|0\rangle}\equiv\hat{\mathcal{T}}|\phi_k\rangle}$ related by the **time reversal** transformation $\hat{\mathcal{T}}$

For instance: $\left\{\begin{matrix}|+\vec{p},\uparrow\rangle & \leftrightarrow & |-\vec{p},\downarrow\rangle\\ |n,l,j,+m_j\rangle & \leftrightarrow & |n,l,j,-m_j\rangle\end{matrix}\right.$ electron states in metals / nucleon states in nuclei

▶ Simplified Hamiltonian

The above approximation is represented by so-called **monopole pairing** interaction:

$$\boxed{\hat{V}_{\rm pair}\approx -G\sum_{k,l}{}'\,\hat{a}_{\bar{k}}^\dagger\hat{a}_k^\dagger\hat{a}_l\hat{a}_{\bar{l}}}$$

$G\equiv$ pairing interaction strength

$\sum_{k,l}'\equiv$ sum over the states close to the **Fermi energy** $\varepsilon_{\rm F}$: $\boxed{|\varepsilon_k-\varepsilon_{\rm F}|<S}$

(with $\varepsilon_{\rm F}$ taken now as the *energy of the highest occupied orbital* in $|\Psi_{\rm HB}\rangle$)

This interaction can be expressed with the aid of the following bifermion operators: $\boxed{\hat{V}_{\rm pair}\approx -Gn\,\hat{P}^\dagger\hat{P}}$

$\boxed{\hat{P}^\dagger\equiv\frac{1}{\sqrt{n}}\sum_k{}'\,\hat{a}_{\bar{k}}^\dagger\hat{a}_k^\dagger}$ $\boxed{\hat{P}\equiv\frac{1}{\sqrt{n}}\sum_l{}'\,\hat{a}_l\hat{a}_{\bar{l}}}$ where $n\equiv$ number of levels ε_k in the $|\varepsilon_k-\varepsilon_{\rm F}|<S$ interval around $\varepsilon_{\rm F}$

If the $k,\bar{k}$ states correspond to $|n,l,j,\pm m_j\rangle$, the $\hat{P}^\dagger$ operator creates a pair with zero total angular momentum (hence the term "monopole")

Boson-like commutator: $\big[\hat{P},\hat{P}^\dagger\big]=1-\frac{1}{n}\sum_k{}'\underbrace{(\hat{a}_k^\dagger\hat{a}_k+\hat{a}_{\bar{k}}^\dagger\hat{a}_{\bar{k}})}_{\hat{n}_k\in[0,2]}\quad\in[-1,+1]$

The full Hamiltonian:

$$\boxed{\hat{H}=\underbrace{\sum_k\varepsilon_k(\hat{a}_k^\dagger\hat{a}_k+\hat{a}_{\bar{k}}^\dagger\hat{a}_{\bar{k}})}_{\hat{T}+\hat{V}_{\rm HF}}\underbrace{-G\sum_{k,l}{}'\,\hat{a}_{\bar{k}}^\dagger\hat{a}_k^\dagger\hat{a}_l\hat{a}_{\bar{l}}}_{\hat{V}_{\rm pair}}=\sum_k\varepsilon_k\hat{n}_k-Gn\,\hat{P}^\dagger\hat{P}}$$

▶ The BCS approach

Splitting of the full Hamiltonian into $\left\{\begin{matrix}\hat{H}_0=\hat{T}+\hat{V}_{\rm HF}+\hat{V}'_{\rm pair} & \text{(the main part)}\\ \hat{V}''_{\rm pair} & \text{(the rest)}\end{matrix}\right.$

$$\hat{H} = \overbrace{E_0 + \underbrace{\sum_k \varepsilon_k(\hat{a}_k^\dagger \hat{a}_k + \hat{a}_{\bar{k}}^\dagger \hat{a}_{\bar{k}})}_{\hat{T}+\hat{V}_{\rm HF}} \underbrace{-\Delta \sum_k{}' (\hat{a}_{\bar{k}}^\dagger \hat{a}_k^\dagger + \hat{a}_k \hat{a}_{\bar{k}})}_{\hat{V}'_{\rm pair}}}^{\hat{H}_0}$$

$$\overbrace{+\Delta \sum_k{}' (\hat{a}_{\bar{k}}^\dagger \hat{a}_k^\dagger + \hat{a}_k \hat{a}_{\bar{k}}) - G \sum_{k,l}{}' \hat{a}_{\bar{k}}^\dagger \hat{a}_k^\dagger \hat{a}_l \hat{a}_{\bar{l}} - E_0}^{\hat{V}''_{\rm pair}}$$

$[\hat{H},\hat{N}]=0$
$[\hat{H}_0,\hat{N}]\neq 0 \neq [\hat{V}''_{\rm pair},\hat{N}]$

Here, Δ is a so far undetermined parameter called **pairing gap** (see below). It is believed that $\hat{V}'_{\rm pair}$ included in $\hat{H}_0$ represents "a larger part" of the full pairing interaction $\hat{V}_{\rm pair}$, while the rest $\hat{V}''_{\rm pair}$ is "small".

The subsequent procedure consists of **2 steps**:
(1) The ground state of $\hat{H}_0$ found analytically $\Rightarrow$ wavefunction $|\Psi_{\rm BCS}(\Delta)\rangle$
(2) $|\Psi_{\rm BCS}(\Delta)\rangle$ is used as the ansatz wavefunction for the variational procedure using the full Hamiltonian $\Rightarrow$ minimization of $\mathcal{E}(\Delta) = \langle\Psi_{\rm BCS}(\Delta)|\hat{H}|\Psi_{\rm BCS}(\Delta)\rangle$ determines the value of parameter Δ.

The **idea behind**:

$$\hat{P}^\dagger\hat{P} = \overbrace{\left[\hat{P}^\dagger - \langle\hat{P}^\dagger\rangle_\Psi\right]\left[\hat{P} - \langle\hat{P}\rangle_\Psi\right]}^{\text{small contribution} \to 0} + \overbrace{\langle\hat{P}\rangle_\Psi\hat{P}^\dagger + \langle\hat{P}^\dagger\rangle_\Psi\hat{P}}^{\text{the main part} \to \hat{V}'_{\rm pair}} - \overbrace{\langle\hat{P}^\dagger\rangle_\Psi\langle\hat{P}\rangle_\Psi}^{\text{const.} \to E_0}$$

The gap can be identified with: $G\sqrt{n}\langle\hat{P}^\dagger\rangle_\Psi = G\sqrt{n}\langle\hat{P}\rangle_\Psi \approx \Delta$

▶ **Bogoljubov transformation** (a toy form)

Spin states $\begin{cases} |\uparrow\rangle \equiv \hat{a}_\uparrow^\dagger|0\rangle \\ |\downarrow\rangle \equiv \hat{a}_\downarrow^\dagger|0\rangle \end{cases}$

quadraticHamiltonian

$$\hat{h}_0 = \varepsilon_0 + \varepsilon\left(\hat{a}_\uparrow^\dagger\hat{a}_\uparrow + \hat{a}_\downarrow^\dagger\hat{a}_\downarrow\right) + \delta\hat{a}_\downarrow\hat{a}_\uparrow + \delta\hat{a}_\uparrow^\dagger\hat{a}_\downarrow^\dagger$$

Eigenproblem of $\hat{h}_0$ in the 3D Hilbert space (spanned by states $|N_a\rangle$ with particle numbers N_a=0,1,2) can be solved analytically via Bogoljubov transform.:

$$\left.\begin{matrix} \hat{a}_\uparrow, \hat{a}_\uparrow^\dagger \\ \hat{a}_\downarrow, \hat{a}_\downarrow^\dagger \end{matrix}\right\} \mapsto \begin{cases} \hat{\alpha}_\uparrow = u\hat{a}_\uparrow + v\hat{a}_\downarrow^\dagger & \hat{\alpha}_\uparrow^\dagger = u\hat{a}_\uparrow^\dagger + v\hat{a}_\downarrow & u,v \in \mathbb{R} \\ \hat{\alpha}_\downarrow = u\hat{a}_\downarrow - v\hat{a}_\uparrow^\dagger & \hat{\alpha}_\downarrow^\dagger = u\hat{a}_\downarrow^\dagger - v\hat{a}_\uparrow & u^2+v^2=1 \end{cases}$$

particles **quasiparticles**

Quasiparticles are fermions (the transformation is "canonical"):

$$\{\hat{\alpha}_\uparrow, \hat{\alpha}_\uparrow\} = \{\hat{\alpha}_\uparrow^\dagger, \hat{\alpha}_\uparrow^\dagger\} = \{\hat{\alpha}_\downarrow, \hat{\alpha}_\downarrow\} = \{\hat{\alpha}_\downarrow^\dagger, \hat{\alpha}_\downarrow^\dagger\} = \{\hat{\alpha}_\uparrow, \hat{\alpha}_\downarrow\} = \{\hat{\alpha}_\uparrow^\dagger, \hat{\alpha}_\downarrow^\dagger\} = 0$$
$$\{\hat{\alpha}_\uparrow, \hat{\alpha}_\downarrow^\dagger\} = \{\hat{\alpha}_\downarrow, \hat{\alpha}_\uparrow^\dagger\} = 0 \qquad \{\hat{\alpha}_\uparrow, \hat{\alpha}_\uparrow^\dagger\} = \{\hat{\alpha}_\downarrow, \hat{\alpha}_\downarrow^\dagger\} = u^2+v^2 = 1$$

Coefficients u, v are determined by the required form of Hamiltonian after the transformation, which is:

$$\hat{h}_0 \mapsto \hat{h}'_0 = e_0 + e\,\underbrace{(\hat{\alpha}_\uparrow^\dagger\hat{\alpha}_\uparrow + \hat{\alpha}_\downarrow^\dagger\hat{\alpha}_\downarrow)}_{\hat{N}_\alpha}$$

This Hamiltonian is solvable: eigensolutions identified with the states having fixed numbers of quasiparticles: $|N_\alpha\rangle \equiv |0_\alpha\rangle, |1_\alpha\rangle, |2_\alpha\rangle$
The ground state is the quasiparticle vacuum: $|\psi_0\rangle \equiv |0_\alpha\rangle$

Amplitudes u, v & constants e, e_0 (together 4 real variables) obtained from the condition $\hat{h}'_0 = \hat{h}_0$, yielding together with the normalization constraint 4 real equations:

$$\hat{h}'_0 = \overbrace{e_0 + 2ev^2}^{=\varepsilon_0} + \overbrace{e\left(u^2 - v^2\right)}^{=\varepsilon} (\hat{a}^\dagger_\uparrow \hat{a}_\uparrow + \hat{a}^\dagger_\downarrow \hat{a}_\downarrow) + \overbrace{euv}^{=\delta}\, \hat{a}_\downarrow \hat{a}_\uparrow + \overbrace{euv}^{=\delta}\, \hat{a}^\dagger_\uparrow \hat{a}^\dagger_\downarrow = \hat{h}_0$$

▶ Solving the main part of the pairing Hamiltonian

The part $\hat{H}_0$ of the total pairing Hamiltonian is quadratic ⇒ solvable

Bogoljubov transformation (the full form):

$$\hat{\alpha}_k = u_k \hat{a}_k + v_k \hat{a}^\dagger_{\bar{k}} \quad \hat{\alpha}^\dagger_k = u_k \hat{a}^\dagger_k + v_k \hat{a}_{\bar{k}} \qquad \hat{a}_k = u_k \hat{\alpha}_k - v_k \hat{\alpha}^\dagger_{\bar{k}} \quad \hat{a}^\dagger_k = u_k \hat{\alpha}^\dagger_k - v_k \hat{\alpha}_{\bar{k}}$$

$$\hat{\alpha}_{\bar{k}} = u_k \hat{a}_{\bar{k}} - v_k \hat{a}^\dagger_k \quad \hat{\alpha}^\dagger_{\bar{k}} = u_k \hat{a}^\dagger_{\bar{k}} - v_k \hat{a}_k \qquad \hat{a}_{\bar{k}} = u_k \hat{\alpha}_{\bar{k}} + v_k \hat{\alpha}^\dagger_k \quad \hat{a}^\dagger_{\bar{k}} = u_k \hat{\alpha}^\dagger_{\bar{k}} + v_k \hat{\alpha}_k$$

$$u_k, v_k \in \mathbb{R} \qquad u_k^2 + v_k^2 = 1$$

$$\{\hat{\alpha}_k, \hat{\alpha}_l\} = 0 = \{\hat{\alpha}^\dagger_k, \hat{\alpha}^\dagger_l\} \quad \{\hat{\alpha}_k, \hat{\alpha}^\dagger_l\} = \delta_{kl}$$
$$\{\hat{\alpha}_{\bar{k}}, \hat{\alpha}_{\bar{l}}\} = 0 = \{\hat{\alpha}^\dagger_{\bar{k}}, \hat{\alpha}^\dagger_{\bar{l}}\} \quad \{\hat{\alpha}_{\bar{k}}, \hat{\alpha}^\dagger_{\bar{l}}\} = \delta_{kl}$$
$$\{\hat{\alpha}_k, \hat{\alpha}_{\bar{l}}\} = 0 = \{\hat{\alpha}^\dagger_k, \hat{\alpha}^\dagger_{\bar{l}}\} \quad \{\hat{\alpha}_k, \hat{\alpha}^\dagger_{\bar{l}}\} = 0 = \{\hat{\alpha}_{\bar{k}}, \hat{\alpha}^\dagger_l\}$$

Remarks:
(a) We assume $(u_k, v_k) = (1, 0)$ for levels "far from" the Fermi level: $|\varepsilon_k - \varepsilon_{\rm F}| > S$
(b) Instead of $\hat{H}_0$ we consider $\hat{\mathbf{H}}_0 = \hat{H}_0 - \mu\hat{N}$, where μ will become a Lagrange multiplier for fixing the average particle number (⇒ chemical potential)
The transformed $\hat{\mathbf{H}}_0$ reads as:

$$\hat{\mathbf{H}}'_0 = \underbrace{2\sum_k \left[(\varepsilon_k - \mu)v_k^2 - \Delta u_k v_k\right]}_{E_0} + \Bigg(\sum_k \underbrace{\left[2(\varepsilon_k - \mu)u_k v_k - \Delta(u_k^2 - v_k^2)\right]}_{0} \hat{\alpha}^\dagger_k \hat{\alpha}_k + \text{H.c.}\Bigg) + \sum_k \overbrace{\left[(\varepsilon_k - \mu)(u_k^2 - v_k^2) + 2\Delta u_k v_k\right]}^{e_k} \overbrace{\left(\hat{\alpha}^\dagger_k \hat{\alpha}_k + \hat{\alpha}^\dagger_{\bar{k}} \hat{\alpha}_{\bar{k}}\right)}^{\hat{\mathfrak{n}}_k}$$

Solution of the diagonalization condition:

$$2(\varepsilon_k - \mu)u_k v_k - \Delta(u_k^2 - v_k^2) = 0 \quad \Rightarrow \quad 2(\varepsilon_k - \mu)u_k\sqrt{1 - u_k^2} = \Delta(2u_k^2 - 1) \quad \Rightarrow$$

$$4\left[\Delta^2 + (\varepsilon_k - \mu)^2\right]u_k^4 - 4\left[\Delta^2 + (\varepsilon_k - \mu)^2\right]u_k^2 + \Delta^2 = 0$$

$$\Rightarrow \quad \boxed{e_k = \sqrt{\Delta^2 + (\varepsilon_k - \mu)^2}} \qquad \Rightarrow \qquad \boxed{u_k^2 = \frac{1}{2}\left[1 + \frac{\varepsilon_k - \mu}{\sqrt{\Delta^2 + (\varepsilon_k - \mu)^2}}\right] \quad v_k^2 = \frac{1}{2}\left[1 - \frac{\varepsilon_k - \mu}{\sqrt{\Delta^2 + (\varepsilon_k - \mu)^2}}\right]}$$

▶ Ground-state wavefunction

The ground state of $\hat{\mathbf{H}}'_0$ ≡ **vacuum of quasiparticles** (⇒ $\hat{\mathfrak{n}}_k = 0$). Written in terms of creation/annihilation operators of the original particles and their vacuum, this state has the following form:

$$\boxed{|\Psi_{\rm BCS}\rangle = \prod_k \left(u_k + v_k \hat{a}^\dagger_{\bar{k}} \hat{a}^\dagger_k\right)|0\rangle}$$

Proof:

$$\hat{\alpha}_l|\Psi_{\rm BCS}\rangle = \overbrace{\left(u_l\hat{a}_l + v_l\hat{a}^\dagger_{\bar{l}}\right)}^{\hat{\alpha}_l}\prod_k \overbrace{\left(u_k + v_k\hat{a}^\dagger_{\bar{k}}\hat{a}^\dagger_k\right)}^{\hat{\beta}_k}|0\rangle = \left\{\left[\hat{\alpha}_l, \prod_k \hat{\beta}_k\right] + \left(\prod_k \hat{\beta}_k\right)\hat{\alpha}_l\right\}|0\rangle =$$

$$\left\{u_l v_l \underbrace{\left[\hat{a}_l, \hat{a}^\dagger_{\bar{l}}\hat{a}^\dagger_l\right]}_{-\hat{a}^\dagger_{\bar{l}}}\prod_{k\neq l}\left(u_k + v_k\hat{a}^\dagger_{\bar{k}}\hat{a}^\dagger_k\right) + \prod_{k\neq l}\left(u_k + v_k\hat{a}^\dagger_{\bar{k}}\hat{a}^\dagger_k\right)\underbrace{\left(u_l + v_l\hat{a}^\dagger_{\bar{l}}\hat{a}^\dagger_l\right)\left(u_l\hat{a}_l + v_l\hat{a}^\dagger_{\bar{l}}\right)}_{+u_l v_l \hat{a}^\dagger_{\bar{l}}|0\rangle}\right\}|0\rangle$$

$$\Rightarrow \boxed{\hat{\alpha}_l|\Psi_{\rm BCS}\rangle = 0} \quad \text{similarly:} \quad \boxed{\hat{\alpha}_{\bar{l}}|\Psi_{\rm BCS}\rangle = 0}$$

The solution $|\Psi_{\rm BCS}\rangle$ approximates the **superconducting state** at $T=0$

▶ **Interpretation**

(a) $|\Psi_{\rm BCS}\rangle$ is a state with **undetermined particle number**

(b) The **average** $\boxed{\langle N\rangle_{\rm BCS} = \sum_k \langle\Psi_{\rm BCS}|\underbrace{\left(\hat{a}^\dagger_k\hat{a}_k + \hat{a}^\dagger_{\bar{k}}\hat{a}_{\bar{k}}\right)}_{\hat{n}_k}|\Psi_{\rm BCS}\rangle \overset{!}{=} N}$ fixed by μ

(c) The **dispersion** $\langle\langle N^2\rangle\rangle_{\rm BCS} = \langle N^2\rangle_{\rm BCS} - \langle N\rangle^2_{\rm BCS}$ is beyond the control (for small systems like nuclei this is a drawback)

(d) u_k and v_k represent probability amplitudes for the pair of states $|\phi_k\rangle, |\phi_{\bar{k}}\rangle$ being empty and occupied:

$$\boxed{p_k^{\rm empty} = |u_k|^2 \quad \text{and} \quad p_k^{\rm occup} = |v_k|^2}$$

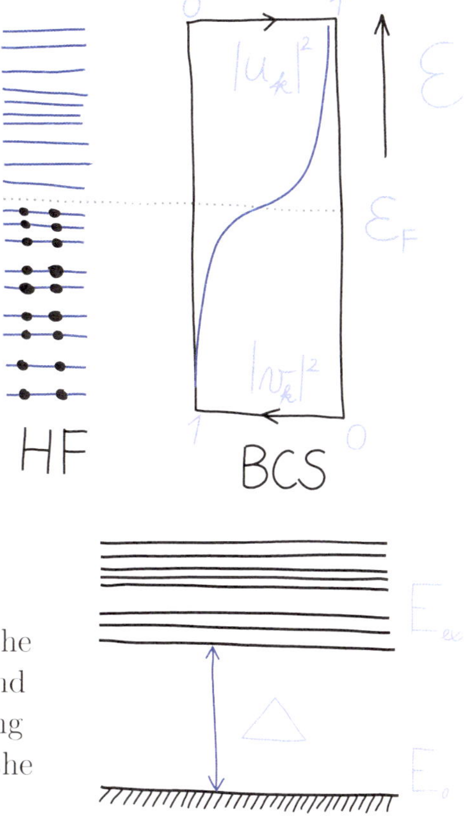

(e) The occupation probability $|v_k|^2$ as a function of ε_k is **smeared** around the value μ. The smearing width $\sim\Delta$. For $\Delta=0$ we get:

$$|v_k|^2 = 1 - |u_k|^2 = \begin{cases} 1 \text{ for } \varepsilon_k \leq \mu \\ 0 \text{ for } \varepsilon_k > \mu \end{cases}$$

$$\Rightarrow \boxed{\mu \equiv \varepsilon_{\rm F}}$$

(f) Excited states (with $\hat{\mathfrak{n}}_k \geq 1$) have energies $\boxed{E_{\rm exc} \geq {\rm Min}\{e_k\} \geq \Delta}$

⇒ **energy gap** above the ground state in the spectrum is a typical signature of pairing and one of the origins of the superconducting behavior (the friction is suppressed due to the difficulty to excite the system)

▶ **Determination of the gap**

(a) **Variational approach**: $\mathcal{E}'(\Delta) = \langle\Psi_{\rm BCS}(\Delta)|(\hat{H} - \mu\hat{N})|\Psi_{\rm BCS}(\Delta)\rangle =$

$$\underbrace{\langle\Psi_{\rm BCS}|\hat{\mathbf{H}}_0'|\Psi_{\rm BCS}\rangle+\Delta{\sum_k}'\langle\Psi_{\rm BCS}|(\hat{a}_k^\dagger\hat{a}_{\bar{k}}^\dagger+\hat{a}_k\hat{a}_{\bar{k}})|\Psi_{\rm BCS}\rangle}_{2\sum_k'(\varepsilon_k-\mu)v_k(\Delta)^2}-G\underbrace{\langle\Psi_{\rm BCS}|{\sum_{k,l}}'\hat{a}_k^\dagger\hat{a}_{\bar{k}}^\dagger\hat{a}_l\hat{a}_{\bar{l}}|\Psi_{\rm BCS}\rangle}_{\left[\sum_k'u_k(\Delta)v_k(\Delta)\right]^2}$$

Minimization of $\mathcal{E}'(\Delta)$: $\quad \frac{\partial}{\partial\Delta}\left\{2{\sum_k}'(\varepsilon_k-\mu)v_k(\Delta)^2-G\left[{\sum_k}'u_k(\Delta)v_k(\Delta)\right]^2\right\}=0$

(b) Derivation from **expectation values of pair operators** $\hat{P}$ or $\hat{P}^\dagger$:

$\Delta=G\sqrt{n}\langle\Psi_{\rm BCS}(\Delta)|\hat{P}|\Psi_{\rm BCS}(\Delta)\rangle=$

$$=G\langle 0|\prod_{k'}(u_{k'}+v_{k'}\hat{a}_{k'}\hat{a}_{\bar{k}'})\underbrace{\left({\sum_l}'\hat{a}_l\hat{a}_{\bar{l}}\right)\prod_k(u_k+v_k\hat{a}_k^\dagger\hat{a}_{\bar{k}}^\dagger)}_{\sum_l'v_l\underbrace{(1-\hat{n}_l)}_{1}\prod_{k\neq l}(u_k+v_k\hat{a}_k^\dagger\hat{a}_{\bar{k}}^\dagger)}|0\rangle=G\underbrace{{\sum_l}'u_l(\Delta)v_l(\Delta)}_{\frac{G}{2}\sum_l'\sqrt{1-\frac{(\varepsilon_l-\mu)^2}{\Delta^2+(\varepsilon_l-\mu)^2}}}$$

Both derivations equivalent $\Rightarrow$ **gap equation**: $\Delta\left(1-\frac{G}{2}{\sum_k}'\frac{1}{\sqrt{\Delta^2+(\varepsilon_k-\mu)^2}}\right)=0$

$\Rightarrow$ $\exists$ a **critical value** $G_{\rm c}$ of pairing strength: $\boxed{\frac{2}{G_{\rm c}}={\sum_k}'\frac{1}{|\varepsilon_k-\mu|}}$

$\Rightarrow$ Solutions:

(1) $G\leq G_{\rm c}$:	$\Delta=0$	(normal solution)
(2) $G>G_{\rm c}$:	$\frac{2}{G}={\sum_k}'\frac{1}{\sqrt{\Delta^2+(\varepsilon_k-\mu)^2}}\Rightarrow\Delta\neq 0$	(superconducting solution)

◀ **Historical remark**

1947: N. Bogolyubov introduces the transformation to quasiparticles

1957: J. Bardeen, L.N. Cooper & J.R. Schrieffer formulate the BCS method

■ Quantum gases

At last we turn to systems of non-interacting indistinguishable particles at a nonzero temperature. Generalizing the concept of a canonical ensemble (see Sec. 1.7), we will point out some crucial differences in thermodynamic properties of bosons and fermions. The respective grand-canonical partition functions will be evaluated and shown to carry universal (not only thermodynamical) information on many-body systems.

▶ Grand-canonical ensemble

Consider a gas of indistinguishable particles at temperature $T=(k\beta)^{-1}$ (with $k\equiv$ Boltzmann const., $\beta\equiv$ inverse temperature) in a finite volume V. Assuming an **exchange** of both **energy** & **particles** between the system and a bath, we fix *neither* the total energy E, *nor* the actual number of particles N in the system, but only the averages $\langle E\rangle$ and $\langle N\rangle$. The most likely choice of the system's

density operator follows from the **maximum entropy** principle. The resulting grand-canonical ensemble generalizes the canonical ensemble (Sec. 1.7) by taking into account also the effects of particle exchange.
Hamiltonian $\hat{H}$ commutes with the particle-number operator $\hat{N}$. For each particle number N, the system has a discrete energy spectrum $\{E_{Ni}\}$. The equilibrium density operator $\hat{\rho}$ is diagonal in the common eigenbasis of $\hat{H}, \hat{N} \Rightarrow$ diagonal matrix elements (probabilities) $\boxed{\rho(N, E_{Ni}) \equiv \rho_{Ni}}$

Constraints induced by the normalization and fixed averages:

$$\sum_{N=0}^{\infty}\sum_{i=1}^{\infty}\rho_{Ni}=1 \qquad \sum_{N=0}^{\infty}\sum_{i=1}^{\infty}\rho_{Ni}N=\langle N\rangle \qquad \sum_{N=0}^{\infty}\sum_{i=1}^{\infty}\rho_{Ni}E_{Ni}=\langle E\rangle$$

Entropy $S=-k\sum_{N,i}\rho_{Ni}\ln\rho_{Ni}$ to be maximized with the above constraints:

$$f=-\sum_{N,i}\rho_{Ni}\ln\rho_{Ni}+(\alpha+1)\sum_{N,i}\rho_{Ni}-\beta\sum_{N,i}\rho_{Ni}E_{Ni}+\gamma\sum_{N,i}\rho_{Ni}N$$

$$\frac{\partial f}{\partial\rho_{Ni}}=-\ln\rho_{Ni}-1+(\alpha+1)-\beta E_{Ni}+\gamma N=0 \quad\Rightarrow\quad \ln\rho_{Ni}=\alpha-\beta E_{Ni}+\gamma N$$

This leads to the **grand-canonical** form of the density operator, which describes an equilibrium state of a many-particle system exchanging energy & particles with the environment:

$$\boxed{\rho_{Ni}=\frac{1}{Z(\beta,\mu)}\,e^{-\beta\left(E_{Ni}-\mu N\right)}} \quad \text{where} \quad \begin{cases} \mu=\frac{\gamma}{\beta}\equiv \textbf{chemical potential} \\ \boxed{Z(\beta,\mu)=\sum_{N,i}e^{-\beta(E_{Ni}-\mu N)} \quad \textbf{partition function}} \end{cases}$$

▶ Quantities derived from the partition function

(a) Energy & particle number averages:

$$\langle E\rangle_{\beta,\mu}=\sum_{N,i}\rho_{N,i}E_{Ni}=\frac{1}{Z(\beta,\mu)}\sum_{N,i}E_{Ni}\,e^{-\beta(E_{Ni}-\mu N)}=-\frac{1}{Z(\beta,\mu)}\frac{\partial Z(\beta,\mu)}{\partial\beta}=-\frac{\partial}{\partial\beta}\ln Z(\beta,\mu)$$

$$\langle N\rangle_{\beta,\mu}=\sum_{N,i}\rho_{N,i}N=\frac{1}{Z(\beta,\mu)}\sum_{N,i}N\,e^{-\beta(E_{Ni}-\mu N)}=\frac{1}{\beta Z(\beta,\mu)}\frac{\partial Z(\beta,\mu)}{\partial\mu}=+\frac{1}{\beta}\frac{\partial}{\partial\mu}\ln Z(\beta,\mu)$$

(b) Energy & particle number dispersions (cf. Sec. 1.7):

$$\langle\langle E^2\rangle\rangle_{\beta,\mu}=+\frac{\partial^2}{\partial\beta^2}\ln Z(\beta,\mu) \qquad \langle\langle N^2\rangle\rangle_{\beta,\mu}=+\frac{1}{\beta^2}\frac{\partial^2}{\partial\mu^2}\ln Z(\beta,\mu)$$

(c) Density of states for a fixed particle number: $\boxed{\varrho(N,E)=\sum_i\delta(E-E_{Ni})}$

Density with a *continuous* variable $N\equiv\bar{N}$ is defined by:

$$\boxed{\varrho(\bar{N},E)=\sum_N\sum_i\delta(\bar{N}-N)\delta(E-E_{Ni})}$$

$$\Rightarrow \int_{N-\epsilon}^{N+\epsilon}\varrho(\bar{N},E)d\bar{N}=\varrho(N,E)$$

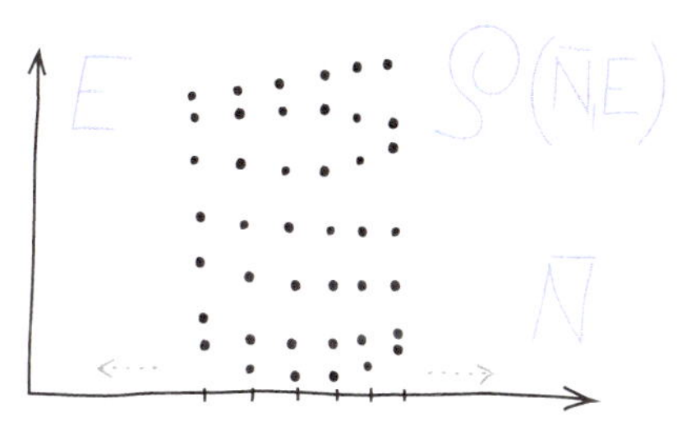

$$Z(\beta,\mu)=\sum_N\sum_i e^{-\beta(E_{Ni}-\mu N)} \quad = \quad \iint \varrho(\bar N,E)\,e^{-\beta(E-\mu\bar N)}d\bar N\,dE$$

partition function — 2D Laplace transform. of state density

$$\Rightarrow \quad \boxed{\begin{aligned}\varrho(\bar N,E) &= \left(\tfrac{1}{2\pi i}\right)^2\iint_{-i\infty}^{+i\infty} Z(\beta,\mu)\,e^{+\beta(E-\mu\bar N)}\beta\,d\mu\,d\beta\\ &= \left(\tfrac{1}{2\pi i}\right)^2\iint_{-i\infty}^{+i\infty} e^{\ln Z(\beta,\mu)+\beta(E-\mu\bar N)}\beta\,d\mu\,d\beta\end{aligned}}$$

state density — inverse 2D Laplace transform. of partition function

Evaluating the grand-canonical partition function of the many-particle system, one can determine the density of energy eigenstates for each particle number

▶ **Partition function of the Bose gas**

Bose gas ≡ ensemble of **non-interacting bosons**

⇒ total energy = sum of single-particle energies: $\boxed{E_{Ni}=\sum_{k=1}^{\infty}n_{ik}\varepsilon_k}$

total number of particles: $\boxed{N=\sum_{k=1}^{\infty}n_{ik}}$ $\boxed{n_{ik}=0,1,2,3,\ldots}$ occup. numbers

⇒ partition function:

$$Z(\beta,\mu)=\sum_N\sum_i e^{-\beta(E_{Ni}-\mu N)} \overset{*}{=} \sum_{\{n_{ik}\}} e^{-\beta\left(\sum_k n_{ik}\varepsilon_k-\mu\sum_k n_{ik}\right)}=\prod_k\underbrace{\sum_{n_{ik}=0}^{\infty}e^{-\beta\left(n_{ik}\varepsilon_k-\mu n_{ik}\right)}}_{\frac{1}{1-e^{-\beta(\varepsilon_k-\mu)}}}$$

* the sum $\sum_{\{n_{ik}\}}$ goes over all sets of occup. numbers

$$\boxed{\ln Z(\beta,\mu)=-\sum_k\ln\left[1-e^{-\beta(\varepsilon_k-\mu)}\right]}$$

For Bose gas in volume V we can change the sum into an integral over the single-particle phase space, using substitutions:

$$\begin{cases}\varepsilon_k \mapsto \frac{p^2}{2M}\\ \sum_k \mapsto \frac{4\pi V}{(2\pi\hbar)^3}\int_0^\infty p^2dp\end{cases}$$

▶ **Partition function of the Fermi gas**

Fermi gas ≡ ensemble of **non-interacting fermions**

$$\Rightarrow \quad E_{Ni}=\sum_{k=1}^{\infty}n_{ik}\varepsilon_k \qquad N=\sum_{k=1}^{\infty}n_{ik} \qquad \text{with occup. numbers } \boxed{n_{ik}=0,1}$$

⇒ partition function:

$$Z(\beta,\mu)=\sum_N\sum_i e^{-\beta(E_{Ni}-\mu N)} = \sum_{\{n_{ik}\}} e^{-\beta\left(\sum_k n_{ik}\varepsilon_k-\mu\sum_k n_{ik}\right)}=\prod_k\underbrace{\sum_{n_{ik}=0,1}e^{-\beta\left(n_{ik}\varepsilon_k-\mu n_{ik}\right)}}_{1+e^{-\beta(\varepsilon_k-\mu)}}$$

$$\boxed{\ln Z(\beta,\mu)=+\sum_k\ln\left[1+e^{-\beta(\varepsilon_k-\mu)}\right]}$$

the sum can be replaced by the same phase-space integral as for bosons